D1420635

MULTI-PROBLEM VIOLENT YOUTH

NATO Science Series

A series presenting the results of scientific meetings supported under the NATO Science Programme.

The series is published by IOS Press and Kluwer Academic Publishers in conjunction with the NATO Scientific Affairs Division.

Sub-Series

I.	Life and Behavioural Sciences	IOS Press
II.	Mathematics, Physics and Chemistry	Kluwer Academic Publishers
III.	Computer and Systems Sciences	IOS Press
IV.	Earth and Environmental Sciences	Kluwer Academic Publishers
V.	Science and Technology Policy	IOS Press

The NATO Science Series continues the series of books published formerly as the NATO ASI Series.

The NATO Science Programme offers support for collaboration in civil science between scientists of countries of the Euro-Atlantic Partnership Council. The types of scientific meeting generally supported are "Advanced Study Institutes" and "Advanced Research Workshops", although other types of meeting are supported from time to time. The NATO Science Series collects together the results of these meetings. The meetings are co-organized by scientists from NATO countries and scientists from NATO's Partner countries – countries of the CIS and Central and Eastern Europe.

Advanced Study Institutes are high-level tutorial courses offering in-depth study of latest advances in a field.
Advanced Research Workshops are expert meetings aimed at critical assessment of a field, and identification of directions for future action.

As a consequence of the restructuring of the NATO Science Programme in 1999, the NATO Science Series has been re-organized and there are currently five sub-series as noted above. Please consult the following web sites for information on previous volumes published in the series, as well as details of earlier sub-series:

http://www.nato.int/science
http://www.wkap.nl
http://www.iospress.nl
http://www.wtv-books.de/nato_pco.htm

Series I: Life and Behavioural Sciences - Vol. 324   ISSN: 1566-7693

Multi-Problem Violent Youth

A Foundation for Comparative Research on Needs,
Interventions and Outcomes

Edited by

Raymond R. Corrado

School of Criminology, Simon Fraser University, Burnaby, BC, Canada

Ronald Roesch

Department of Psychology, Simon Fraser University, Burnaby, BC, Canada

Stephen D. Hart

Department of Psychology, Simon Fraser University, Burnaby, BC, Canada

and

Jozef K. Gierowski

*Department of Forensic Psychology, Institute of Forensic Research, Cracow,
Poland*

IOS
Press

Ohmsha

Amsterdam • Berlin • Oxford • Tokyo • Washington, DC

Published in cooperation with NATO Scientific Affairs Division

Proceedings of the NATO Advanced Research Workshop on
Multi-Problem Violent Youth: A Foundation for Comparative Research on Needs, Interventions and Outcomes
31 August – 3 September 2000
Cracow, Poland

© 2002, IOS Press

ISBN 1 58603 071 X (IOS Press)
ISBN 4 274 90484 9 C3045 (Ohmsha)
Library of Congress Control Number: 2001097033

Publisher
IOS Press
Nieuwe Hemweg 6B
1013 BG Amsterdam
Netherlands
fax: +31 20 620 3419
e-mail: order@iospress.nl

Distributor in the UK and Ireland
IOS Press/Lavis Marketing
73 Lime Walk
Headington
Oxford OX3 7AD
England
fax: +44 1865 75 0079

Distributor in the USA and Canada
IOS Press, Inc.
5795-G Burke Centre Parkway
Burke, VA 22015
USA
fax: +1 703 323 3668
e-mail: iosbooks@iospress.com

Distributor in Germany, Austria and Switzerland
IOS Press/LSL.de
Gerichtsweg 28
D-04103 Leipzig
Germany
fax: +49 341 995 4255

Distributor in Japan
Ohmsha, Ltd.
3-1 Kanda Nishiki-cho
Chiyoda-ku, Tokyo 101
Japan
fax: +81 3 3233 2426

PRINTED IN THE NETHERLANDS

Introduction

Raymond R. Corrado, Ronald Roesch, Stephen D. Hart and Jozef K. Gierowski

This book is the first product of an Advanced Research Workshop (ARW), entitled *Multi-problem Violent Youth: A Foundation for Comparative Research on Needs, Interventions and Outcomes*, that was funded by the Scientific Affairs Division of the North Atlantic Treaty Organization (NATO) and held in Cracow, Poland. The primary focus of the workshop was the development of a procedure for assessing risk and needs factors for serious antisocial behavior—and especially violence—in children and adolescents. This book contains chapters prepared by speakers and participants at the workshop and also the conceptual framework for the risk-needs instrument that eventually will become the basis of ongoing, international collaborative research projects.

This workshop resulted from our belief that research on the correlates of youth violence had reached the point where it could be linked to relatively more recent research on the correlates of violence in childhood and infancy, including prenatal and perinatal risk factors. Although several risk instruments were already available, in our view none was comprehensive in nature or scope. A more comprehensive instrument could assist government and community agencies to develop individual, family, and community interventions designed to reduce the violence risk in children and adolescents. An important assumption that guided our work was the belief that early intervention is an essential, and potentially more effective, approach to helping families and children who are at risk for violence. Early intervention may reduce the use of punitive alternatives, such as incarceration of adolescents and young adults, that are costly and often ineffective, in large part because the interventions take place long after problem behaviors have developed.

In our original application for funding submitted to NATO, we referred to the growing policy concern about youth violence in Europe and North America. In many countries, the media, sections of the public and certain politicians promoted the view that youth violence had escalated. They pointed to "unprecedented" acts of random and brutal violence in cities, the emergence of gang violence and immigrant youth-organized crime associations involved in drug trafficking, prostitution and robbery. The media sensationalizing of brutal acts of youth violence created the image within the public that their safety and public order was threatened. Some politicians, in turn, reacted to public anger by promoting or introducing harsher criminal justice responses to violent youth and young adults. In North America and Britain, these policy responses typically included raising violent accused youth to adult courts and long terms of incarceration. The increase in transfers to the adult system is attributed to a justice philosophy that holds that some youth are too violent to be managed in the juvenile justice system, that some youth are beyond rehabilitation, and that a more punishment oriented approach, as reflected by the adult criminal justice system, would provide better protection for the public. Empirical research has not shown this to be the case. Indeed, the research suggests that raising youth to adult court and incarceration in prisons rather than keeping them in the juvenile system appears to have effects that run counter to the expectations, as some studies have shown that recidivism of transferred youth is higher than comparable youths not transferred (Bishop, Frazier, Lanza-Kaduce, & Winner, 1996; Snyder, Sickmund, & Poe-Yamagata, 2000). Neither do "get tough" approaches to youth violence appear to have reduced homicide or overall crime rates. Further, concerns have been raised that ethnic and cultural minorities are overrepresented among youths transferred to the adult sentence, and that once sentenced in adult court, youth are at greater risk of victimization in prison. We believe that this policy diverts resources from early intervention programs that might provide a more effective means to achieve the goals of violence reduction and protection of the public. Whether youth violence escalated in frequency or severity is an important empirical issue (data from a number of countries suggest that rates of violent crime among youth may be stable or even declining), but we believe that the more immediate policy concern should be how best to react to the correlates of serious youth violence. The assumption that violence is caused by the absence of punishment and retribution requires ignores the impressive body of research that has identified a far more complex array of risk factors (see chapters by Farrington; Lösel; Reppucci, Fried, & Schmidt, this book).

We initially discussed the idea of applying for the NATO ARW with our colleagues, Professor Giovanni Traverso from the University of Siena and, later, Professor David Farrington from the University of Cambridge. We agreed there was sufficient comparative research to begin developing procedures to assist the assessment and management of risk and need factors that would reflect research on the correlates of violence at various stages of development, from the prenatal period through early adulthood. We formed a research committee to draft the application to NATO. Professors Ronald Roesch and Stephen Hart from Simon Fraser University and Professor Jozef Gierowski and Teresa Jaskiewicz-Obydinska from the Institute of Forensic Research, Cracow, joined this committee. Our Polish colleagues were instrumental in developing the list of key researchers from NATO partner countries and non-NATO countries in Eastern Europe. As well, Dr. Aleksander Glazek graciously offered to host the ARW at the Institute of Forensic Research. Reflecting the NATO goal of promoting collaborative partnerships, the research committee invited attendees from Eastern Europe, Northern Europe, Southern Europe, and North America who were considered to be experts on some aspect of youth violence, and particularly those working within a developmental framework.

The ARW application eventually was submitted to NATO with Professors Corrado and Gierowski named as Principal Investigators. Shortly after it received funding, a preliminary meeting was held in Vancouver, funded jointly by NATO and by Simon Fraser University. The meeting included a day of lectures and a day of meetings, which permitted the research committee to receive advice from North American researchers regarding current research and practice — as well as beliefs about best practice — with respect to the assessment and management of violence risk violence in youths. Following the Vancouver meeting, several graduate students in criminology and psychology from Simon Fraser University assisted the research committee in pursuing some of the topics identified. Candice Odgers, Gina Vincent, Rebecca Dempster, and Andrea McEachran worked throughout the summer of 2000 to review and summarize the relevant literature.

The ARW was held in Cracow from August 31 to September 3, 2000. It was attended by 37 researchers from 16 countries. (A full list of participants appears at the end of this chapter.) Focused presentations were followed by intensive discussions aimed at developing a conceptual framework for the proposed risk-needs management instrument. The ARW achieved two important goals. First, and most important, a solid network of connections was established among researchers. There were constant exchanges of information and ideas regarding policy and research issues. Second, the conceptual framework of the risk-needs instrument was developed.

This book summarizes the proceedings and the results of the ARW. It is divided into four sections. The chapters in Section I review risk factors for violence associated with different developmental stages and in specific groups or populations. The chapters in Section II focus on assessment issues, including reviews of existing assessment instruments. The chapters in Section III describe risk-needs research projects undertaken in several countries, including France, Italy, Poland, and The Netherlands. Finally, the chapters in Section IV present the conceptual framework for the new risk-needs instrument and provide details about its content.

We would like to thank all the people who attended the ARW for making presentations, leading discussions, sharing their ideas and views during the discussions, and submitting papers summarizing their work. We hope that this book captures for the major ideas put forward at the meeting in Cracow, and also conveys to readers something of the sense of fellowship, excitement, and purpose felt by those of us who were there.

References

Bishop, D. M., Frazier, C. E., Lanza-Kaduce, L., & Winner, L. (1996). The transfer of juveniles to criminal court: Does it make a difference? *Crime & Delinquency, 42,* 171-191.

Snyder, H. N., Sickmund, M., & Poe-Yamagata, E. (2000). *Juvenile transfers to criminal court in the 1990's: Lessons learned from four studies.* Rockville, MD: National Institute of Justice.

Appendix

Multi-problem Violent Youth: A Foundation for Comparative

Research on Needs, Interventions and Outcomes

An Advanced Research Workshop

funded by
The North Atlantic Treaty Organization

Co-Directors

Raymond Corrado
School of Criminology
Simon Fraser University
8888 University Drive
Burnaby BC
CANADA V5A 1S6
corrado@sfu.ca

Jozef Gierowski
Department of Forensic Psychology
Institute of Forensic Research
u. Westerplatte 9
31-033
Cracow POLAND

Organizing Committee

David Farrington
Institute of Criminology
Cambridge University
7 West Road
Cambridge CB3 DT9
ENGLAND
dpf1@cus.cam.ac.uk

Ronald Roesch
Department of Psychology
Simon Fraser University
8888 University Drive
Burnaby BC
CANADA V5A 1S6
rroesch@arts.sfu.ca

Giovanni Traverso
Department of Forensic Sciences
University of Siena
Policlinico "Le Scotte"
53100 Siena
ITALY
traverso@unisi.it

Research Staff

Rebecca Dempster
Department of Psychology
Simon Fraser University
8888 University Drive
Burnaby BC
CANADA V5A 1S6
rdempster@arts.sfu.ca

Andrea McEachran
Department of Psychology
Simon Fraser University
8888 University Drive
Burnaby BC
CANADA V5A 1S6
amceachran@sfu.ca

Candice Odgers
School of Criminology
Simon Fraser University
8888 University Drive
Burnaby BC
CANADA V5A 1S6
codgers@sfu.ca

Gina Vincent
Department of Psychology
Simon Fraser University
8888 University Drive
Burnaby BC
CANADA V5A 1S6

Participants

Canada

Stephen Hart
Department of Psychology
Simon Fraser University
8888 University Drive
Burnaby BC
CANADA V5A 1S6
shart@arts.sfu.ca

Sheilagh Hodgins
Department of Psychology
Université de Montréal
C.P. 6128, succ. Centre-ville
Montréal, Québec
CANADA H3C 3J7
sheilagh.hodgins@umontreal.ca

Marc Le Blanc
Ecole de criminologie
Universite de Montreal
C.P. 6128, succ. Centre Ville
Montreal, Quebec
CANADA H3C 3J7
marc.leblanc@umontreal.ca

Marlene Moretti
Department of Psychology
Simon Fraser University
8888 University Drive
Burnaby BC
CANADA V5A 1S6
moretti@sfu.ca

Pratibha Reebye
Infant Psychiatry Clinic, Room 414A
B. C. Childrens' Hospital
4480 Oak Street
Vancouver, BC
CANADA V6H3V4
reebye@home.com

Czech Republic

Vojtech Sedlacek
Department of Criminology
Lhotecka 559/7
PO Box 54
143 01 Prague 4
CZECH REPUBLIC

Finland

Jaana Haapasalo
Department of Psychology
University of Jyvaskyla
P. O. Box 35
FIN-40351 Jyvaskyla
FINLAND
jhaapasa@psyka.jyu.fi

France

Catherine Blatier
Les Cadorats
3, Allée des géraniums
38650 Avignonet
FRANCE
Catherine.Blatier@upmf-grenoble.fr

Germany

Doris Bender
Department of Psychology
University of Erlanger-Nuremberg
Bismarckstr. 1
D-91054 Erlanger
GERMANY
dsbender@phil.uni-erlangen.de

Friedrich Lösel
Institute fur Psychologie
Univ. Erlanger-Nuremberg
Lehrstahl, Bismarchster 1
D-91054 Erlanger
GERMANY
fhloesel@phil.uni-erlangen.de

Anneliese Pontius
Waldschmidt St 6
60316 Frankfurt
GERMANY
apontius@hms.harvard.edu

Italy

Luca Iani
Department of Social Psychology, University of Rome 'La Sapienza'
Via dei Marsi, 78, 00185 Rome
ITALY
iani.l@libero.it

Lithuania

Vanda Vaitekoniene
National Child Department Center
Vytavto 15
Vilnius 2004
LITHUANIA
vaitekoniene@takas.lt

The Netherlands

Eric Blaauw
Clinical Psychology
Vrije Universiteit Amsterdam
De Boelelaan 1109
1081 HV Amsterdam
THE NETHERLANDS
E.Blaauw@psy.vu.nl

Frans Winkel
Clinical Psychology
Vrije Universiteit Amsterdam
De Boelelaan 1109
1081 HV Amsterdam
THE NETHERLANDS
FW.Winkel@psy.vu.nl

Poland

Teresa Jaskiewicz-Obydzinska
Department of Forensic Psychology
Institute of Forensic Research
u. Westerplatte 9 u.
31-033 Crakow
POLAND
tjaskiew@ies.krakow.pl

Michal Slawik
Department of Forensic Psychology
Institute of Forensic Research
Westerplatte 9
31-033 Crakow
POLAND

Maciej Szaszkiewicz
Department of Forensic Psychology
Institute of Forensic Research
u. Westerplatte 9
31-033 Crakow
POLAND

Ewa Wach
Department of Forensic Psychology
Institute of Forensic Research
u. Westerplatte 9
31-033 Crakow
POLAND

Portugal

Antonio Castro Fonseca
Faculdade de psicologia e de Ciencias da Educacao
Universidade de Coimbra
Rua do Colegio Novo
3000 Coimbra
PORTUGAL
fonseca@ci.uc.pt

Romania

Gheorghe Ghidanac
High Security Penitary PhD.
Sos. Alexandrei 154
Bucharest
ROMANIA

Bonatiu Mihaela
High Security Penitary PhD.
Sos. Alexandrei 154
Bucharest
ROMANIA

Russia

Vladislav Ruchkin (from Russia but currently on a post-doctoral fellowship at Yale University)
230 South Frontage Road
New Haven, CT 06520-7900
USA
vladislav.ruchkin@yale.edu

Sweden

Martin Grann
Division of Forensic Psychiatry
Karolinska Institute
P.O. Box 4044
SE-14104 Huddinge
SWEDEN
Martin.Grann@neurotec.ki.se

Niklas Langstrom
Division of Forensic Psychiatry
Karolinska Institute
P.O. Box 4044
SE-14104 Huddinge
SWEDEN
Niklas.Langstrom@neurotec.ki.se

Britt af Klinteberg
Department of Psychology
Stockholm University
S-106 91 Stockholm
SWEDEN
bkg@psychology.su.se

Ukraine

Volodymyr Pyatokha
Volyn Regional Hospital
Av. President Grushevski 21
Lutsk. 43007
UKRAINE
pyatokha@lutsk.ukrpack.net

United Kingdom

Danny Clark
Directorate of Regimes,
Sentence Management Group
Abel House, Room 717
John Islip Street
London SW1P 4LH
ENGLAND
Danny.Clark@homeoffice.gsi.gov.uk

David Cooke
Douglas Inch Centre
2 Woodside Terrace
Glasgow G3 7UY
SCOTLAND
djcooke@rgardens.u-net.com

United States

Robert Barnoski
Washington State Institute for Public Policy
PO Box 40999
110 East Fifth Avenue, Suite 214
Olympia, WA 98504-0999
USA
barney@wsipp.wa.gov

Edward Mulvey
Law and Psychiatry Program
WPIC
3811 O'Hara Street
Pittsburgh, PA 15213
USA
mulveyep@MSX.UPMC.EDU

Trish Beuhring
University of Minnesota
Division of General Pediatrics and Adolescent Health
University Gateway #260
Minneapolis, MN 55455
USA
Beuhr001@tc.umn.edu

N. Dickon Reppucci
Department of Psychology
P.O. Box 400400
University of Virginia
Charlottesville, VA 22904-4400
USA
ndr@virginia.edu

DEDICATION

To Penny, Chris, and Mark for your understanding and love. It's a long way from Trail to here. Raymond R. Corrado

To all my sons, Stefan, Michael, David, and Jeremy. Ronald Roesch

To Mackenzie, with all my love. Stephen D. Hart

Contents

I

Risk Factors

Multi-Problem Violent Youth
R.R. Corrado et al. (Eds.)
IOS Press, 2002

Youth Violence: Risk and Protective Factors*

N. Dickon Reppucci, Carrie S. Fried and Melinda G. Schmidt

Although reality is certainly exaggerated by the mass media, there is no doubt that violence by and against youths is an enormous social problem both in the United States and in Europe.. Nearly one in six of all violent crime arrests in the U.S. in 1997 involved a juvenile under 18 years of age (FBI, 1997). Compared with an increase of 19% among persons 18 years of age and over, arrest rates for violent crimes among juveniles increased 49% between 1988 and 1997. Many European countries also experienced increases in their rates of juvenile violent crime (Pfeiffer, 1998). Comparison of rates across countries is complicated because legal definitions of crime vary from country to country and because crime is measured differently in each country. For example in the US and Great Britain, numbers of arrests are frequently used to measure crime; in comparison, in Italy and Germany, the number of cases solved by police, regardless of whether apprehension of the offender occurs is the measure used (Pfeiffer, 1998). Although recent data suggest that juvenile arrests for murder and other violent crimes may have peaked in the US in 1994 (Sickmund, Snyder, & Poe-Yamagata, 1997), violent crime rates in the US have been consistently higher than in Canada and European countries (Hagan and Foster, 2000). Still, Pfeiffer (1998) reports that violent crimes by juveniles in Germany, England and Wales remain substantial. In contrast, countries such as Canada, England and Wales have higher property crime rates than the US (Hagan & Foster, 2000).

This conference is focused on developing a unified instrument to be used to collect data on youthful offenders in all NATO countries so that differences and similarities among the youth of these nations may be compared and contrasted. This is not an easy task! What information should be uniformly collected that would help to understand juvenile offending in a fashion that would allow comparisons while at the same time being helpful to the development of various prevention and treatment interventions? In an attempt to provide some sort of delineation of important background information on youth, we outline briefly what we have learned over the past several decades about risk and protective factors, with special emphasis on violent youth. We then proceed in the following fashion. We first summarize various risk factors that are associated with an increased probability of violence, then comment very briefly on developmental trajectories of violence, finally turning to protective factors that are associated with a decreased probability of delinquency and violence. The former have been the major focus of research, while the latter have the potential to be the most provocative in terms of helping to tailor new interventions. Our goal is to suggest that whatever instrument is devised, it is crucial to decide what information is absolutely essential to obtain on each youth, e.g., age, gender, ethnicity, crime, but then from the myriad possibilities, what other information should be collected. One caveat: Because of space constraints and the enormity of available information, what is presented is severely truncated; for an exhaustive review, see Loeber and Farrington (1998).

1. Definitional Issues

The difficulty of defining violence complicates the identification of risk factors for violent behavior among youth, as well as efforts to prevent youth violence. The National Center for Health Statistics (1993) defines violence as the threatened or actual use of physical force or power that either results in or has a high likelihood of resulting in death, injury or deprivation. However, studies of violence tend to employ either an overly broad definition that blurs the distinction between aggression and violence or an overly narrow definition that focuses on specific types of violence, such as robbery, rape, murder, and assault. If researchers rely on the use of official records (e.g., arrests, convictions) to measure outcomes, many violent acts that never lead to an arrest or a conviction are excluded (Farrington, 1997). Moreover, the relative levels of violence among groups that are more likely to be arrested and convicted (e.g., the poor, minorities, especially African Americans) as a result of institutionalized racism and possibly lower quality legal representation may be exaggerated.

The terms *violence* and *aggression* are often used interchangeably because many violent acts are aggressive in nature and vice versa. However, in this article, aggression refers to the intent to hurt or gain advantage over others, without necessarily involving physical attack, while violence involves the use of physical force against another individual (Megargee, 1982). Several researchers have advocated the need to make distinctions among different types of violence and aggression by treating them as heterogeneous categories. Such distinctions could lead to a more heterogeneous treatment of violent youth, which would have implications for prevention, treatment, and processing by the legal system.

2. Risk Factors as Predictors of Violence

Numerous studies of youth violence have examined the predictors or factors that put youth at increased risk for exhibiting violent behavior. These predictors can be categorized using three levels of analysis: A) the individual level, including biological, cognitive, and emotional variables; B) the immediate systems level, including family, peer, school, and neighborhood influences; and C) the cultural and societal level including poverty and racism, media portrayals of violence, accessibility of firearms, and societal drug and alcohol use.

A caution is in order before proceeding. Although risk factors may help to figure out which youth are most likely in need of preventive intervention, they can not pinpoint which individual children will become serious or violent offenders. Most youth with risk factors do not become chronic offenders. Lipsey & Derson (1998) point out that if we rely on any one of the strongest risk factors, it would suggest 3 to 6 children who would not become serious offenders for every child who would. They further emphasize that for a risk to be a good predictor of individual outcome, it must be both that most individuals with the outcome have the risk and most individuals without the risk do not have the outcome. Another way of measuring risk is how much exposure to the risk increases the probability of the outcome. However, being exposed to the risk, e.g., child abuse, does not mean that every or even most youth will develop the delinquent behavior. Most victims of child abuse do not become violent offenders, and many violent offenders have never been abused. A third measure is how much the outcome would decrease if the risk were eliminated. For example, although poverty is a major factor in delinquency, it is not powerful in prediction by itself because so many youth grow up in poverty, (e.g., in the United States, more than 20% of youth live in poverty). Still it can contribute to a substantial amount of the delinquent behavior. With this stated, it is widely accepted that the more risk factors experienced by a youth, the higher the risk for delinquent behavior. However, we have not yet been able to weigh the various factors in terms of negative impact, either individually or in conjunction with one another, even though the development of such a calculus would be of immense value. Finally, it is worth noting that most youth commit some illegal behavior; in fact, Moffitt (1993) claims that, "it is statistically aberrant to refrain from crime during adolescence (p. 685-86)," and Farrington's (1989) famous Cambridge study found that "96 percent of a sample of inner-city London males admitted to committing at least one of ten common offenses (including theft, burglary, violence, vandalism and drug abuse) at some time between the ages of ten and thirty-two."

3. Individual Level Factors

At the individual level, biological, cognitive, and emotional factors contribute to the development of violence and aggression. An accumulation of individual risk factors is presumed to increase the likelihood of eliciting violence or aggression, especially when coupled with multiple immediate systems and societal and cultural risk factors.

3.1 Biological Bases of Violence

Although a biological predisposition may make it more or less likely that an individual will exhibit violent behaviors in certain contexts, data suggest that biological factors account for a small proportion of the variance in violent behavior (Miczek, Mirsky, Carey, DeBold, & Raine, 1994). Violent and aggressive behavior have been linked to heritability, hormones, neurotransmitters, the autonomic nervous system, and neurological functioning. With regards to heritability, the consensus is that genes may influence other physiological processes that may predispose an individual to aggressive behavior (Gottesman & Goldsmith, 1994).

Testosterone is the hormone that is most often implicated in influencing levels of aggression and violence. Among adults, substantial evidence links aggression and testosterone levels, although research on adolescents has yielded mixed results (see Coie & Dodge, 1998). Most likely, testosterone and aggression influence each other in a reciprocal fashion, such that testosterone could be considered both a cause and consequence of aggression (Archer, 1994).

Autonomic nervous system research has consistently found that low resting heart rate is related to violent behavior (Farrington, 1997). Raine, Venables, and Williams (1990) used resting heart rate and skin conductance measures of males at age 15 to predict criminality at age 24. They were able to classify 75 of 101 participants accurately. Low resting heart rate may be an indication of fearlessness or autonomic under-arousal (Farrington, 1997). Fearless individuals are more likely to engage in violent behaviors than fearful individuals, and autonomic under-arousal leads to sensation-seeking, and thus, participation in risky behaviors.

Several neurotransmitters have been linked to the development of aggression. For example, low levels of serotonin (5-HT) activity consistently correlate with aggression in both children and adults. Experimental studies have demonstrated that changes in serotonin (5-HT) functioning are inversely related to aggressive behavior, especially among high-trait aggressive individuals (e.g., Cleare & Bond, 1995). Dopamine and norepinephrine have also been implicated in the development of aggression, but the relationships are not clearly established.

Studies of neurological functioning have consistently found that violence is related to deficits in the frontal cortex. According to Raine's (1998) prefrontal dysfunction model, damage to the prefrontal cortex results in cognitive, personality, arousal and anticipatory fear deficits, which in turn predispose individuals to aggressive behavior. Prenatal exposure to environmental toxins (e.g., lead, in utero alcohol and drugs) and perinatal complications may result in neurological damage that prevents the inhibition of violent behavior (Reiss & Roth, 1993). Moffitt, Lynam, and Silva (1994) found that neurological deficits at age 13 were correlated with delinquency at age 18.

3.2 Cognitive Factors

Low IQ, reading problems, attention deficits, and hyperactivity are among the cognitive precursors of aggression and violence. Low Verbal IQ is consistently predictive of aggressive and delinquent behavior among adolescents even after controlling for race and SES (Lynam, Moffitt & Stouthamer-Loeber, 1993). Furthermore, many studies have found that when compared to their non-delinquent peers, both non-violent and violent delinquents demonstrate relatively lower scores on verbal IQ than on performance IQ measures (e.g., Cornell & Wilson, 1992). Examining several theories that have been used to explain the relationship between low IQ and delinquency, Lynam and colleagues (1993) found support for only one of them, the indirect effect model, which holds that IQ causes delinquency through its effect on school performance; youth with cognitive deficits experience school failure, which in turn contributes to delinquency. Several longitudinal studies demonstrate that childhood hyperactivity and attention deficits are also strongly associated with both self-reported violence and arrests for violent offenses in late adolescence and early adulthood (e.g., Farrington, 1989; Hechtman & Weiss, 1986). However, without hyperactivity, attention deficits appear to be only weakly associated with later aggression (Loeber, 1988).

Research also indicates that aggressive adolescents have more social cognitive deficiencies (e.g., Crick & Dodge, 1994), lower moral reasoning maturity (e.g., Arbuthnot, Gordon, & Jurkovic, 1987), and poorer abstract reasoning and problem-solving skills than non-aggressive youths (e.g., Seguin, Pihl, Harden, Tremblay, & Boulerice, 1995). An individual's cognitive appraisal of an event is a major factor in determining the subsequent behavioral response. Crick & Dodge (1994) found that aggressive children selectively attend to aggressive cues, are more likely to attribute hostile intent to others, generate fewer solutions to problems and select action-oriented rather than reflective solutions.

3.3 Temperament and Emotional Regulation

Based on maternal reports, children with a difficult temperament as infants are more likely to be aggressive and have other behavior problems in later childhood (e.g., Kingston & Prior, 1995). However, maternal reports may be inaccurate because: (a) the mother's attitude toward the child, rather than the child's actual behavior may be driving the reports (e.g., Bates, Maslin & Frankel, 1985); and (b) the interaction between the child's temperament and the quality of the mother-child relationship may be responsible for the correlation between temperament and later aggressive behavior. Even so, longitudinal studies demonstrate consistently high stability in aggression from early childhood into young adulthood (Olweus, 1979; Farrington, 1994), and that early aggressive behavior is predictive of later arrests for violent crimes. Early onset of violent behavior and delinquency is also predictive of more serious and chronic violence (Tolan & Thomas, 1995).

Although there are no studies that have specifically examined the relationship between empathy and violence, several studies support the notion that a lack of empathy may be a risk factor for participation in aggressive and antisocial behaviors. A meta-analysis of 22 studies determined that empathy measured by self-report questionnaires was moderately correlated with aggressive (pooled correlation = -.18) and antisocial behavior (pooled correlation = -.26) (Miller & Eisenberg, 1988). Recently, Cohen and Strayer (1996) found that conduct disordered youth scored lower than non-conduct disordered youth on three separate measures of empathy. Feshbach (1997) suggests that empathic reactions serve to inhibit aggression because individuals who vicariously experience the pain and distress of others would be less likely to engage in aggressive behaviors.

4. Immediate Systems Level Factors

Immediate systems are those with which the individual has direct contact, and include neighborhood, school, peer group and family factors.

4.1 Neighborhoods

Sociocultural and neighborhood characteristics are strongly related to levels of violence in a community. Poverty, economic inequality, racial composition, geographic region, institutionalized racism, population density, residential mobility and family disruption are just some of the factors that have been implicated as causes of high rates of violence in some communities. It is beyond the scope of this article to comment on all of these. Rather it is important to note that economically poor neighborhoods tend to have higher rates of violence, even though all such neighborhoods do not experience high rates of violence. They also share other characteristics that make violence more prevalent. Poor neighborhoods tend to be inhabited by a more transient population and to be more disorganized (Hawkins, Herrenkohl, Farrington, Brewer, Catalano, & Harachi, 1998). High levels of mobility make it more difficult for residents to develop support networks and feel a sense of attachment to the community (American Psychological Association Commission on Violence, 1993). Also, the types of public and social services (e.g., recreation programs) that are available in more affluent neighborhoods are often limited in low-income neighborhoods (National Research Council, 1993).

In violent neighborhoods children are more likely to witness shootings, stabbings, and killings (e.g., Jenkins & Bell, 1994), which has an effect on the development of violence and aggression (Widom, 1989). A recent study found that exposure to violence was the strongest predictor of adolescents' self-reported use of violence (DuRant, Cadenhead, Pendergrast, Slavens & Linder, 1994). Neighborhoods also appear to affect early onset of violent behavior, such that in the most disadvantaged neighborhoods more 10 to 12-year-old boys were engaging in acts of violence. Yet, among 13 to 15-year-old boys violence was fairly evenly distributed across neighborhoods (Loeber & Wikstrom, 1993).

4.2 Family

The family has been cited as the most powerful single influence on development of aggression and violence (McGuire, 1997). Higher levels of aggression and violence are found among children of low socioeconomic status (SES), poorly educated, highly mobile, single parent families (Gagnon, Craig, Tremblay, Zhou & Vitaro, 1995; Kupersmidt, Briesler, DeRosier, Patterson, & Davis, 1995; Tremblay, Masse, Kurtz & Vitaro, 1997; Saner & Ellickson, 1996). However, Kupersmidt and colleagues (1995) note that single parenthood is associated with neighborhood type and it may be the interaction between these factors that accounts for some of the variance in aggression. The criminal behavior of parents is also strongly and consistently associated with violence in youths (e.g., Baker & Mednick, 1984).

The influence of child-rearing practices on the development of violence has received considerable attention. Coercive interactions (Capaldi & Patterson, 1996), lax and ineffective parental discipline (Snyder & Patterson, 1987; Weiss, Dodge, Bates, & Pettit, 1992), poor parental monitoring (Gorman-Smith, Tolan, Zelli, & Huesmann, 1996), physical punishment (Farrington, 1978; Eron, Huesmann, & Zelli, 1991), and child physical abuse (Manly, Cicchetti, & Barnett, 1994) are all associated with higher rates of aggression and violence among children. However, it is unclear whether parenting practices cause aggressive and violent behavior, or whether these behaviors lead to harsh and ineffective parenting styles.

Longitudinal research seems to indicate that there is a reciprocal relationship between child-rearing practices and child behavior (Vuchinich, Bank, & Patterson, 1992). The effects of parenting strategies appear to be mediated through the quality of the parent-child relationship. Lack of maternal warmth (e.g., Booth, Rose-Krasnor, McKinnon, & Rubin (1994), insecure attachment (e.g., Erickson, Sroufe, & Egeland, 1985), and parental indifference and rejection (e.g., Farrington, 1991) have all been associated with increased levels of aggression and violence in childhood and adolescence. However, studies (e.g., Booth et al., 1994; Bates, Bayles, Bennett, Ridge, & Brown, 1991) examining the link between attachment and later antisocial and violent behavior have produced mixed results.

Children who grow up in homes characterized by high levels of family conflict and family violence are at increased risk for becoming aggressive and violent. Exposure to parents' marital conflict is predictive of violent criminal convictions (Farrington, 1989; McCord, 1979) and self-reported violence in adolescence (Elliott, 1994). The evidence also seems to indicate that children who have been physically abused or neglected are more likely to be arrested for a violent crime (e.g., Widom, 1989).

4.3 Peer Groups

Unlike adults, juveniles commit most violent and serious delinquent acts in the company of peers rather than alone (Zimring, 1998). The process of association with aggressive and antisocial peers begins at a young age with rejection by prosocial peers. There is substantial evidence that aggressive children are likely to be rejected by their peers (for a review see Coie, Dodge, & Kupersmidt, 1990). Peer rejection leads to increased aggression and antisocial activity, and promotes association with antisocial peers (Dodge, Bates, &, Pettit, 1990; Dishion, Patterson, Stoolmiller, & Skinner, 1991; Patterson & Bank, 1989). Deviant, rejected children gravitate toward each other and form their own "coercive cliques" (Cairns, Cairns, Neckerman, Gest, & Gariepy, 1988), where aggression is valued and deviant behaviors are promoted. Other non-deviant adolescents may be attracted to the glamour of these delinquent peer groups, which may contribute to the late onset of aggression in adolescents with no prior history of delinquency or aggression (Moffitt, 1993). Several longitudinal studies have found strong positive correlations between association with delinquent peers in early adolescence and self-reported violence in late adolescence or early adulthood (e.g., Farrington, 1989; Maguin, et al., 1995; Saner & Ellickson, 1996). These findings were confirmed by a recent meta-analysis that concluded that lack of positive social ties and association with antisocial peers at age 12-14 were powerful predictors of later violent or serious delinquency (Lipsey & Derzon, 1998).

4.4 Schools

Rates of violence are higher in schools with greater percentages of students who do not place a high value on good grades, do not believe the curricula is relevant, and do not think their school experience will have a positive influence on their lives (National Research Council, 1993). Higher rates of violence are also found in schools with low student attendance, high student-teacher ratios, instability in the school population, and poor academic quality of the school (Hellman & Beaton, 1986). Although violence in schools is a reflection of the community where the school is located, relations between school characteristics and aggressive behaviors exist even after controlling for neighborhood crime rates. Aggressive and violent behavior is more likely to occur in schools with lax enforcement of rules and undisciplined classrooms (National Research Council, 1993). It is unclear whether high rates of school violence and aggression cause or are caused by the discipline practices of the school. Probably an individual youth's academic achievement and attitude toward school interact with the school environment to predict aggression and violence. A recent meta-analysis of longitudinal research found that school attitude and performance from age 6 to 14 were significant predictors of violence at age 15-25 (Lipsey & Derzon, 1998).

5. Societal and Cultural Level Factors

Although each NATO country may have cultural and societal factors that contribute to youth violence, the role of several societal-level factors are often critical. These factors include poverty and racism, the media, accessibility of firearms, and societal abuse of drugs and alcohol.

5.1 Poverty and Racism

Poverty is one of the strongest predictors of violence (Centerwall, 1984). However, poverty alone does not account for high rates of violence. Rather violence is likely to occur when poverty is combined with discrimination, prejudice, and structural inequities, resulting in injustice, deprivation, frustration, anger and hopelessness (Hill, Soriano, Chen, & LaFromboise, 1994). The effect of poverty on violence is conditional on numerous other variables, from parental stress, to lack of social networks, to joblessness. Two of the most salient variables may be racism and residential segregation. In the United States, statistics from the three most common measurement approaches, --- official arrest and conviction records, self-report surveys, and victimization surveys--- all show high rates of serious offending among African American adolescents, i.e., they are over-represented among both victims and perpetrators of violent crimes (Reiss & Roth, 1993). However, these youth are more likely than any other racial or ethnic group to grow up in segregated neighborhoods of concentrated urban poverty (Wilson, 1987). In fact, studies that control for socioeconomic status and concentrated neighborhood poverty demonstrate that the disparity between the percentage of African-Americans and the percentage of the general population in the United states who are perpetrators of violence is quite small (Centerwall, 1984; National Research Council, 1993; Reiss & Roth, 1993). Moreover, as shifts in the American economy resulted in the movement of industry out of the cities (Wilson, 1987), African Americans have suffered disproportionately, because youth who grow up in these communities are likely to live in impoverished neighborhoods, where parents have fewer social supports and are more likely to be stressed, which decreases their ability to parent effectively (Hill et al., 1994; McLoyd, 1990). Poverty appears to influence youth aggressiveness in part through its effect on parenting (Guerra, Huesmann, Tolan, Van Acker, & Eron, 1995).

5.2 The Media

Based on the average number of hours that children in the United States spend watching television and content analyses of the programs they watch, it is estimated that by the time a child leaves elementary school, he or she has seen 8,000 murders and more than 100,000 other acts of violence (Huston et al., 1992). Decades of research confirm the relationship between exposure to television violence and levels of childhood and adolescent aggression and violence (Eron, Gentry, & Schlegel, 1994). Although the magnitude of the effect, and the processes by which viewing violent television contributes to violence are disputed, the relationship between exposure to violent media and aggressive behavior is no longer controversial (Fried, Reppucci, and Woolard, 2000).

Experimental and correlational studies provide evidence of the deleterious influence of media violence on children. A recent meta-analysis (Wood, Wong, & Chachere, 1991) indicates that the effects of TV violence account for about 10% of the variance in child aggression. The results of hundreds of experimental studies demonstrate that increases in aggressive behavior can be produced by exposing children to either extended or brief amounts of television violence in either laboratory or natural settings (e.g., Bandura, Ross, & Ross, 1963; Berkowitz & Rawlings, 1963; Josephson, 1987). Numerous correlational studies confirm that more aggressive children watch more violent television (e.g., Huesmann, Lagerspetz, & Eron, 1984; Milavsky, Kessler, Stipp, & Rubens, 1982). Huesmann and Eron (1984) followed a cohort of 8-year-old children for 22 years, and found that at 30 years of age, those boys who watched more violent television at age 8 were more likely to have been arrested for a violent crime.

5.3 Accessibility of Firearms

The growth in juvenile homicide in the United States between 1987 and 1994 was entirely accounted for by the growth in handgun-related homicides (Zimring, 1998). Likewise, the 17% decline in juvenile homicides in 1995 was accounted for by the decline in homicides committed with a firearm (Sickmund et al., 1997) and decreases in weapons arrests in 1996 and 1997 tend to coincide with decreases in homicides (FBI, 1997; FBI, 1996).

Guns are being carried by large numbers of American youth. As many as 11.4% of high school males possess a gun (Callahan & Rivera, 1992) and 15% of high school students carried a handgun to school in the previous year (O'Donnell, 1995). Youth who dropout of school and youth who admitted having been expelled or suspended from school, having sold drugs, and/or having engaged in assault and battery are more likely to own handguns (Callahan & Rivara, 1992). Self-protection (e.g., Sheley & Wright, 1993) and a desire for respect and power are the main reasons that youths report for carrying weapons (Wilkinson & Fagan, 1996).

Much political debate exists regarding the ability of tougher gun control laws to control access to firearms. One longitudinal study used a multiple time series design to compare changes in crime rates in countries experiencing the introduction of new gun control legislation to changes in crime rates in neighboring countries where no firearms laws were introduced during the same time period. The results indicate that major legal changes may be effective in reducing certain types of violent crimes, e.g., homicide, but minor changes are ineffective in reducing any type of violent crime (Podell & Archer, 1994). It should also be noted that the rates of homicide by juveniles in Canada and many European countries, although lower than the U. S. rate, followed a similar pattern of rising in the late 1980s and early 1990s, but declining in the mid-1990s. However, handgun use did not play a part in the homicide trends in Canada (Hagan & Foster, 2000).

5.4 Societal Drug and Alcohol Abuse

Substance abuse is correlated with youth violence, but the associations are complicated and a clear causal link has not been established. Substance abuse is theorized to have an influence on violent behavior in three ways: (a) the direct effect of alcohol and drugs on an individual's propensity to commit a violent act; (b) the violence that is committed as a result of trying to obtain money to purchase drugs; and (c) the systemic effects related to the violence that arises from the association with drug distribution systems (Goldstein, 1985). The research regarding the direct influence of drugs and alcohol on violent behavior is inconclusive. Although there is a moderately strong correlation between substance use and violence (Elliott, Huizinga, & Ageton, 1985), substance use usually follows the onset of violent behavior

(Osgood, 1995; Tolan & Loeber, 1993). One study that examined the relationship between substance use and violence demonstrated that aggressive behavior at age 12 predicted alcohol use at age 18, but alcohol use at age 12 did not predict aggression at age 18 (White, Brick, & Hansell, 1993). Although a relationship between drug and alcohol use and criminal behavior exists, not all delinquents use alcohol and/or drugs, and many alcohol and drug users never commit serious delinquent or violent acts.

Empirical research on the violence resulting from efforts to obtain money to purchase drugs and systemic violence resulting from the sale and distribution of drugs is lacking. Blumstein (1995) suggests that youth violence has increased because youth have been recruited into the drug industry. The picture is complicated by potentially confounding variables (e.g., prior histories of violence, neighborhood factors) that link violence and drug distribution networks (Reiss & Roth, 1993). Despite the intuitive appeal of a theory that links drug trafficking and youth violence, more evidence is needed before a clear connection can be made.

Certainly, the risk factors for both youth violence and substance abuse are similar and we cannot ignore the high rate of co-occurrence between them. Some researchers believe that both sets of behaviors are caused by the same underlying factors (Snyder & Sickmund, 1995), while others believe that substance abuse is a direct risk factor contributing to youth violence (Greenwood, 1995). However, in the absence of controlled experimental studies, it is difficult to determine the causal pathway between substance abuse and violence, although statistical models offer promise in disentangling the relationship.

In sum, numerous risk factors at all levels have been identified as being related to delinquent, violent and/or aggressive behavior in youth. However, no single factor, in and of itself, even comes close to predicting violent behavior accurately.

6. Developmental Trajectories of Violence and Aggression

Developmental trajectories of violence suggest that behavior develops in a time ordered fashion. In order to study the developmental pathways of violence and aggression it is necessary to measure intra-individual change by measuring the same individual repeatedly over time. The ultimate goal is to be able to identify the serious and violent offenders, who make up a small proportion of the population. Approximately 6%-8% of the male population commits about 60%-85% of the serious violent crimes (see Tolan & Gorman-Smith, 1998) and among frequent offending juveniles, only 29% have ever been arrested for a violent offense (Snyder, 1998). Even among the most aggressive children, only a small proportion will ever become a violent offender. While early aggression predicts later violence these predictions are far from perfect and will yield false-positives (youth who are at risk for violence, but who are not violent) and false-negatives (individuals who were not highly aggressive as children, but who become violent) (Loeber & Hay, 1997).

Patterson, Capaldi, and Bank (1991) and Moffitt (1993) have suggested that age of onset can be used to distinguish early starters or life-course persistent offenders from late starters or adolescent-limited offenders. Moffitt (1993) hypothesizes that the majority of adolescent offenders fall into the adolescent-limited category, and that as these juveniles mature into adult roles, they weigh the costs of delinquency against the rewards of adult opportunities and privileges and desist from delinquent behaviors. Several longitudinal studies indicate that earlier onset and early escalation of the offense seriousness seem to be risk factors for serious and violent offending that all too often do not abate.

Aside from age of onset, the other developmental trajectory that has proven to be most effective at categorizing aggressive and delinquent behavior among youths distinguishes between overt and covert behaviors. Loeber and Hay (1994) describe three developmental pathways of delinquency and problem behaviors: 1) The Overt Pathway, characterized by aggression (e.g., bullying) as the first stage, followed by physical fighting, and finally violence (e.g., assault, rape); 2) The Covert Pathway, characterized by an escalation of clandestine, concealing behaviors (e.g., shoplifting, vandalism, fraud); and 3) The Authority Conflict Pathway, characterized by conflicts and avoidance of authority figures. Of the three pathways only the Overt Pathway leads to violent behaviors (Tolan & Gorman-Smith, 1998; Tolan, Gorman-Smith, Huesmann, & Zelli, 1997). This theory has been found to fit better for life-course persistent offenders than for adolescence-limited offenders (Loeber, Keenan, & Zhang, 1997). In contrast, Capaldi and Patterson (1996) argue that the violent acts committed by adolescents are part of a general involvement in criminal behavior and that distinct pathways for violent versus non-violent offenders do not exist.

Understanding the risk factors and the developmental trajectories that lead to violence, help to determine the most effective strategies and the most appropriate time for prevention and intervention efforts. Intervention planning should be theory driven and informed by the study of risk factors for violence to ensure that at least one, but preferably multiple, risk factors are targeted (Fried, Reppucci, & Woolard, 2000). We now move to the issue of protective factors as perhaps the key to preventive and treatment interventions.

7. Protective Factors

Responses to increases in youth violence have generally come in two forms. The first, all too prevalent response, at least in the United States, has been an effort to protect society from the young "superpredators" by enacting tougher juvenile crime legislation (Reppucci, 1999). In the 1990s, most states toughened their juvenile justice system by enacting laws that make it easier to prosecute juveniles as adults, creating minimum sentencing requirements and blended sentencing provisions (Sickmund et al., 1997). Some proponents of these policies believe that getting tough on crime will have a deterrent effect on potential violent offenders. However, harsher sentences do not appear to be effective in deterring violent criminal behavior (Gottfredson & Hirschi, 1995), and these sentences may actually have the effect of prolonging the criminal careers of otherwise adolescence-limited offenders.

The second response is the prevention of youth violence through interdisciplinary programmatic intervention. Specifically, recent intervention research has focused on the identification of important risk and protective factors, adopted an epidemiological and developmental framework, and targeted interventions at multiple levels (e.g., individual, family, school, neighborhood) (Reppucci, Woolard, & Fried, 1999). The task for researchers and policymakers is not only to understand the impact of various risk factors, but also to identify protective factors that are associated with a decreased probability of violence. Farrington (1998) has stated that "protective factors may have more implications than risk factors for prevention and treatment (p. 451)." In comparison to the study of risk factors, relatively few investigations have explored potential protective factors (Farrington, 1998). The remainder of this paper focuses on identified protective factors on individual and immediate systems levels that may result in a pathway to desistance from serious delinquency and violence. However, we caution that few studies of protective factors, in contrast to risk factors, have focused on violent behavior per se. Also, we are not aware of any specific empirical studies of societal/cultural variables as protective factors.

The search for protective factors against serious delinquency and behavioral problems was initiated by the groundbreaking work of Garmezy and Neuchterlein (1972), who coined the term "invulnerables" to describe a group of African-American children raised in a ghetto in the face of poverty and prejudice who grew up to be highly competent despite the odds against them. Werner and Smith (1982) later studied a group of children they labeled "resilient" who possessed four or more risk factors for delinquency by the age of two, but refrained from developing delinquent and behavioral difficulties in adolescence. These early studies of children at risk resulted in the identification of three general types of protective factors: 1) a positive disposition or likeable temperament, 2) a warm, emotionally supportive family, and 3) a source of external adult support (e.g., a teacher, a neighbor) who rewards the child's competencies and determination (Garmezy, 1993; Werner, 1989).

These advances brought to light the need to clarify and define a protective factor. Farrington (1998) has provided three general definitions: 1) "a protective factor is merely the opposite of a risk factor, e.g., if low intelligence is a risk factor, high intelligence may be a protective factor (p. 451)"; 2) a protective factor may be free-standing, having "no corresponding, symmetrically opposite, risk factor (p.452)", i.e., such a factor while serving a protective benefit, would not constitute a risk if it were absent, (e.g., nervousness or withdrawal in childhood; Hawkins et al., 1998); and 3) a protective factor may interact with (i.e., buffer) risk factors to minimize their negative effects (e.g., having a positive adult figure who cares may buffer having delinquent peers). In addition, protective factors may be more or less influential at different levels of delinquency, e.g., non-delinquency, minor delinquency, serious delinquency (Stouthamer-Loeber, Loeber, Farrington, Zhang, VanKammen, and Maguin, 1993). For example, a good relationship with parents may have a stronger positive effect in the initial stages of delinquency (e.g., promoting non-delinquency as opposed to minor delinquency) and a less strong effect in later stages of delinquency (e.g., suppressing serious delinquency). These definitions have helped researchers successfully identify protective factors in existence on several ecological levels.

8. Individual Level Protective Factors

Biological Factors. Although little research on potential biological protective factors for delinquency exists, a prospective investigation of 101 15-year-old male school children in England by Raine, Venables, & Williams (1996) found psychophysiological differences between the youth who displayed antisocial behaviors in adolescence but desisted from crime by age 29, and those who did not desist and were later convicted as adult criminals. Desistors had better classical conditioning (i.e., learning) and faster fear dissipation. Raine and colleagues use the context of Moffitt's theory of adolescence-limited and life-course persistent criminal behavior to explain these findings. "Good conditioners are well-behaved in the prevailing prosocial environments they experience in early development, but they may for a temporary period become easily conditioned into the antisocial mores that predominate only during adolescence...A change back to a prosocial life norm in early adulthood when the participant leaves these antisocial peer influences (e.g., starting work, setting up a home) may see a return to prosocial behavior in the good conditioner (p. 628)."

Psychosocial Factors. Jessor, VanDenBos, Vanderryn, Costa, and Turbin (1995) investigated the role of psychosocial risk and protective factors in promoting or insulating adolescents against engaging in alcohol and drug abuse, delinquency and sexual precocity. Over the course of four years, 1486 7th, 8th, and 9th grade students filled out questionnaires annually that allowed for evaluation of seven potential protective factors: positive orientations to school and health, positive relationships with adults, negative attitudes toward deviant behavior, perceptions of strong social controls or punishment for transgressions, awareness of friends' modeling of prosocial behaviors, and personal involvement in prosocial activities; and six risk variables: low expectations for success, low self-esteem, feelings of hopelessness, low school achievement, awareness of friends' modeling antisocial behavior, and greater orientation towards friends than parents. Both risk and protective factor indices were calculated and used to predict the problem behaviors as indicated by scores on the Multiple Problem Behavior Index (MPBI). The researchers found that both the risk and protective factors accounted for unique variance in the problem behavior score. Moreover, the risk-protection interaction significantly increased the predictive ability of the equation with regard to the MBPI score. They concluded, "High risk is associated with high involvement in problem behavior when protection is absent or low but not when protection is high. In fact, under the condition of highest protection, the predicted MPBI score for high risk is not much higher than the predicted score for low risk (p.928)." The most powerful protective factors were intolerant attitudes toward deviant behavior, positive orientation to school, perceived regulatory controls, and perceptions of friends' modeling of prosocial behaviors. In addition, the researchers found that protective factors in place early in adolescence were more influential than were risk factors: "the greater the earlier protection, the greater the reduction in MBPI in subsequent years (p.930)."

Born, Chevalier, and Humblet (1997) focused on understanding desistance from delinquency and studied 363 incarcerated, Belgium youths ages 12 - 18 across 7 time periods from childhood to after the youth's 18th birthday. For youths to be considered high risk or not, they had to meet four of the following five criteria: 1) living in an unstable family, 2) persistent financial trouble, 3) low sociocultural background, 4) living in an environment or neighborhood favoring delinquency and deviant behavior, and 5) having delinquent family members. To be deemed resilient, they had to be categorized as high risk, but have shown only low level, small variety, low gravity delinquent acts even after incarceration. Thus, resilient youth were high risk but maintained a relatively low level of delinquency (about 7% of total sample). The most powerful protective factors for these resilient youth were largely individual personal variables rather than demographic variables such as socioeconomic class or family structure. The protective factors included having more personal resources, greater self-control, and more maturity; being less aggressive and more compassionate toward others and feeling greater attachment to others; being less likely to have some sort of psychopathology; and possessing a greater ability to adapt to detention which yielded a better prognosis following incarceration.

One further study deserves mention. Bliesener and Loesel (1992) conducted a cross-sectional study in Northwest Germany. They examined 66 "resilients" and 80 "deviants", both boys and girls, ages 14-17, from 60 residential homes for children and adolescents. All had high, cumulative risk loads in childhood, but resilients had attained adolescence with their mental health relatively intact, while deviants were showing signs of behavioral and emotional disorders. Resilient adolescents were more intelligent, had better coping skills, higher self-esteem, lower feelings of helplessness, and a more flexible temperament than the deviant youth. Finally, other researchers have found other protective individual factors, including low disruptive behavior and positive motivation (Loeber, Stouthamer-Loeber, VanKammen, & Farrington, 1991) and high self-efficacy (Rutter, 1985).

9. Immediate Systems Level Protective Factors

Family. In contrast to studies of family variables as risk factors for serious delinquency and violence in adolescence, only a few investigations have examined potential protective effects of the family environment. However, these studies clearly support the notion that positive family relations tend to mitigate the negative effects of deviant peers (Borduin & Schaeffer, 1998). The Seattle Social Development Project found that positive parent-child communication and high parental involvement during adolescence served to protect against violent behavior in later adolescence (Williams, 1994). Hawkins and Catalano (1992) reported that warm relationships with family members could serve as a protective factor against serious delinquency, and Deater-Deckard and Dodge (1997) found that maternal warmth played a buffering role in the relationship between harsh discipline and aggression. Bliesener and Loesel (1992) reported that their resilient German children had been raised in a less conflictual, more autonomous environment. Finally, Rutter (1985) highlighted positive parental modeling and child-rearing behavior as one of five general protective findings in his 1985 review of studies of resilience. Even more impressively, Poole and Rigoli (1979) found that if conditions of low family support existed, then involvement with delinquent peers was strongly predictive of antisocial behavior, but under conditions of high family support, such involvement with delinquent peers was only slightly predictive. Boys who had highly delinquent friends and nonsupportive family relations reported 500% more criminal activity than did boys with highly delinquent friends and supportive family relations. Similar results were found by Dishion, Patterson, Stoolmiller, and Skinner (1991) in a community sample of 10-year-old boys in that high levels of parental discipline skill and monitoring buffered the negative effects of child involvement with deviant peers.

Peer Group. Several studies have examined the protective effects associated with peer group membership. Bender and Loesel (1997), using the same sample of German youth as Bliesener and Loesel (1992), found that clique membership and satisfaction with social support could fill either protective or risk functions depending on the peer group in that they fostered behavioral continuity. As an aside, it is worth noting that a lack of social embeddedness was a protective factor for their deviant youth sample. Ageton (1983) and Elliott (1994) have also found that associating with peers who are non-approving of delinquent behavior protects against serious delinquency in adolescence.

Hoge, Andrews, and Lescheid (1996) examined the main effects and interactions of four protective variables related to peers and social factors (positive peer relations, good school performance, participation in organized leisure activities, and positive response to authority) and three composite risk factors related to family (family relationship problems, e.g., low family cohesion; inadequate parenting, e.g., inconsistent discipline; and parental personal problems, e.g., mental illness) on two outcome measures–recidivism rate and overall adjustment–in delinquent youths. All four protective factors were associated with lower recidivism and better overall adjustment. Somewhat surprisingly though, no interactive effects were found between the risk and protective factors; when protective factors were effective, they had positive effects at both low and high levels of risk variables.

10. Future Directions for Protective Factors

The question remains: How can we put protective factors to use in preventing serious juvenile delinquency? Given what we know from the limited research on protective factors and this somewhat narrow overview, we might choose to focus on malleable protective factors such as positive orientation to school (Jessor et al., 1995), compassion and attachment to others (Born et al., 1997), and family factors such as family support and positive parental modeling. Interventions having these variables as cornerstones could include programs aimed at engaging youth in school, empathy training and conflict resolution programs, and parenting and family intervention programs.

In order to capitalize most effectively on the benefits of protective factors in the prevention of juvenile delinquency, however, we must do more than simply identify which protective variables matter most. Successful interventions will be those based on research which considers the complex questions of when and for whom the identified protective factors have the most positive impact. Several researchers have begun to find differing effects of protective variables based on age, gender, and race. For example, the moderating effects of some protective factors have been found to disappear as adolescents move into high school (Jessor et al, 1995), while others, such as peer-related protective factors, tended to operate more strongly to buffer risk factors for older juveniles (ages 15-17) than for younger juveniles (Hoge et al., 1996). Similarly, Hawkins et al. (1998) described age differences in effectiveness of the protective factor of commitment to school in that commitment at mid-adolescence was a greater protection against violence at age 18 than commitment to school at age 10. Moreover, Stouthamer-Loeber et al (1993) demonstrated that the magnitude of both risk and protective effects increased with age.

With regard to gender and race, Jessor and his colleagues (1995) found no moderating effects of individual-level protective factors on risk factors for either boys or African Americans. In contrast, Williams (1994) found that commitment or attachment to school resulted in a greater reduction in violence for African American students and boys than for European American students and girls. Bender and Loesel (1997) found that girls were more influenced (both positively and negatively) by social resources than were boys, e.g., the presence of a boyfriend or a small social group served as a protective factor against serious delinquency for the most aggressive and antisocial girls. It is also important to note that girls and boys appear to differ in their development of aggression, and that girls' self-reported violence during adolescence peaks about two years earlier (around age 14) than does boys' (Loeber & Stouthamer-Loeber, 1998). Although some girls may follow the same trajectory from childhood into serious and violent behavior in adolescence as do boys, adolescent-limited girls may interact differently with their peers than do boys, thereby resulting in somewhat different pathways to delinquency (Caspi, Lynam, Moffitt, & Silva, 1993).

11. Conclusion

What we have summarized is only a smattering of what is in the literature. This selective review has largely emphasized studies that have been focused on the risk and protective factors for adolescent delinquency and violence, and not for problem behaviors of adolescence in general. Its goal has been to suggest that as we strive to develop a systematic measuring instrument, we must keep the concepts of both risk and protective factors clearly in mind. Although we are not yet at a point of understanding where we can formulate a calculus of risk and protective factors that could be used to develop interventions with definitive positive results, this is a direction that holds promise. As we have already indicated, although the identification of risk and protective factors and pathways to serious delinquency is well underway, it is imperative that we continue to struggle with the developmental implications for the timing and process of intervention (Coie, et al., 1993) and to define populations that could benefit from enhanced protection. In other words, risk factors need to be mapped to the appropriate developmental periods because youth can move in and out of risk depending upon development and changing environmental and family circumstances. However, many risks faced by juveniles are not readily amenable to change. This suggests that the identification of malleable protective factors and pathways to desistance could be more fruitful for designing successful preventive interventions for serious delinquency and violence. Moreover, examination of factors related to desistance from serious delinquency can also play an integral role in further distinguishing life-course-persistent offenders from adolescent-limited offenders. This distinction is necessary in order to design and implement separate prevention programs to cater to these two types of offenders, and to keep from ensnaring adolescent-limited offenders into prolonged criminal careers (Howell & Hawkins, 1998). Hopefully, the instrument that is developed will provide needed information that will contribute to the solution of these problems.

Notes

[*] Some portions of this paper are based on C.S. Fried & N. D. Reppucci (in press), Youth violence and the law. In B. Bottoms, M. Kovera, & B. McAuliff (Eds.), *Children and the law: Research and social policy,* Oxford University Press.

References

Ageton, S.S. (1983). *Sexual assault among adolescents.* Lexington, MA: Lexington Books.

American Psychological Association Commission on Violence. (1993). *Violence and youth: Psychology's response* (Vol. 1). Washington, DC: American Psychological Association.

Arbuthnot, J., Gordon, D.A., & Jurkovic, G.J. (1987). Personality. In H.C. Quay (Ed.), *Handbook of juvenile delinquency* (pp. 139-183). New York: Wiley.

Archer, J. (1994). Testosterone and aggression: A theoretical review. *Journal of Offender Rehabilitation, 21,* 3-39.

Baker, R.L.A., & Mednick, B.R. (1984). *Influences on human development: A longitudinal perspective.* Boston: Kluwer-Nijhoff.

Bandura, A., Ross, D., & Ross, S.H. (1963). Imitation of film-mediated aggressive models. *Journal of Abnormal and Social Psychology, 66,* 3-11.

Bates, J.E., Bayles, K., Bennett, D.S., Ridge, B., & Brown, M.M. (1991). Origins of externalizing behavior problems at eight years of age. In D.J. Pepler & K.H. Rubin (Eds.), *The development and treatment of childhood aggression* (pp. 93-120). Hillsdale, NJ: Erlbaum.

Bates, J.E., Maslin, C.A., & Frankel, K.A. (1985). Attachment security, mother-child interaction, and temperament as predictors of behavior problem ratings at age three years. *Monographs of the Society for Research in Child Development, 50,* 167-193.

Bender, D. & Loesel, F. (1997). Protective and risk effects of peer relations and social support on antisocial behaviour in adolescents from multi-problem mileus. *Journal of Adolescence, 20,* 661-78.

Berkowitz, L., & Rawlings, E. (1963). Effects of film violence on inhibitions against subsequent aggression. *Journal of Abnormal and Social Psychology, 66,* 405-412.

Bliesener, T., & Loesel, F. (1992). Resilience in juveniles with high risk of delinquency. In F. Loesel, D. Bender, et al. (Eds.), *Psychology and law: International perspectives* (pp. 62-75). Berlin, Germany: Walter De Gruyter.

Blumstein, A. (August, 1995). Violence by young people: Why the deadly nexus? *National Institute of Justice Journal,* 2-9.

Booth, C.L., Rose-Krasnor, L, McKinnon, J., & Rubin, K.H. (1994). Predicting social adjustment in middle childhood: The role of preschool attachment security and maternal style. From family to peer group: Relations between relationship systems [Special issue]. *Social Development, 3,* 189-204.

Borduin, C.M., & Schaeffer, C.M. (1998). Violent offending in adolescence: Epidemiology, correlates, outcomes, and treatment. In T.P. Gullotta, G.R. Adams, & R. Montemayor, *Delinquent violent youth: Theory and interventions* (pp. 144-174). Thousand Oaks: Sage.

Born, M., Chevalier, V., & Humblet, I. (1997). Resilience, desistance and delinquent career of adolescent offenders. *Journal of Adolescence, 20,* 679-694.

Cairns, R.B., Cairns, B.D., Neckerman, H.J., Gest, S.D., & Gariepy, J.L. (1988). Social networks and aggressive behavior: Peer support or peer rejection? *Developmental Psychology, 24,* 815-823.

Callahan, C.M., & Rivera, F.P. (1992). Urban high school youth and handguns: A school-based survey. *Journal of the American Medical Association, 267,* 3038-3042.

Capaldi, D.M., & Patterson, G.R. (1996). Can violent offenders be distinguished from frequent offenders?: Prediction from childhood to adolescence. *Journal of Research in Crime and Delinquency, 33,* 206-231.

Caspi, A., Lynam, D., Moffitt, T.E., & Silva, P.A. (1993). Unraveling girls' delinquency: Biological, dispositional, and contextual contributions to adolescent misbehavior. *Developmental Psychology, 29,* 19-30.

Centerwall, B.S. (1984). Race, socioeconomic status, and domestic homicide, Atlanta, 1971-2. *American Journal of Public Health, 74,* 813-815.

Cleare, A.J., & Bond, A.J. (1995). The effect of tryptophan depletion and enhancement on subjective and behavioral aggression in normal male subjects. *Psychopharmacolgy, 118,* 72-81.

Cohen, D., & Strayer, J. (1996). Empathy in conduct-disordered and comparison youth. *Developmental Psychology, 32,* 988-998.

Coie, J.D., & Dodge, K.A. (1998). Aggression and antisocial behavior. In W. Damon & N. Eisenberg (Eds.), *Handbook of child psychology: Vol. 3. Social, emotional and personality development* (pp. 779-862). New York: John Wiley & Sons.

Coie, J.D., Dodge, K.A., & Kupersmidt, J. (1990). Peer group behavior and social status. In S.R. Asher & J.D. Coie (Eds.), *Peer rejection in childhood* (pp. 17-59). New York: Cambridge University Press.

Coie, J.D., Watt, N.F., West, S.G., Hawkins, J.D., Asarnow, J.R., Markman, H.J., Ramey, S.L., Shure, M.B., & Long, B. (1993). The science of prevention: A conceptual framework and some directions for a national research program. *American Psychologist, 48,* 1013-1022.

Cornell, D. G., & Wilson, L. A. (1992). The PIQ>VIQ discrepancy in violent and non-violent delinquents. Journal of Clinical Psychology, 48, 256-261.

Crick, N.R., & Dodge, K.A. (1994). A review and reformulation of social information processing mechanisms in children's social adjustment. *Psychological Bulletin, 115*, 74-101.

Deater-Deckard, K., & Dodge, K.A. (1997). Externalizing behavior problems and discipline revisited: Nonlinear effects and variation by culture, context, and gender. *Psychological Inquiry, 8,* 161-175.

Dishion, T.J., Paterson, G.R., Stoolmiller, M., & Skinner, M.L. (1991). Family, school, and behavioral antecedents to early adolescent involvement with antisocial peers. *Developmental Psychology, 27,* 172-180.

Dodge, K.A., Bates, J.E., & Pettit, G.S. (1990). Mechanisms in the cycle of violence. *Science, 250,* 1678-1683.

DuRant, R.H., Cadenhead, C., Pendergrast, R.A., Slavens, G., & Linder, C.W. (1994). Factors associated with the use of violence among urban black adolescents. *American Journal of Public Health, 84,* 612-617.

Elliott, D.S. (1994). Serious violent offenders: Onset, developmental course, and termination—The American Society of Criminology 1993 presidential address. *Criminology, 32,* 1-21.

Elliott, D.S., Huizinga, D., Ageton, S.S. (1985). *Explaining delinquency and drug use.* Thousand Oaks, CA: Sage.

Erickson, M.F., Sroufe, L.A., & Egeland, B. (1985). The relationship between quality of attachment and behavior problems in preschool in a high-risk sample. *Monographs of the Society for Research in Child Development, 50,* 147-186.

Eron, L.D., Gentry, J.H., & Schlegel, P. (1994). *Reason to hope: A psychosocial perspective on violence and youth.* Washington, DC: American Psychological Association.

Eron, L.D., Huesmann, L.R., & Zelli, A. (1991). The role of parental variables in the learning of aggression. In D.J. Pepler & K.H. Rubin (Eds.), *The development and treatment of childhood aggression* (pp. 169-189). Hillsdale, NJ: Erlbaum.

Farrington, D.P. (1978). The family background of aggressive youths. In L.A. Hersov, M. Berger, & D. Schaffer (Eds.), *Aggression and antisocial behavior in childhood and adolescence* (p.73-93). Oxford: Pergamon.

Farrington, D.P. (1989). Early predictors of adolescent aggression and adult violence. *Violence and Victims, 4,* 79-100.

Farrington, D.P. (1991). Childhood aggression and adult violence: Early precursors and later-life outcomes. In D.J. Pepler & K.H. Rubin (Eds.), *The development and treatment of childhood aggression* (pp. 5-29). Hillsdale, NJ: Lawrence Erlbaum.

Farrington, D.P. (1994). Childhood, adolescent, and adult features of violent males. In L.R. Huesmann (Ed.), *Aggressive behavior: Current perspectives* (pp. 215-240). New York: Plenum.

Farrington, D.P. (1997). The relationship between low resting heart rate and violence. In A. Raine, P.A., Brennan, D.P. Farrington, & S.A. Mednick (Eds.), *Biosocial bases of violence* (pp. 89-105). New York: Plenum.

Farrington, D.P. (1998). Predictors, causes, and correlates of male youth violence. In M. Tonry & M.H. Moore, (Eds.), *Crime and justice: A review of research: Vol. 24. Youth violence* (pp. 421-476). Chicago: University of Chicago Press.

Federal Bureau of Investigation (FBI). (1996). *Uniform Crime Reports.* Washington, DC: U.S. Department of Justice.

Federal Bureau of Investigation (FBI). (1997). *Uniform Crime Reports.* Washington, DC: U.S. Department of Justice.

Feshbach, N. D. (1997). Empathy: The formulative years. In A. C. Bohart & L. S. Greenberg (Eds.), *Empathy reconsidered: New directions in psychotherapy* (pp. 33-59). Washington, DC: American Psychological Association.

Fried, C.S., Reppucci, N.D., & Woolard, J.L. (2000). Youth violence. In J. Rappaport, & E. Seidman (Eds.), *Handbook of community psychology.* New York: Plenum.

Gagnon, C., Craig, W.M., Tremblay, R.E., Zhou, R.M., & Vitaro, F. (1995). Kindergarten predictors of boys' stable behavior problems at the end of elementary school. *Journal of Abnormal Child Psychology, 23,* 751-766.

Garmezy, N. & Neuchterlein, K. (1972). Invulnerable children: The fact and fiction of competence and disadvantage. *American Journal of Orthopsychiatry, 42,* 328-329.

Garmezy, N. (1993). Children in poverty: Resilience despite risk. *Psychiatry, 56,* 127-136.

Goldstein, P.J. (1985). The drugs/violence nexus: A tripartite conceptual framework. *Journal of Drug Issues, 15,*

493-506.

Gorman-Smith, D., Tolan, P.H., Zelli, A., & Huesmann, L.R. (1996). The relation of family functioning to violence among inner-city minority youth. *Journal of Family Psychology, 10,* 115-129.

Gottesman, I.I., & Goldsmith, H.H. (1994). Developmental psychopathology of antisocial behavior: Inserting genes into its ontogenesis and epigenesis. In C.A. Nelson (Ed.), *Threats to optimal development: Integrating biological, psychological, and social risk factors* (pp. 69-104). Hillsdale, NJ: Erlbaum.

Gottfredson, M.R., & Hirschi, T. (1995). National crime control policies. *Society, 32,* 30-36.

Greenwood, P. (1995). Juvenile crime and juvenile justice. In J. Wilson & J. Petersilia (Eds.), *Crime* (pp.15-38). San Francisco: ICS.

Guerra, N.G., Huesmann, L.R., Tolan, P.H., Van Acker, R., & Eron, L.D. (1995). Stressful events and individual beliefs as correlates of economic disadvantage and aggression among urban children. *Journal of Consulting and Clinical Psychology, 63,* 518-528.

Hagan, J., & Foster, H. (2000). Making corporate and criminal America less violent: Public norms and structural reforms. *Contemporary Sociology, 29,* 44-53.

Hawkins, J.D., & Catalano, R.F. (1992). *Communities that care: Action for drug abuse prevention.* San Francisco, CA: Jossey-Bass.

Hawkins, J.D., Herrenkohl, T., Farrington, D.P., Brewer, D., Catalano, R.F., & Garachi, T.W. (1998). A review of predictors of youth violence. In R. Loeber & D.P. Farrington (Eds.), *Serious and violent juvenile offenders: Risk factors and successful interventions* (pp. 106-146). Thousand Oaks, CA: Sage.

Hechtman, L. & Weiss, G. (1986). Controlled prospective fifteen year follow-up of hyperactives as adults: Non-medical drug and alcohol use and antisocial behaviour. *Canadian Journal of Psychiatry, 31,* 557-567.

Hellman, D.A., & Beaton, S. (1986). The pattern of violence in urban public schools: The influence of school and community. *Journal of Research in Crime and Delinquency, 23,* 102-127.

Hill, H.M, Soriano, F.I, Chen, S.A., LaFromboise, T.D. (1994). Sociocultural factors in the etiology and prevention of violence among ethnic minority youth. In L.D. Eron, J.H. Gentry, & P. Schlegel (Eds.), *Reason to hope: A psychosocial perspective on violence and youth* (pp. 59-97). Washington, DC: American Psychological Association.

Hoge, R.D., Andrews, D.A., & Lescheid, A.W. (1996). An investigation of risk and protective factors in a sample of youthful offenders. *Journal of Child Psychology & Psychiatry & Allied Disciplines, 37,* 419-424.

Howell, J.C. & Hawkins, J.D. (1998). Prevention of youth violence. In M. Tonry & M.H. Moore, (Eds.), *Crime and justice: A review of research: Vol 24. Youth Violence* (pp.263-315). Chicago: University of Chicago Press.

Huesmann, L.R., & Eron, L.D. (1984). Cognitive processes and the persistence of aggressive behavior. *Aggressive Behavior, 10,* 243-251.

Huesmann, L.R., Lagerspetz, K., & Eron, L.D. (1984). Intervening variables in the TV violence-aggression relation: Evidence from two countries. *Developmental Psychology, 20,* 746-775.

Huston, A.C., Donnerstein, E., Farichild, H., Feshbach, N.D., Katz, P.A., Murray, J.P., Rubinstein, E.A., Wilcox, B.L., & Zuckerman, D. (1992). *Big world, small screen: The role of television in American society.* Lincoln: University of Nebraska Press.

Jenkins, E.J., & Bell, C.C. (1994). Violence exposure, psychological distress, and high risk behaviors among inner-city high school students. In S. Friedman (Ed.), *Anxiety disorders in African-Americans* (pp. 76-88). New York: Springer.

Jessor, R., VanDenBos, J., Vanderryn, J., Costa, F.M., & Turbin, M.S. (1995). Protective factors in adolescent problem behavior: Moderator effects and developmental change. *Developmental Psychology, 31,* 923-933.

Josephson, W.L. (1987). Television violence and children's aggression: testing the priming, social script, and disinhibition predictions. *Journal of Personality and Social Psychology, 53,* 882-890.

Kingston, L., & Prior M. (1995). The development of patterns of stable, transient, and school-age aggressive behavior in young children. *Journal of the American Academy of Child and Adolescent Psychiatry, 34,* 348-358.

Kupersmidt, J.B., Briesler, P.C., DeRosier, M.E., Patterson, C.J., & Davis, P.W. (1995). Childhood aggression and peer relations in the context of family and neighborhood factors. *Child Development, 66,* 360-375.

Lipsey, M.W., & Derzon, J.H. (1998). Predictors of violent or serious delinquency in adolescence and early adulthood: A synthesis of longitudinal research. In R. Loeber & D.P. Farrington (Eds.), *Serious and violent juvenile offenders: Risk factors and successful interventions* (pp. 86-105). Thousand Oaks, CA: Sage.

Loeber, R. & Stouthamer-Loeber, M. (1998). Development of juvenile aggression and violence. Some common misconceptions and controversies. *American Psychologist, 53,* 242-259.

Loeber, R. (1988). Behavioral precursors and accelerators of delinquency. In W. Buikhuisen & S.A. Mednick (Eds.), *Explaining delinquency* (pp.51-67). Leiden, Holland: Brill.

Loeber, R., & Farrington, D.P. (1998). *Serious and violent juvenile offenders: Risk factors and successful interventions*. Thousand Oaks, CA: Sage.

Loeber, R., & Hay, D. (1994). Developmental approaches to aggression and conduct problems. In M. Rutter & D.F. Hay (Eds.), *Development through life: A handbook for clinicians* (pp. 488-515). Oxford: Blackwell Scientific.

Loeber, R., & Hay, D. (1997). Key issues in the development of aggression and violence from childhood to early adulthood. *Annual Review of Psychology, 48*, 371-410.

Loeber, R., & Wikstrom, P. (1993). Individual pathways to crime in different types of neighborhoods. In D.P. Farrington, R.J. Sampson, & P. Wikstrom (Eds.), *Integrating individual and ecological aspects of crime* (pp. 169-204). Stockholm: Liber-Verlag.

Loeber, R., Keenan, K., & Zhang, Q. (1997). Boys' experimentation and persistence in developmental pathways toward serious delinquency. *Journal of Child and Family Studies, 6,* 321-357.

Loeber, R., Stouthamer-Loeber, M., VanKammen, W.B., & Farrington, D.P. (1991). Initiation, escalation and desistance in juvenile offending and their correlates. *Journal of Criminal Law and Criminology, 82,* 36-82.

Lynam, D.R., Moffitt, T.E., & Stouthamer-Loeber, M. (1993). Explaining the relation between IQ and delinquency: Class, race, test motivation, school failure, or self-control? *Journal of Abnormal Psychology, 102,* 187-196.

Maguin, E., Hawkins, J.D., Catalano, R.F., Hill, K., Abbott, R., & Herrenkohl, T. (1995, November). *Risk factors measured at three ages for violence at age 17-18*. Paper presented at the American Society of Criminology, Boston.

Manly, J.T., Cicchetti, D., & Barnett, D. (1994). The impact of subtype, frequency, chronicity, and severity of child maltreatment on social competence and behavior problems. *Developmental Psychology, 7,* 121-143.

McCord, J. (1979). Some child-rearing antecedents of criminal behavior in adult men. *Journal of Personality and Social Psychology, 37,* 1477-1486.

McGuire, J. (1997). Psychosocial approaches to the understanding and reduction of violence in young people. In V. Varma (Ed.), *Violence in children and adolescents* (pp.65-83). Bristol, PA: Jessica Kingsley.

McLoyd, V.C. (1990). The impact of economic hardship on black families and their children: Psychological distress, parenting and socioemotional development. *Child Development, 61,* 311-346.

Megargee, E.I. (1982). Psychological determinants and correlates of criminal violence. In M.E. Wolfgang & N.A. Weiner (Eds.), *Criminal violence*. Beverly Hills, CA: Sage.

Miczek, K.A., Mirsky, A.F., Carey, R., DeBold, J., & Raine A. (1994). An overview of biological influences on violent behavior. In A.J. Reiss, K.A. Miczek, & J.A. Roth (Eds.), *Understanding and preventing violence: Vol. 3. Social influences* (pp. 377-570). Washington, DC: National Academy Press.

Milavsky, J.R., Kessler, R., Stipp, H., & Rubens, W.S. (1982). Television and aggression: Results of a panel study. In D. Pearl, L. Bouthilet, & J. Lazar (Eds.), *Television and behavior: Ten years of scientific progress and implications for the 80's* (Vol 2). Washington, DC: Government Printing Office.

Miller, P. A., & Eisenberg, N. (1988). The relation of empathy to aggressive and externalizing/antisocial behavior. *Psychological Bulletin, 103,* 324-344.

Moffitt, T. E., & Lynam, D. R., Silva, P.A. (1994). Neuropsycholgical tests predict persistent male delinquency. *Criminology, 32,* 101-124.

Moffitt, T.E. (1993). Adolescence-limited and life-course persistent antisocial behavior: A developmental taxonomy. *Psychological Review, 100,* 674-701.

National Center for Health Statistics. (1993). *Health, United States, 1992*. Hyattsville, MD: Public Health Service.

National Research Council. (1993). *Losing generations: Adolescents in high-risk settings*. Washington, DC: National Academy Press.

O'Donnell, C.R. (1995). Firearm deaths among children and youth. *American Psychologist, 50,* 771-776.

Olweus, D. (1979). Stability of aggressive reaction patterns in males: A review. *Psychological Bulletin, 86,* 852-875.

Osgood, D.W. (1995). *Drugs, alcohol, and adolescent violence*. Boulder, CO: The Center for the Study and Prevention of Violence.

Patterson, G.R., & Bank, C.L. (1989). Some amplifying mechanisms for pathological processes in families. In M. Gunnar & E. Thelen (Eds.), *Systems and development: Symposia on child psychology* (pp. 167-210). Hillsdale, NJ: Erlbaum.

Patterson, G.R., Capaldi, D., & Bank, L. (1991). An early starter model for predicting delinquency. In D.J. Pepler & K.H. Rubin (Eds.), *The development and treatment of childhood aggression* (pp. 139-168). Hillsdale, NJ: Erlbaum.

Pfeiffer, C. (1998). Juvenile crime and juvenile violence in European countries. In M. Tonry (Ed.) *Crime and justice: A review of research* (Vol. 23, pp. 255-328). Chicago: University of Chicago Press.

Podell, S., & Archer, D. (1994). Do legal changes matter? The case of gun control laws. In M. Costanzo & S. Oskamp (Eds.), *Violence and the law* (pp. 37-60). Thousand Oaks, CA: Sage.

Poole, E.D., & Rigoli, R.M. (1979). Parental support, delinquent friends, and delinquency: A test of interaction effects. *Journal of Criminal Law and Criminology, 70,* 188-193.

Raine, A. (1998). Antisocial behavior and psychophysiology: A biosocial perspective and a prefrontal dysfunction hypothesis. In D.M. Stoff, J. Breiling, & J.D. Maser (Eds.), *Handbook of antisocial behavior.* New York: Wiley.

Raine, A., Venables, P.H., & Williams, M. (1990). Relationships between central and autonomic measures of arousal at age 15 years and criminality at age 24 years. *Archives of General Psychiatry, 47,* 1003-1007.

Raine, A., Venables, P.H., & Williams, M. (1996). Better autonomic conditioning and faster electrodermal half-recovery time at age 15 years as possible protective factors against crime at age 29 years. *Developmental Psychology, 32,* 624-630.

Reiss, A.J., & Roth, J.A. (1993). *Understanding and preventing violence: Panel on the understanding and control of violent behavior* (Vol. 1). Washington, DC: national Academy Press.

Reppucci, N.D. (1999). Adolescent development and juvenile justice. *American Journal of Community Psychology, 27,* 307-326.

Reppucci, N.D., Woolard, J.L., & Fried, C.S. (1999). Social, community and preventive interventions. *Annual Review of Psychology, 50,* 387-418.

Rutter, M. (1985). Resilience in the face of adversity: Protective factors and resistance to psychiatric disorder. *British Journal of Psychiatry, 147,* 598-611.

Saner, H., & Ellickson, P. (1996). Concurrent risk factors for adolescent violence. *Journal of Adolescent Health, 19,* 94-103.

Seguin, J.R., Pihl, R.O., Harden, P.W., Tremblay, R.E., & Boulerice, B. (1995). Cognitive and neuropsychological characteristics of physically aggressive boys. *Journal of Abnormal Psychology, 104,* 614-624.

Sheley, J.F., & Wright, J.D. (1993). Motivations for gun possession and carrying among serious juvenile offenders. *Behavioral Science Law, 11,* 375-388.

Sickmund, M., Snyder, H.N., & Poe-Yamagata, E. (1997). *Juvenile offenders and victims: 1997 update on violence.* Washington, DC: Office of Juvenile and Delinquency Prevention.

Snyder, H.N. (1998). Appendix: Serious, violent, and chronic juvenile offenders – An assessment of the extent of and trends in officially recognized serious criminal behavior in a delinquent population. In R. Loeber & D.P. Farrington (Eds.), *Serious and violent juvenile offenders: Risk factors and successful interventions* (pp. 428-444). Thousand Oaks, CA: Sage.

Snyder, H.N., & Sickmund, M. (1995). *Juvenile offenders and victims: 1995 update on violence.* Pittsburgh: U.S. Department of Justice, Office of Juvenile Justice and Delinquency Prevention, National Center for Juvenile Justice.

Snyder, J.N., & Patterson, G.R. (1987). Family interaction and delinquent behavior. In H.C. Quay (Ed.), *Handbook of juvenile delinquency* (pp. 216-243). New York: John Wiley.

Stouthamer-Loeber, M., Loeber, R., Farrington, D.P., Zhang, Q., VanKammen, W., & Maguin, E. (1993). The double edge of protective and risk factors for delinquency: Interrelations and developmental patterns. *Development and Psychopathology, 5,* 683-701.

Tolan, P.H., & Gorman-Smith, D. (1998). Development of serious and violent offending careers. In R. Loeber & D.P. Farrington (Eds.), *Serious and violent juvenile offenders: Risk factors and successful interventions* (pp. 68-85). Thousand Oaks, CA: Sage.

Tolan, P.H., & Loeber, R. (1993). Antisocial behavior. In P.H. Tolan & B.J. Cohler (Eds.), *Handbook of clinical research and practice* (pp. 307-331). New York: Wiley.

Tolan, P.H., & Thomas, P. (1995). The implications of age of onset for delinquency risk: II. Longitudinal data. *Journal of Abnormal Child Psychology, 23,* 157-181.

Tolan, P.H., Gorman-Smith, D., Huesmann, L.R., & Zelli, A. (1997). Assessment of family relationship characteristics: A measure to explain risk for antisocial behavior and depression in youth. *Psychological Assessment, 9,* 212-223.

Tremblay, R.E., Masse, L.C., Kurtz, L., & Vitaro, F. (1997). From childhood physical aggression to adolescent maladjustment: The Montreal Prevention Experiment. In R.D. Peters & R.J. McMahon (Eds.), *Childhood disorders, substance abuse, & delinquency: Prevention and early intervention approaches* (pp. 1-62). Thousand Oaks, CA: Sage.

Vuchinich, S., Bank, L., & Patterson, G.R. (1992). Parenting, peers, and the stability of antisocial behavior in preadolescent boys. *Developmental Psychology, 28,* 510-521.

Weiss, B., Dodge, K.A., Bates, J.E., & Pettit, G.S. (1992). Some consequences of early harsh discipline: Child aggression and a maladaptive social information processing style. *Child Development, 63,* 1321-1335.

Werner, E.E. (1989). High-risk children in young adulthood: A longitudinal study from birth to 32 years. *American Journal of Orthopsychiatry, 59,* 72-81.

Werner, E.E., & Smith, R.S. (1982). *Vulnerable but invincible: A study of resilient children.* New York: McGraw Hill.

White, H.R., Brick, J., & Hansell, S. (1993). A longitudinal investigation of alcohol use and aggression in adolescence. *Journal of Studies on Alcohol, Suppl 11*, 62-77.

Widom, C.S. (1989). Does violence beget violence? A critical review of the literature. *Psychological Bulletin, 106*, 3-28.

Wilkinson, D., & Fagan, J. (1996). The role of firearms in violence "scripts": The dynamics of gun events among adolescent males. *Law and Contemporary Problems, 59*, 55-89.

Williams, J.H. (1994). Understanding substance abuse, delinquency involvement, and juvenile justice system involvement among African-American and European-American adolescents. Unpublished dissertation, University of Washington, Seattle.

Williams, K.R. (1984). Economic sources of homicide: Reestimating the effects of poverty and inequality. *American Sociological Review, 49*, 283-289.

Wilson, W.J. (1987). *The truly disadvantaged.* Chicago: University of Chicago Press.

Wood, W., Wong, F., & Chachere, J. (1991). Effects of media violence on viewer's aggression in unconstrained social interaction. *Psychological Bulletin, 109*, 371-383.

Zimring, F.E. (1998). *American youth violence.* New York: Oxford University Press.

Multi-Problem Violent Youth
R.R. Corrado et al. (Eds.)
IOS Press, 2002

Multiple Risk Factors for Multiple Problem Violent Boys

David P. Farrington

The main aim of this chapter is to investigate childhood risk factors for multiple problem violent boys in the Cambridge Study in Delinquent Development (described in more detail below), which is a prospective longitudinal survey of 411 South London males from age 8 to age 46.

1. Risk Factors for Multiple Problem Boys

Loeber, Farrington, Stouthamer-Loeber, and van Kammen (1998b) carried out the most extensive previous investigation of risk factors for multiple problem boys in the Pittsburgh Youth Study, which is a prospective longitudinal survey of three samples (total $N = 1517$) of inner city males originally aged 7, 10 and 13. They investigated 8 types of child problems: delinquency (according to boys, mothers and teachers), substance use and depressed mood (self-reported), conduct problems and attention deficit (based on the Diagnostic Interview Schedule for Children), physical aggression, covert behavior and shy/withdrawn behavior (based on mother and teacher ratings). They investigated a very wide range of risk factors: individual (e.g., low guilt, low school achievement), family (e.g., poor parental supervision, mother's physical punishment) and macro (e.g., low socio-economic status, bad neighborhood). However, their analysis was correlational, not predictive: risk factors and problem behaviors were measured at the same time.

All eight types of child problem behavior were significantly intercorrelated, and all had substantial positive weightings on the first factor in a principal component analysis. This suggested that all eight reflected the same underlying construct of problem behavior or general deviance (Jessor, Donovan & Costa, 1991). On the other hand, the second factor contrasted the externalizing problems of delinquency and physical aggression (high negative weightings) with the internalizing problems of depressed mood and shy/withdrawn behavior (high positive weightings), which had the lowest weightings on the first factor. Thus, this analysis also supports the common distinction between externalizing and internalizing problem behaviors.

The present chapter is concerned with externalizing or antisocial behavior, and Loeber et al. (1998b) clearly showed that delinquency, physical aggression, conduct problems, attention deficit and covert (untrustworthy or manipulating) behavior were strongly intercorrelated. They found that risk factors tended to be similar for all these outcomes. For example, of the significant risk factors for physical aggression, 84% were also significant risk factors for delinquency in the youngest sample, 82% were significant for delinquency in the middle sample, and 68% were significant for delinquency in the oldest sample.

A number of risk factors predicted most of these child problems in all three samples: low guilt, low school achievement, poor parental supervision, mother's physical punishment, parental stress, parental anxiety or depression, a broken family, low socio-economic status and the family on welfare. Loeber et al. (1998b) identified the multiple problem boys (those showing 4 or more of the 8 types of problem behavior) and found that these risk factors were among the most important predictors. Interestingly, when the multiple problem boys were eliminated from the analyses, the relationships between risk factors and outcomes were generally attenuated but not eliminated, suggesting that all the significant relationships were not being generated entirely by the multiple problem boys. Loeber et al. (1998b) also developed risk scores for multiple problem boys based on the most predictive risk factors.

The focus in the present chapter is on multiple problem violent boys: that is, boys who are *both* violent *and* show multiple problem (non-violent) behavior. In the Cambridge Study, a number of previous analyses have investigated risk factors for violent behavior and risk factors for multiple problem (antisocial) behavior, but no previous analysis has focussed on multiple problem violent boys.

2. Risk Factors for Violence

Prospective longitudinal surveys are needed to investigate the childhood predictors of youth violence. The best of these surveys follow up large community samples of several hundreds, include interview as well as record data, and span a follow-up period of at least five years. Farrington (1998) and Hawkins et al. (1998) have provided extensive literature reviews of longitudinal surveys of youth violence and of results obtained in them. These surveys have been carried out in many different countries, including the United Kingdom (e.g., Wadsworth, 1979), the United States (e.g., Loeber, Farrington, Stouthamer-Loeber & van Kammen, 1998a), Canada (e.g., LeBlanc & Frenchette, 1989), Sweden (e.g., Stattin & Magnusson, 1989), Finland (e.g., Pulkkinen & Pitkanen, 1993), Denmark (e.g., Hogh & Wolf, 1983) and New Zealand (e.g., Henry, Caspi, Moffitt & Silva, 1996). The most important risk factors are similar in different countries (e.g., Farrington & Loeber, 1999). Results obtained in the Cambridge Study will now be summarized (see also Farrington, 2000a).

The first analysis of predictors of self-reported violence (at age 10-14) was published by Farrington and West (1971). Significant predictors at age 8-10 included childhood misconduct, high daring, low non-verbal IQ, large family size and low family income. Farrington and West concluded that self-reported violent boys were similar in many respects to early delinquents (those convicted at age 10-14) but that the delinquents were more deprived in regard to low income and poor housing.

Farrington (1978) carried out a more extensive study of predictors of aggression at age 8-10 (difficult to discipline, rated by teachers), 12-14 (aggression rated by teachers), 16-18 (self-reported violence) and convictions for violence between ages 10 and 20. The most important independent predictors of convictions for violence were harsh or erratic parental discipline, a convicted parent, poor parental supervision, a broken family, high daring, and low nonverbal IQ. There was considerable continuity from aggression at age 8-10 to violence at age 16-18 (see also Farrington, 1982).

Farrington (1989a) investigated the predictors of aggression at age 12-14 (rated by teachers), self-reported violence at age 16-18, self-reported fights at age 32, and convictions for violence between ages 10 and 32. Regression analyses showed that the best predictors of convictions included high daring, authoritarian parents, a convicted parent, low verbal IQ, and harsh parental discipline. The best predictors of self-reported fights at age 32 included the father rarely joining in the boy's leisure activities at age 12, high self-reported delinquency at age 18, and high daring at age 8-10. Farrington (1991b) showed that the predictors of convictions for violence up to age 32 and of frequent non-violent offenders were very similar, a result later replicated by Capaldi and Patterson (1996). Generally, violent offenders were frequent offenders.

Farrington (1997) investigated the ability of various combined scales to predict convictions for violence between ages 10 and 20 and self-reported violence at age 15-18, and Farrington (1998) reviewed a variety of risk factors for these two outcomes. The most recent analysis of violence (Farrington, 2001) compared predictors of adult convictions for violence (between ages 21 and 40) with predictors of self-reported violence between ages 27 and 32. Interestingly, adolescent convictions for non-violent crimes predicted adult violence better than did adolescent convictions for violent crimes, because offenders were versatile rather than specialized. Boys who were heavy drinkers, drug users, aggressive and hostile to the police as adolescents tended to be violent as adults.

As adolescents, adult violent offenders tended to be restless or lacking in concentration, daring, extraverted or impulsive. They often had convicted parents or young mothers and were exposed to harsh or erratic parental discipline. They tended to experience parental conflict and broken families, and their fathers tended not to join in their leisure activities. They tended to come from low income families, and in turn they tended to have unskilled manual jobs themselves when they grew up. All of these factors help to identify children with a high potential for adult violence.

3. Risk Factors for Antisocial Behavior

It is surprising that, apart from the seminal work of Robins (1966, 1979), "we have relatively few studies that have measured the effects of these [child and family] risks, prospectively measured, on adult personality disorder symptoms" (Cohen, 1996, p. 126). There are many studies showing that childhood conduct disorder symptoms predict adult antisocial personality disorder (APD) symptoms (e.g., Offord & Bennett, 1994; Robins & Ratcliff, 1978; Zoccolillo, Pickles, Quinton, & Rutter, 1992). There are very few studies of explanatory risk factors (i.e. risk factors that are not measuring some kind of underlying antisocial behavior construct) as predictors of later antisocial behavior. As a rare example, Luntz and Widom (1994) found that 20% of boys in Indianapolis who were abused or neglected up to age 11 showed APD (on DSM-IIIR) 20 years later, compared with10% of control boys.

In the Cambridge Study, scales were constructed at different ages as the best measures available in this project of the personality and behavioral features included in the (DSM-IIIR) definitions of conduct disorder and antisocial personality disorder. These scales were termed "antisocial personality" (ASP) scales (Farrington, 1991a). The ASP scale at age 18 included 14 variables: convicted between ages 15 and 18, self-reported delinquency and self-reported violence at age 15-18, antisocial group activity (including violence, vandalism and hanging about on the streets), drug taking, heavy smoking, heavy drinking, drunk driving, irresponsible sex (having intercourse without using contraceptives or ensuring that the girl was using contraceptives), heavy gambling, an unstable job record (including periods of unemployment, several short-lived jobs, and getting fired), anti-establishment attitudes (negative to police, bosses, school, rich people and civil servants) tattooed, and impulsiveness (based on items such as "I generally do and say things quickly without stopping to think"). The reliability of the ASP scale at age 18 was 0.74, and the worst quarter were the 88 males with six or more adverse features out of 14.

Farrington (2000b) reported the most extensive study of childhood psychosocial risk factors in relation to later teenage and adult antisocial behavior (or symptoms of antisocial personality). The best predictors at age 8-10 of ASP at age 18 were a convicted parent, large family size, a nervous or depressed mother, high neuroticism, poor parental child rearing behavior (harsh or erratic discipline, parental conflict) and low school attainment. A risk score based on these 6 factors was developed. Only 9% of 213 boys with 0-1 risk factors became antisocial at age 18, compared with 67% of 43 boys with 4 or more risk factors.

Summarizing, the main questions addressed in this chapter are:

1. How is youth violence related to other types of problem behaviors measured at age 15-18?
2. What are the most important childhood risk factors at age 8-10 for multiple problem violent youth?
3. How accurately can multiple problem violent youth be predicted using a risk score based on multiple childhood risk factors?

4. Method

4.1 Design of the Survey

The Cambridge Study in Delinquent Development is a prospective longitudinal survey of the development of offending and antisocial behavior in 411 males. At the time they were first contacted in 1961-62, these males were all living in a working-class inner-city area of South London. The sample was chosen by taking all the boys who were then aged 8-9 and on the registers of 6 state primary schools within a one-mile radius of a research office that had been established. Hence, the most common year of birth of these males was 1953. In nearly all cases (94%), their family breadwinner at that time (usually the father) had a working-class occupation (skilled, semi-skilled or unskilled manual worker). Most of the males were white (97%) and of British origin. The study was originally directed by Donald J. West, and it has been directed since 1982 by David P. Farrington, who has worked on it since 1969. It has been mainly funded by the Home Office. The major results can be found in four books (West, 1969, 1982; West & Farrington, 1973, 1977) and in summary papers by Farrington and West (1990) and Farrington (1995). These publications should be consulted for more details about the childhood risk factors and teenage problem behaviors discussed here.

A major aim in this survey was to measure as many factors as possible that were alleged to be causes or correlates of offending. The males were interviewed and tested in their schools when they were aged about 8, 10, and 14, by male or female psychologists. They were interviewed in a research office at about 16, 18 and 21, and in their homes at about 25 and 32, by young male social science graduates. They are currently being interviewed at age 46. At all ages except 21 and 25, the aim was to interview the whole sample, and it was always possible to trace and interview a high proportion: 389 out of 410 still alive at age 18 (95%) and 278 out of 403 still alive at age 32 (94%), for example. The tests in schools measured individual characteristics such as intelligence, attainment, personality, and psychomotor impulsivity, while information was collected in the interviews about such topics as living circumstances, employment histories, relationships with females, leisure activities such as drinking and fighting, and offending behavior.

In addition to interviews and tests with the males, interviews with their parents were carried out by female social workers who visited their homes. These took place about once a year from when the male was about 8 until when he was aged 14-15 and was in his last year of compulsory education. The primary informant was the mother, although many fathers were also seen. The parents provided details about such matters as family income, family size, their employment histories, their child-rearing practices (including attitudes, discipline, and parental conflict), their degree of supervision of the boy, and his temporary or permanent separations from them. The teachers completed questionnaires when the boys were aged about 8, 10, 12 and 14. These furnished data about troublesome and aggressive school behavior, restlessness and poor concentration, school attainment and truancy. Ratings were also obtained from the boys' peers when they were in the primary schools, about such topics as daring, dishonesty, troublesomeness and popularity.

4.2 Information on Offending

Searches were carried out in the central Criminal Record Office (National Identification Service) in London to try to locate findings of guilt of the males, of their parents, of their brothers and sisters, and (in recent years) of their wives and female partners. The latest search of conviction records took place in the summer of 1994, when most of the males were aged 40. Up to this age, 164 males (40%) were convicted (Farrington, Barnes, & Lambert, 1996; Farrington, Lambert, & West, 1998). Convictions were only counted if they were for offenses normally recorded in the Criminal Record Office, thereby excluding minor crimes such as common assault, traffic infractions and drunkenness. The most common offenses included were thefts, burglaries and unauthorized takings of vehicles, although there were also quite a few crimes of violence, vandalism, fraud and drug abuse.

Of the 760 recorded offenses, 119 were classified as violent: 52 assaults causing bodily harm, 28 offences of threatening behavior, 18 robberies, 18 offensive weapon crimes and three sex offenses. Assaults had to be relatively serious (involving visible damage worse than bruises, swelling, or a black eye) to be counted as indictable bodily harm offenses rather than common assault. Threatening behavior offences usually involved a serious threat of violence. Offensive weapon crimes usually involved sheath knives, flick knives, crowbars, shotguns and the like. Half of those convicted for possessing an offensive weapon also had a conviction for some other violent crime. Three out of 10 sex offenses involved violence: one alleged rape which led to a conviction for unlawful sexual intercourse, one violent indecent assault on a female, and one man (aged 33) who used a 13-year-old boy for homosexual buggery and then passed the boy on to other homosexuals for the purpose of buggery.

Up to age 40, 65 of the males were convicted for violence: 16% of 404 at risk, excluding seven not convicted for violence who died up to age 32. The vast majority of these males (54 out of 65) also had convictions for nonviolent offenses. The 36 males (9%) convicted for violence between ages 10 and 20 (committing 53 violent crimes) are the major focus of this chapter; these boys are the officially violent offenders. Only three boys were convicted for violence between ages 10 and 13, and only 11 between ages 14 and 16, but 26 were convicted between ages 17 and 20.

In order not to rely on official records for information about offending, self-reports were obtained from the males at every age from 14 to 32 (Farrington, 1989b). The advantage of self-reports is that they reveal many more offenses than official records, which to some extent show only the tip of the iceberg of offending. Also, official records may be limited by biases in police or court processing. Self-reports, however, may be distorted by concealment, exaggeration or forgetting. Since official records and self-reports have somewhat different strengths and weaknesses, it is important to establish which results are replicated in both methods. If a childhood risk factor predicts both self-reported and official violence, that risk factor is probably related to violent behavior rather than to any biases in measurement.

At age 15-18, the self-reported violence measure was the sum of four items, each scored 1 - 4: the number of fights in the previous three years, number of fights started, number of times carried a weapon in case it was needed in a fight, and number of times used a weapon in a fight. The 79 boys (20%) with the highest scores (10 or more out of 16) were defined as the self-reported violent boys and contrasted with the remaining 310 out of 389 interviewed (West & Farrington, 1977, p.82).

5. Results

5.1 Correlates of Youth Violence

Table 1 summarizes the most important 10 types of problem behavior at age 15-18 that were correlated with convictions and self-reports of violence. The "combined violence" boys were the 95 who were identified either by convictions or by self-reports or by both. The main measure of strength of relationship used here is the Odds Ratio (OR). Essentially, the OR indicates the increase in risk associated with a risk factor; ORs of 2 or greater, showing a doubling of the risk, suggest strong predictive relationships (Cohen, 1996).

Table 1: Correlates of Youth Violence (Odds Ratios)

	Violence Convictions	Self-Reported Violence	Combined Violence
Non-violence convictions	15.5	6.6	9.2
Self-reported delinquency	2.7	11.3	8.3
Drug use	5.3	3.3	3.9
Heavy drinking	4.5	1.7	2.4
Drunk driving	1.9*	2.1	2.0
Irresponsible sex	3.5	2.7	2.9
Heavy gambling	1.8*	2.3	2.7
Unstable job record	4.8	3.3	3.6
Anti-establishment attitude	2.7	2.9	2.4
Impulsive	1.4*	2.0	1.8

* $p > .05$

Convictions for violent offenses between ages 10 and 20 were strongly related to convictions for non-violent offenses in the same age range (OR = 15.5); 30 of the 36 violent offenders also had convictions for non-violent offenses. Similarly, self-reported violence (at age 15-18) was strongly related to self-reported delinquency in the same age range (OR = 11.3); 52 of the 79 self-reported violent boys were also self-reported delinquents (among the 97 boys who admitted most acts of burglary, taking vehicles, stealing from vehicles, shoplifting, stealing from automatic machines, vandalism and receiving stolen property).

Apart from non-violent offending, Table 1 shows that youth violence (combined) was most strongly related to drug use, an unstable job record, irresponsible sex, heavy drinking and an anti-establishment attitude. However, it was significantly related to all 10 types of problem behavior.

A multiple problem score was developed at age 18 by counting the number of types of problem behavior (out of 10) shown by each boy. Table 2 shows the number of boys with each score who were violent (on the combined measure). Not surprisingly, the percentage that were violent increased with the number of problems, from 2.9% of these with no problems to 85.7% of those with 8-9 problems (6 out of 7). The violent boys had more problems on average than the nonviolent boys (4.4 as opposed to 1.9; $t = 11.5$, p<.0001).

Table 2: Multiple Problem Score vs. Youth Violence

Score	Non-Violent (294)	Violent (95)	Total (389)	Percent Violent
0	68	2	70	2.9
1	76	10	86	11.6
2	59	11	70	15.7
3	43	10	53	18.9
4	21	13	34	38.2
5	15	14	29	48.3
6	7	19	26	73.1
7	4	10	14	71.4
8-9	1	6	7	85.7
Mean Score	1.9	4.4	2.5	

The 110 boys with 4 or more problems were defined as the "multiple problem (MP) boys". Of these 62 were violent, and hence were multiple problem violent (MPV) boys, while the remaining 48 were MP non-violent boys. Conversely, there were 33 non-MP violent boys and 246 non-MP non-violent boys.

5.2 Risk Factors for Multiple Problem Violent Boys

Table 3 shows how far 21 important psychosocial risk factors predicted multiple problem violent boys. Risk factors were chosen for inclusion in this analysis on the basis of their importance in previous analyses. As far as possible, all factors were dichotomized into the "worst" quarter versus the remainder (see Farrington & Loeber, 2000, for a discussion about the advantages and problems of dichotomization). Thus, variables that could not be dichotomized to identify about 15-35% as a risk category were not included. Also, variables with considerable missing data (more than about 10%) were excluded. The average fraction of cases missing on the 21 variables shown in Table 3 was 2.8%.

Also, since regression analyses were planned to investigate the extent to which each risk factor predicted MPV boys independently of other risk factors, highly correlated risk factors were not both included. For example, low school attainment was chosen in preference to low school track. Five of the original 26 risk factors chosen did not significantly predict MPV boys and are not shown in Table 3: nervousness of the boy, a nervous father, unpopularity of the boy, and neuroticism and extraversion of the boy. Nonsignificant ORs are indicated with an asterisk.

The first column of Table 3 shows how far each risk factor predicts multiple problem boys compared with the remainder, while the second column shows how far each risk factor predicts violent boys compared with the remainder. For example, 46.1% of low-income boys were MP, compared with 23% of the remainder (OR = 2.9); 41.6% of low-income boys were violent, compared with 19.3% of the remainder (OR = 3.0). The third column contrasts the MP non-violent boys with the non-MP non-violent boys to investigate how far each risk factor predicts MP boys among the non-violent boys. The fourth column contrasts the violent non-MP boys with the non-violent non-MP boys, to investigate how far each risk factor predicts violent boys among the non-MP boys. To the extent that these comparisons are nonsignificant (as was the case with low family income), this suggests that the predictions of MP and violent boys was largely driven by characteristics of the MPV boys.

Table 3: Risk Factors for Multiple Problem Violent Boys
(Odds Ratios)

Variable at age 8-10 (%)	MP vs. Rest	V vs. Rest	MP vs. Rest (Non-V)	V vs. Rest (Non-MP)	MPV vs. Rest (Non-MPV)
Socio-economic					
Low family income (23)	2.9	3.0	1.5*	1.3*	4.7
Poor housing (37)	1.5*	2.0	1.0*	1.7*	2.2
Low social class (19)	1.6*	1.5*	0.9*	0.4*	2.1
Large family size (24)	3.3	2.7	2.9	2.0*	4.4
Family					
Convicted parent (27)	4.0	2.8	2.5	1.0*	5.6
Nervous mother (32)	2.5	1.6*	1.6*	0.6*	3.0
Young mother (22) ·	1.6*	1.9	1.2*	1.8*	2.1
Harsh discipline (28)	2.4	2.3	2.0*	1.7*	3.2
Poor supervision (19)	2.1	2.6	1.1*	1.6*	3.4
Parental conflict (24)	2.5	2.0	1.6*	0.8*	3.4
Disrupted family (22)	2.0	2.1	1.0*	0.8*	3.1
School					2.1
Delinquent school (21)	2.0	1.4*	1.7*	0.7*	3.1
Low nonverbal IQ (25)	2.9	1.8	2.6	1.0*	2.2
Low verbal IQ (25)	1.9	1.5*	1.6*	1.0*	2.7
Low attainment (23)	2.7	1.8	3.1	1.5*	
					2.1
Individual					2.7
Small (18)	1.9	1.7*	1.8*	1.4*	1.9
Lacks concentration (20)	2.1	2.1	1.9*	1.8*	7.0
High impulsivity (25)	1.9	1.3*	2.0	1.0*	2.3
Daring (30)	3.6	4.7	2.4	3.7	6.8
Dishonest (25)	2.1	1.9	2.3	2.0*	
Troublesome (22)	4.7	3.8	4.7	3.4	

Notes: MP = Multiple Problem Boys
         V = Violent Boys
         * p>.05

The fifth column contrasts the MPV boys with the non-MP non-violent boys. For example, 42.3% of low income boys were MPV, compared with 13.5% of the remainder (OR = 4.7). The fact that relationships between low income and MP boys and violence are largely driven by the MPV boys can also be demonstrated from the retrospective percentages who were low income boys: this was true of 48.4% of the MPV boys, compared with 21.2% of the non-MP violent boys, 22.9% of the MP non-violent boys, and 16.7% of the non-MP nonviolent boys.

In Table 3, almost all relationships between risk factors and violence (all except daring and troublesomeness) were largely driven by the MPV boys. This was also true of most relationships between risk factors and MP boys. The best childhood predictors of the MPV boys were daring (taking risks), troublesomeness, a convicted parent, low family income and large family size (four or more siblings). Other significant socio-economic predictors were poor housing and low social class (a manual job of the family breadwinner, usually the father). Other significant family predictors were a nervous or depressed mother, a young mother (aged 19 or less at the time of her first birth), harsh or erratic parental discipline, poor parental supervision, parental conflict and a disrupted family (a separation of the boy from at least one of his parents, usually the father). Significant school predictors were attending a high delinquency rate school, low nonverbal IQ (on the Progressive Matrices), low verbal IQ (on the Mill Hill Vocabulary test) and low school attainment. Other significant individual predictors were small height, poor concentration/restlessness, high impulsivity on psychomotor tests and peer-rated dishonesty.

5.3 Risk Scores for Multiple Problem Violent Boys

In order to investigate which risk factors predicted MPV boys independently of other risk factors, forward stepwise multiple regression and logistic regression analyses were carried out. Strictly speaking, logistic regression analysis should be used with dichotomous data. However, a problem with logistic regression is that a case that is missing on any one variable has to be deleted from the whole analysis, sometimes causing a considerable loss of data. Fortunately, with dichotomous data, ordinary least squares (OLS) regression produces very similar results to logistic regression (Cleary & Angel, 1984), and the results obtained by the two methods are mathematically related (Schlesselman, 1982). Missing data were handled in the OLS regression by pairwise deletion, using all the available data for estimation of the relationships between all possible pairs of variables. In these analyses, the 62 MPV boys were contrasted with the remaining 327 interviewed at age 18.

Table 4 shows the results of the regression analyses. One-tailed statistical tests were used because all predictions were directional. The most important independently predictive childhood risk factors were daring, a convicted parent, low family income, a nervous or depressed mother, troublesomeness, a disrupted family, low nonverbal IQ and parental conflict. A risk score was constructed, based on these 8 independently important risk factors. The risk score for each boy was simply the number of risk factors that he possessed, from 0 to 8. Where a boy was missing on a risk factor, his score was pro-rated accordingly.

Table 4: Regression Analyses Predicting Multiple Problem Violent Boys

	F change	OLS p	LRCS Change	Logistic p
Daring	35.26	.0001	38.19	.0001
Convicted parent	25.63	.0001	19.50	.0001
Low family income	9.89	.0009	8.46	.002
Nervous mother	6.12	.007	5.25	.011
Troublesome	4.20	.021	4.29	.019
Disrupted family	2.73	.050	1.65	.099
Low nonverbal IQ	2.61	.053	---	---
Parental conflict	2.04	.077	---	---

Notes: OLS = Ordinary Least Squares Regression
         LRCS = Likelihood Ratio Chi-Squared
         Order of entry of variables in regression shown in parentheses
         p values one-tailed

Table 5 shows that the percentage of boys who were MPV increased with the risk score, from 1.1% of those with no risk factors to 83.3% of those with 7-8 risk factors (5 out of 6). At the bottom of this table, these boys are divided into three risk categories, low (score 0-2), medium (score 3-4) and high (score 5-8). Prospectively, the percentages of boys who became MPV were 5.8% (of the two-thirds of the sample who were low risk), 26.7% (of the one-fifth of the sample who were medium risk) and 52.2% (of the one-eighth of the sample who were high risk). Retrospectively, 38.7% of MPV boys were high risk,

37.1% were medium risk, and 24.2% were low risk. Focussing on the high risk boys versus the rest, the true positive rate was 52.2%, the true negative rate was 87.5%, the sensitivity (the percentage of MPV boys identified correctly) was 38.7%, the specificity (the percentage of non-MPV boys identified correctly) was 93.3%, and the odds ratio was 8.8.

Table 5: Multiple Risk Factors vs. Multiple Problem Violent Boys

No. Risk Factors	Non-MPV (327)	MPV (62)	Total (389)	Percent MPV
0	89	1	90	1.1
1	95	9	104	8.7
2	58	5	63	7.9
3	36	10	46	21.7
4	27	13	40	32.5
5	13	10	23	43.5
6	8	9	17	52.9
7-8	1	5	6	83.3
Low (0-2)	242	15	257	5.8
Medium (3-4)	63	23	86	26.7
High (5-8)	22	24	46	52.2

Note: MPV = Multiple Problem Violent Boys

Unfortunately, these figures over-estimate the extent to which MPV boys could be predicted in a new sample, because the risk score was chosen according to those variables that were the best predictors of MPV boys. A better indication of predictability can be obtained by using a risk score that was constructed at a very early stage of the project. The "vulnerable background" measure was based on five explanatory risk factors: low family income, large family size, poor parental child-rearing (harsh or erratic discipline, parental conflict), a convicted parent and low nonverbal IQ (West & Farrington, 1973, p.131). The vulnerable boys at age 8-10 were the 15% with three or more of these five risk factors.

Of the vulnerable boys, 44.8% became MPV (true positives), while the true negative rate was 89.1%, the sensitivity was 41.9%, the specificity was 90.2%, and the odds ratio was 6.7. Thus, it is reasonable to conclude that (given a prevalence of MPV boys of 10-15%), a childhood risk score could identify a high risk group of 10-15% of boys, of whom about 40-50% would become MPV boys, conversely identifying about 40-50% of MPV boys. This indicates the possible degree of overlap between multiple risk factors and multiple problem violent youth in a new study of an inner-city sample.

6. Conclusions

Violent youth tend to be versatile in their offending and in their other types of problem behavior. They disproportionately take drugs, get drunk, drive after drinking, engage in irresponsible sex, are heavy gamblers, impulsive, unsatisfactory employees and have anti-establishment attitudes. Many childhood risk factors predict youth who are both violent and show multiple problem behavior: socio-economic factors such as low family income and large family size, family factors such as a convicted parent and a disrupted family, school factors such as low nonverbal IQ, and individual factors such as daring and troublesomeness. A risk score based on 8 childhood factors predicted multiple problem violent youth. About 40-50% of multiple risk factor boys became multiple problem violent youth. These results should be used to inform the development of risk assessment instruments for multiple problem violent youth, and the testing of risk-focused prevention techniques.

References

Capaldi, D. M., & Patterson, G. R. (1996). Can violent offenders be distinguished from frequent offenders? Prediction from childhood to adolescence. *Journal of Research in Crime and Delinquency, 33,* 206-231.

Cleary, P.D., & Angel, R. (1984). The analysis of relationships involving dichotomous dependent variables. *Journal of Health and Social Behavior, 25,* 334-348.

Cohen, P. (1996). Childhood risks for young adult symptoms of personality disorder. *Multivariate Behavioral Research, 31,* 121-148.

Farrington, D. P. (1978). The family backgrounds of aggressive youths. In L. Hersov, M. Berger, & D. Shaffer (Eds.), *Aggression and antisocial behaviour in childhood and adolescence* (pp. 73-93). Oxford: Pergamon.

Farrington, D. P. (1982). Longitudinal analyses of criminal violence. in M. E. Wolfgang & N. A. Weiner (Eds.), *Criminal violence* (pp. 171-200). Beverly Hills, CA: Sage.

Farrington, D. P. (1989a). Early predictors of adolescent aggression and adult violence. *Violence and Victims,* 4, 79-100.

Farrington, D. P. (1989b). Self-reported and official offending from adolescence to adulthood. In M. W. Klein (Ed.), *Cross-national research in self-reported crime and delinquency* (pp. 399-423). Dordrecht, Netherlands:Kluwer.

Farrington, D. P. (1991a). Antisocial personality from childhood to adulthood. *The Psychologist,* 4, 389-394.

Farrington, D. P. (1991b). Childhood aggression and adult violence: Early precursors and later life outcomes. In D. J. Pepler & K. H. Rubin (Eds.), *The development and treatment of childhood aggression* (pp. 5-29). Hillsdale, NJ: Erlbaum.

Farrington, D. P. (1995). The development of offending and antisocial behaviour from childhood: Key findings from the Cambridge Study in Delinquent Development. *Journal of Child Psychology and Psychiatry,* 36, 929-964.

Farrington, D. P. (1997). Early prediction of violent and non-violent youthful offending. *European Journal on Criminal Policy and Research,* 5, 51-66.

Farrington, D. P. (1998). Predictors, causes and correlates of male youth violence. In M. Tonry & M. H. Moore (Eds.) *Youth violence* (pp. 421-475). Chicago: University of Chicago Press.

Farrington, D. P. (2000a). Adolescent violence: Findings and implications from the Cambridge Study. In G. Boswell (Ed.), *Violent children and adolescents: Asking the question why* (pp. 19-35). London: Whurr.

Farrington, D. P. (2000b). Psychosocial predictors of adult antisocial personality and adult convictions. *Behavioral Sciences and the Law,* 18, 605-622.

Farrington, D. P. (2001). Predicting adult official and self-reported violence. In G-F. Pinard & L. Pagani (Eds.) *Clinical assessment of dangerousness: Empirical contributions* (pp. 66-88). Cambridge: Cambridge University Press.

Farrington, D. P., Barnes, G., & Lambert, S. (1996). The concentration of offending in families. *Legal and Criminological Psychology, 1,* 47-63.

Farrington, D. P., Lambert, S., & West, D. J. (1998). Criminal careers of two generations of family members in the Cambridge Study in Delinquent Development. *Studies on Crime and Crime Prevention, 7,* 85-106.

Farrington, D. P., & Loeber, R. (1999). Transatlantic replicability of risk factors in the development of delinquency. In P. Cohen, C. Slomkowski,, & L. N. Robins (Eds.), *Historical and geographical influences on psychopathology* (pp. 299-329). Mahwah, NJ: Erlbaum.

Farrington, D. P., & Loeber, R. (2000). Some benefits of dichotomization in psychiatric and criminological research. *Criminal Behaviour and Mental Health, 10,* 100-122.

Farrington, D. P., & West, D. J. (1971). A comparison between early delinquents and young aggressives. *British Journal of Criminology, 11,* 341-358.

Farrington, D. P., & West, D. J. (1990). The Cambridge Study in Delinquent Development: A long-term follow-up of 411 London males. In H-J. Kerner & G. Kaiser (Eds.) *Kriminalitat: Personlichkeit, lebensgeschichte und verhalten (Criminality: Personality, behavior and life history)* (pp.115-138). Berlin, Germany: Springer-Verlag.

Hawkins, J. D., Herrenkohl, T., Farrington, D. P., Brewer, D., Catalano, R. F., & Harachi, T. W. (1998). A review of predictors of youth violence. In R. Loeber & D. P. Farrington (Eds.), *Serious and violent juvenile offenders: Risk factors and successful interventions* (pp. 106-146). Thousand Oaks, CA: Sage.

Henry, B., Caspi, A., Moffitt, T. E., & Silva, P. A. (1996). Temperamental and familial predictors of violent and non-violent criminal convictions: Age 3 to age 18. *Developmental Psychology, 32,* 614-623.

Hogh, E., & Wolf, P. (1983). Violent crime in a birth cohort: Copenhagen 1953-1977. In K. T. van Dusen & S. A. Mednick (Eds.), *Prospective studies of crime and delinquency* (pp. 249-275). Boston: Kluwer-Nijhoff.

Jessor, R., Donovan, K., & Costa, F. M. (1991). *Beyond adolescence: Problem behavior and young adult development.* Cambridge: Cambridge University Press.

LeBlanc, M., & Frechette, M. (1989). *Male criminal activity from childhood through youth.* New York: Springer-Verlag.

Loeber, R., Farrington, D. P., Stouthamer-Loeber, M., & van Kammen, W. B. (1998a). *Antisocial behavior and mental health problems: Explanatory factors in childhood and adolescence.* Mahwah, NJ: Erlbaum.

Loeber, R., Farrington, D. P., Stouthamer-Loeber, M., & van Kammen, W. B. (1998b). Multiple risk factors for multi-problem boys: Co-occurrence of delinquency, substance use, attention deficit, conduct problems, physical aggression, covert behavior, depressed mood and shy/withdrawn behavior. In R. Jessor (Ed.), *New perspectives on adolescent risk behavior* (pp. 90-149). Cambridge: Cambridge University Press.

Luntz, B. K., & Widom, C. S. (1994). Antisocial personality disorder in abused and neglected children grown up. *American Journal of Psychiatry, 151,* 670-674.

Offord, D. R., & Bennett, K. T. (1994). Conduct disorder: Long-term outcomes and intervention effectiveness. *Journal of the American Academy of Child and Adolescent Psychiatry, 33,* 1069-1078.

Pulkkinen, L., & Pitkanen, T. (1993). Continuities in aggressive behavior from childhood to adulthood. *Aggressive Behavior, 19,* 249-263.

Robins, L. N. (1966). *Deviant children grown up.* Baltimore, MD: Williams and Wilkins.

Robins, L. N. (1979). Sturdy childhood predictors of adult outcomes: Replications from longitudinal studies. In J. E. Barrett, R. M. Rose, & G. L. Kierman (Eds.), *Stress and mental disorder* (pp. 219-235). New York: Raven Press.

Robins, L. N., & Ratcliff, K. S. (1978). Risk factors in the continuation of childhood antisocial behavior into adulthood. *International Journal of Mental Health, 7,* 96-116.

Schlesselman, J. J. (1982). *Case-control studies.* New York: Oxford University Press.

Stattin, H., & Magnusson, D. (1989). The role of early aggressive behavior in the frequency, seriousness, and types of later crimes. *Journal of Consulting and Clinical Psychology, 57,* 710-718.

Wadsworth, M. E. J. (1979). *Roots of delinquency.* London: Martin Robertson.

West, D. J. (1969). *Present conduct and future delinquency.* London: Heinemann.

West, D. J. (1982). *Delinquency: Its roots, careers and prospects.* London: Heinemann.

West, D. J., & Farrington, D. P. (1973). *Who becomes delinquent?* London: Heinemann.

West, D. J., & Farrington, D. P. (1977). *The delinquent way of life.* London: Heinemann.

Zoccolillo, M., Pickles, A., Quinton, D., & Rutter, M. (1992). The outcome of childhood conduct disorder: Implications for defining adult personality disorder and conduct disorder. *Psychological Medicine, 22,* 971-986.

Multi-Problem Violent Youth
R.R. Corrado et al. (Eds.)
IOS Press, 2002

Risk/Need Assessment and Prevention of Antisocial Development in Young People: Basic Issues from a Perspective of Cautionary Optimism

Friedrich Lösel

International longitudinal research in developmental criminology has led to substantial improvements in our knowledge about the risks and needs of children and youth who become seriously delinquent (see Loeber & Farrington, 1998, 2001). It has also revealed the importance of discriminating between different developmental pathways (Loeber & Stouthamer-Loeber, 1998; Moffitt, 1993; Nagin & Tremblay, 1999). Although various models have been proposed, there is widespread consensus over an early-starting, long-term persistent pathway (e.g., Moffitt, 1993; Patterson et al., 1998). Naturally, such a model is a simplified reconstruction of the complexity of individual developments, however, it is of particular interest for policy making and practice. Because the small group of long-term persistent antisocial individuals offends very frequently and seriously, it is responsible for more than one-half of classic criminality (Loeber et al., 1998; Wolfgang et al., 1972). This is why risk/need assessment and related measures of developmental prevention and intervention are a very actual concept of crime policy and social policy.

However, remembering the pioneering work of Glueck and Glueck (1950) or McCord et al. (1959), we should bear in mind that this is not the first time that early prediction and developmental prevention have become a core issue in criminology. Although these studies formed the basis for the later upswing of developmental criminology, the 1960s and early 1970s saw attention being shifted away to other core concepts such as offender treatment and rehabilitation. A few years later, however, many people believed in the ideology of "nothing works" in correctional treatment (Lösel, 1995a). Nowadays, American states that were at the forefront of rehabilitation are among the leading advocates of harsh punishment and incarceration (Haney & Zimbardo, 1998).

That concepts in crime policy swing to and fro like fashion trends is not just due to changes in the political climate of societies. It is also a consequence of the over-exaggerated promises and empirically unfounded expectations that often accompany the proposal of policy concepts. To avoid a similar fate for the current interest in early risk/need assessment and prevention/intervention in the development of antisocial behavior, it is necessary to take a realistic perspective on its possibilities and limitations. Such a perspective goes beyond technical issues of statistical or structured instruments of risk assessment and universal, selective, or indicated prevention programs. It must refer also to basic issues of developmental psychopathology. The present chapter will address a few of these topics illustrated by research examples: (a) Multiple factors and developmental sequences; (b) developmental change and predictive accuracy; (c) multiple-setting-multiple-informant approach; (d) variable- versus person-orientation; (e) protective factors and processes; (f) risk communication; and (g) the efficacy of prevention and treatment programs. Some shorter remarks will address legal, ethical, gender, and cultural issues.

1. Multiple Factors and Developmental Sequences

Much of criminological literature has focused on relatively few and static general causes of delinquency such as social class and deprivation, subcultural normative conflict, differential association, or social bonding. However, developmental criminology clearly shows that serious and violent juvenile offending cannot be explained sufficiently by one or the other single factor. For example, Hawkins et al. (1998) performed a comprehensive review of international longitudinal research on predictors of serious and violent juvenile offending. For each domain of risks such as family, school, peer group, community, personality, and deviant behavior they listed the correlations between predictors and outcome at different ages. Table 1 summarizes their findings across the numerous studies, domains, and variables.

Table 1: Correlations Between Risk Factors and Youth Violence in Hawkins et al. (1998)

r	n	Percent
<.10	108	39.1
.10 - .20	127	46.0
.21 - .30	37	13.4
>.30	4	1.5

According to Cohen's criteria (1988), nearly all coefficients are small (< .30). The majority are even below .20. Medium effect sizes (> .30) are very rare, and when found, they tend to involve predictors that already represent deviant behavior themselves (e.g., child aggressiveness and delinquency).

Lipsey and Derzon's (1998) meta-analysis of the predictors of serious or violent delinquency in adolescence and early adulthood (age 15-25 years) produced similar results: Only two of their predictors at Age 6-11 and Age 12-14 revealed a correlation of >.30 with the criterion. In the 6- to 11-year-olds, these were general offenses and substance use; in the 12- to 14-year-olds, antisocial peers and social ties (few social activities, low popularity). Other risks, such as parent-child relations, antisocial parents, broken home, low SES, ethnicity, school attitudes and performance, noncriminal aggression, or low intelligence, all had correlations of < .30 with behavioral outcome. The predictive validity of several variables was also not the same at different ages. For example, SES and ethnicity were better predictors at the younger ages, whereas peer factors were more relevant in early adolescence. In accordance with such results, developmental approaches refer to numerous biological, psychological, and social risk factors whose predictive values vary over time and circumstances. Different factors and mechanisms may be relevant for processes of onset, persistence, aggravation, or desistance in antisocial developments (Farrington et al., 1990; LeBlanc & Loeber, 1993).

The empirical and probabilistic term of "risk" does not imply that causal theories became less important. Indeed, there are recent examples for their strong impact on research (e.g., Gottfredson & Hirschi, 1990; Moffitt, 1993). In comparison to many other criminological concepts, developmental approaches place more emphasis on the integration of bio-pycho-social factors. Such integrations are particularly necessary to solve the complex problems of prediction and intervention that are typical for applied research. In assessment practice, the problem of integration is often "solved" by forming sum scales of various factors. However, this may be too unspecific for an adequate planning of intervention because it does not take into account the principles of multifinality and equifinality in development (Cicchetti & Rogosch, 1996). Multifinality means that a certain condition may lead to different outcomes in behavior. This is the case when, for example, family stressors enhance the risk of aggressive behavior in some young persons, of depressive problems in others, or even of comorbidities involving both (see Rutter, 1997). Vice versa, according to the principle of equifinality, the origins of a certain behavior problem may vary substantially between different cases. An example of this is Frick's (1998) model of different developmental origins in child conduct disorders. On the one pathway, poor parental socialization and low intelligence are particularly significant. On another, it is low behavioral inhibition and callous, unemotional personality traits that take the central role.

The notion of multifinality and equifinality in development is related to findings on an individual's information processing as a core mediator of external or internal stimuli (e.g., Crick & Dodge, 1994; Huesman, 1997). Within this framework, the young person is not only an object of influences but also - and increasingly as he or she grows older - an active creator of his/her own development. This is in line with transactional models of human development (Bronfenbrenner & Ceci, 1994; Sameroff & Fiese, 2000). For such processes, the assumption of feedback loops may be more adequate than that of linear paths of causality. A typical example is the bidirectional or even more complex relation between poor parenting in the family and antisocial child behavior(e.g., Boyce et al., 1998; Plomin, 1994). Similarly, gang membership is both a result of selective mating and the origin of enhanced seriousness in antisocial behavior (Thornberry, 1998).

Although we still have only a rudimentary causal knowledge about the complicated interplay of multiple factors and domains, the practice of prediction and prevention needs some integrative framework. Currently, social-cognitive and biosocial learning theories on the development of serious antisocial behavior seem to be the most adequate one (e.g., Bandura, 1986; Huesman, 1997; Moffitt, 1993; Raine, 1997). Based on the most consistent results from this area of research, Figure 1 depicts a structure of multiple factors and developmental sequences.

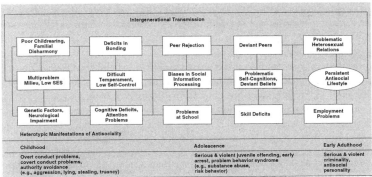

Figure 1: Model of cumulated risks in the development of persistent antisocial behavior
(from Lösel and Bender, 2001)

The model conceptualizes the development of serious and persistent antisocial behavior as a chain reaction of cumulating biological, psychological, and social risk factors. It integrates many of the static and dynamic predictors of persistent antisocial behavior. Each construct represents a number of risks. For example, on the level of family functioning, parental conflicts, lack of emotional warmth, aggressive childrearing, lax, erratic, or harsh discipline, child abuse and neglect are risk factors (e.g., Farrington, 1992; Farrington & Loeber, 2001). On the structural level of the family milieu, risk factors are low income, living on social welfare, divorce, teenage motherhood, parental criminality, alcohol and drug abuse, chronic unemployment, public housing, disintegrated neighborhoods, and availability of drugs (e.g., Hawkins et al., 1998). On the biological level, low resting heart rate, deficits in prefrontal functioning, high testosterone level, low platelet MAO activity, maternal smoking during pregnancy, a fetal alcohol syndrome, premature birth, and perinatal complications are among the risk factors (e.g., Raine et. al, 1997). Whereas some variables represent an independent and general risk, others become important primarily when they interact with further factors (e.g., hormone-context interactions; Susman & Ponirakis, 1997).

Due to the complexity of multiple risks and needs, a dual assessment approach seems to be most adequate. On the one hand, robust omnibus scales or checklists with a few core variables from various domains are useful for screening procedures (e.g., Augimeri et al., 2000; Bartel & Forth, 2000; Farrington, 1997). On the other hand, more specific analyses in relation to specific types of youngsters, programs, and developmental phases require detailed assessments of selected domains. The results of the NATO ARW at Cracow are steps toward developing an integrated framework for such differentiated approaches (see Chapters 25 and 26, this volume).

2. Developmental Change and Predictive Accuracy

The early risk diagnosis and prevention of antisocial behavior focuses on developmental continuity. Systematic prediction instruments are designed to attain the highest possible hit rate. False positives and false negatives are primarily viewed as errors that may be reduced by better instruments. The related technical and statistical issues have been comprehensively discussed already in the 1980s (e.g., Farrington & Tarling, 1985; Gottfredson & Tonry, 1987). However, a primarily technical approach may overshadow the fact that change is just as much a part of human development as continuity. This other side of the coin should sensitize for the limits of even well-constructed prediction instruments. Simultaneously, it may shift the focus toward modeling both developmental stability and flexibility.

For example, Lahey et al. (1995) report that approximately 50% of boys with a DSM diagnosis of conduct disorder did not remain in this category continuously over four years. This is a typical rate of problem stability from preschool to school age (Campbell, 1995; Lavigne et al., 1998). Robins (1978) and Moffitt et al. (1996) have shown that about one half of children with conduct disorders or extreme antisocial behavior did not go on to serious criminal outcomes in adolescence. According to Haapasalo and Tremblay (1994), 8% of boys were stable frequent fighters from kindergarten age to ages 10-12, whereas 12% had desisted by this time. In a study of physical aggression from Age 6 to Age 15, the group of desisters was much larger than the group of boys who remained chronically aggressive (4% vs. 28%; Nagin & Tremblay, 1999). Although Patterson et al. (1998) found a clearly persistent pathway, approximately 50% of children who were high in antisocial behavior at Age 9-10 did not progress to early arrest and chronic offenses by Age 18.

The high stability coefficients for aggressiveness as ascertained in meta-analyses (Olweus, 1979; Zumkley, 1994) also do not rule out relative flexibility in development. Whereas average correlations are approximately $= .70$ after 1 year, they decline as a function of the length of the interval between two measurement times. Furthermore, such correlations document only the similarity of the ranking orders in an inter-individual comparison and not the stability of the behavior itself (Farrington, 1990; Loeber & Stouthamer-Loeber, 1998). The latter may change in frequency and quality. For example, on the one hand, studies show a marked rise in the prevalence of serious and violent offending during early adolescence (Loeber et al., 1998). On the other hand, the frequency of physical forms of aggression declines continuously from preschool age onward (Tremblay, 2000). Although changed modes of aggressive behavior may indicate heterotypic continuity, different types of development may be overseen when behavior is aggregated too sweepingly to form a general syndrome of antisocial behavior.

To avoid misunderstanding; emphasizing not only continuity but also change as a basic principle of development is no argument against early risk/need assessment and management. However, looking at both sides of the coin helps to build up a realistic perspective that does not misinterpret developmental flexibility as a methodological deficit. Although accumulation of risks is a much better predictor than single variables or domains (e.g., Farrington, 1997; LeBlanc, 1998), basic processes of developmental change set a clear limit even for well-constructed instruments.

General experiences in predicting human behavior, the findings on antisocial development, and the intercorrelations of predictors suggest that we should not expect much more than = .40 in long-term predictions. This upper threshold is also realistic in the prediction of recidivism in offenders (e.g., Gendreau et al., 1995). To give an impression of the practical relevance of such a predictive power, Table 2 presents two computational examples. In both scenarios, the predictor-outcome correlation is = .40. As suggested by Lipsey and Derzon (1998), we assume a base rate of 8% serious and violent juvenile offenders. Data are presented for two selection rates: In the first model, the 25% of individuals with the highest predictor scores are identified as being at risk; in the second model, only the highest 10% are selected. To make the examples more concrete, Table 2 does not contain probabilities but absolute figures for a fictitious city with a youth population of = 10,000.

Table 2: Two Examples of Accuracy in Predicting Serious Juvenile Delinquency
Both Examples: n = 10,000; Base Rate = 8%; Predictive Validity: phi = .40

Example 1: Selection Rate = 25%

		Outcome		
		Seriously delinquent	Not seriously delinquent	
Prediction	Seriously deliquent (at risk)	675	1,825	2,500 (25%)
	Not seriously delinquent (not at risk)	125	7,375	7,500 (75%)
		800 (8%)	9,200 (92%)	

Example 2: Selection Rate = 10%

		Outcome		
		Seriously delinquent	Not seriously delinquent	
Prediction	Seriously delinquent (at risk)	405	595	1,000 (10%)
	Not seriously delinquent (not at risk)	395	8,605	9,000 (90%)
		800 (8%)	9,200 (92%)	

The first example and the second example (in brackets) reveal the following hit and error rates:
1. 84% (51%) of those who later become seriously delinquent will be among those who are identified as being at risk (sensitivity);
2. 80% (94%) of those who later do not become seriously delinquent will be identified as being not at risk (specificity).
3. 20% (7%) of those who actually will not become seriously delinquent are mistakenly identified as being at risk (false positives).
4. 16% (49%) of those who will become seriously delinquent are mistakenly identified as being not at risk (false negatives).
5. 27% (41%) of those identified as being at risk will later become seriously delinquent;
6. 98% (96%) of those identified as being not at risk will become not seriously delinquent later;
7. the risk of becoming seriously delinquent for those identified as being at risk is approximately 21-times (15-times) higher than it is for those identified as being not at risk (odds ratio).

These examples demonstrate clearly that advocates as well as critics of risk assessment on juvenile delinquency can select data supporting their position: On the one hand, in both models, those who will not become seriously delinquent are predicted very well. The majority of those who will become seriously delinquent is also identified correctly, particularly when the selection rate is 25%. In the latter case, sensitivity and specificity are in the range of scientifically accepted AIDS tests (Bland, 1996). On the other hand, a majority of those identified as at risk do not become seriously delinquent. The higher the selection rate, the greater the number of children receiving an indication for intervention although they do not need a program. In contrast, the lower the selection rate, the more children are in need who mistakenly get no program.

The question of an optimal cut-off point goes beyond statistical issues. Although measures of Receiver Operating Characteristics (e.g., Douglas et al., 1999) or of Relative Improvement Over Chance (Loeber & Dishion, 1983) can indicate an optimum for different base rates, selection rates, and rates of false positives and negatives, the final decision must be based on reasons of social utility (Blumstein et al., 1985). Optimal payoffs may be rather different for the respective institution, community, society, children, and families. Furthermore, the benefit of the procedure depends on the efficacy, costs, and potential negative side effects of the available prevention measures. If, for example, a program is relatively successful, cheap, and without negative side-effects, a higher selection rate may be chosen. In contrast, a moderately efficient and relatively expensive program with probable negative side effects may be justified only for a small group of very high risks (if there is no better alternative).

3. Multiple-Setting-Multiple-Informant Approach

Risk assessment procedures for children and adolescents have to be economical enough to allow widespread implementation in daily life. If instruments require too much time or are too complicated, there is a risk that they will be either applied inadequately or even not at all. Nonetheless, calls for economic test procedures should not lead to a restriction to only one source of information from only one behavior context. Correlations between reports on the behavior problems of children and adolescents by parents, teachers, psychiatrists, psychologists, social workers, or the young persons themselves generally tend to be rather low. Ratings between different informants reveal particularly low correlations when they do not refer to the same social context (Achenbach et al., 1987). In these cases, they are often lower than = .30, and thereby inadequate for an objective individual diagnosis. When, in contrast, informants refer to the same context (e.g., both parents, or teachers and observers in one classroom), correlations are higher, particularly for easily recognizable externalizing problem behavior (e.g., Lösel et al., 1991).

As with the discussion above on stability, high correlations between judges do not necessarily mean agreement on observed behavior (which would require categorical analyses), but only similarity in the perception of inter-individual differences. Accordingly, the level of agreement also depends on the variance within the group being judged, so that children and adolescents with extreme characteristics may tend to receive the same judgment. Nonetheless, it should not be overlooked that typical assessment instruments with multiple items confound the intensity of the variable and behavior consistency: High or low scores imply that a great number of items on risk behavior are affirmed or denied. At the same time, this indicates consistency across different situations. Medium scale scores, in contrast, show changing behavior depending on the situation.

Low correlations between different informants are in no way just an expression of judgment error, but also reflect valid differences in behavior between settings. This is why a multi-setting-multi-informant approach should be the diagnostic standard in developmental psychopathology. The need for this becomes particularly clear in prognostic issues such as the early detection of multiproblem youth. The repeated restriction to one source of data can lead to an overestimation of the stability of problem behavior. This is indicated in data from several studies on the assessment of child and youth problem behavior that were carried out in our research team (see Tables 3 and 4).

Table 3: Multiple-Informant-Multiple Setting Correlations for Measures of Externalizing, Aggressive, and Delinquent Problem Behavior

Mother (PSBQ) - Educator (PSBQ); Kindergarten; Age 4-5; $n = 482$	.27
Mother (PSBQ) - Father (PSBQ); Kindergarten; Age 4-5; $n = 545$	.58
Mother (ECBI) - Father (ECBI); Kindergarten; Age 4-5; $n = 539$	.60
Mother (CBCL) - Educator (TRF); Clinical setting; Age 6-12; $n = 79$	.32
Observer (DOF) - Self (YSR); School; Age 12-16; $n = 39$	.30
Observer (DOF) - Teacher (TRF); School; Age 12-16; $n = 39$	.59
Self (DBS) - Teacher (Rating); School; Age 14-15; $n = 154$	.43
Teacher (Rating) - Self (BVQ); School; Age 14-15; $n = 1,163$	.29
Educator (TRF) - Self (YSR); Clinical setting; Age 14-17; $n = 142$	.27
Peer (Rating) - Peer (Rating); School; Age 15-17; $n = 102$	.34
Observer (Rating) - Observer (Rating); School; Age 15-17; $n = 102$	.48
Observer (Rating) - Peer (Rating); School; Age 15-17; $n = 102$	.43
Teacher (BVQ) - Self (BVQ); School; Age 15-17; $n = 102$	.57
Teacher (TRF) - Self (YSR); School; Age 15-17; $n = 102$	.52
M (n-weighted)	.40

Note. PSBQ = Preschool Social Behavior Questionnaire (Tremblay), ECBI = Eyberg Child Behavior Inventory, CBCL = Child Behavior Checklist (Achenbach), Teacher Report Form (Achenbach), DOF = Direct Observer Form (Achenbach), YSR = Youth Self report (Achenbach), DBS = Delinquenzbelastungsskala (Lösel), BVQ = Bully/Victim Questionnaire (Olweus).

Table 3 reports studies on cross-sectional assessments of problem behavior in children and adolescents from different informants. The mean correlation is .40. Table 4 reports the results of prospective longitudinal assessments using two measurement waves at intervals ranging from 8 months to 2 years. The mean correlation here was notably higher at .67. Hence, the stability of problem behavior is probably overestimated when information is restricted to one informant.

Table 4: Stability Coefficients (One Informant) for Measures of Problem Behavior

M: Externalizing problems (PSBQ); 1 year; Age 4-6; $n = 252$	.73
F: Externalizing problems (PSBQ); 1 year; Age 4-6; $n = 224$	.77
M: Problem intensity (ECBI); 1 year; Age 4-6; $n = 250$	.75
F: Problem intensity (ECBI); 1 year; Age 4-6; $n = 225$	.75
S: Externalizing problems (YSR); 8 months; Age 11-17; $n = 385$	.48
S: Externalizing problems (YSR); 8 months; Age 11-17; $n = 401$	.70
S: Violent offenses (DBS); 20 months; Age 15-17; $n = 102$	.77
S: Physical aggression (BVQ); 20 months; Age 15-17; $n = 102$	.72
S: Property offenses (DBS); 20 months; Age 15-17; $n = 102$	.55
T: Aggressiveness (rating); 20 months; Age 15-17; $n = 102$	.51
S: Externalizing problems (YSR); 2 years; Age 16-18; $n = 141$	.53
T: Externalizing problems (TRF); 2 years; Age 16-18; $n = 141$	.72
M (n - weighted)	.67

Note. M = Mother, F = Father, S = Self, T = Teacher; PSBQ = Preschool Social Behavior Questionnaire (Tremblay), ECBI = Eyberg Child Behavior Inventory (Eyberg), YSR = Youth Self Report (Achenbach), DBS = Delinquenzbelastungsskala (Lösel), BVQ = Bully/Victim Questionnaire (Olweus), TRF = Teacher Report Form (Achenbach)

Particularly in the risk assessment of large populations of youngsters, however, it is not always possible to draw on multiple sources of information for economic reasons. Nonetheless, multiple gating procedures can be used (LeBlanc, 1998; Loeber & Dishion, 1983). The first stage of risk assessment can, for example, take simple data from school or preschool teachers, and shift to a detailed second stage with further informants only when initial scores are high. A procedure based on simple decision rules seems to be a relatively successful judgment heuristic, particularly when confronted with incomplete information or time pressure (Gigerenzer & Selten, 2001).

4. Variable- and Person-Oriented Approach

Most traditional research on the risks and origins of the development of antisocial behavior is variable-oriented. This means that it addresses relations between manifest variables or latent constructs. The statistical methods applied include cross-sectional or longitudinal correlations, analyses of variance, multiple regressions, path analyses, and LISREL models. However, practical questions in risk/need assessment, prevention, and intervention address individual patterns or configurations of characteristics. Basic research on human development is also increasingly emphasizing the need to supplement the variable-oriented approach with a person-oriented one (e.g., Block, 1971; Magnusson & Bergman, 1988). This is because an exclusively variable-oriented approach has difficulty in detecting complex interaction effects and specific individual courses of development (Magnusson & Cairns, 1996). Longitudinal correlations between variables may also be artificial because they trace back only to subgroups with specific patterns. Bergman and Magnusson (1997), for example, found that aggression and hyperactivity at Age 13 correlated with criminality and alcohol abuse at Age 18-23. However, this general relationship was due only to youngsters with a severe multiproblem syndrome (approximately 10% of the whole sample). Clusters with only some risks at Age 13 were not overrepresented among the criminal and/or alcohol abusing groups in young adulthood.

Consistent interaction effects are still relatively rare in longitudinal studies on criminal behavior (Farrington & Loeber, 2000). Perhaps, this may be due to a paucity of pattern analyses in those domains in which complex interactions would seem to be particularly probable. This applies to, for example, the interplay of biological and social variables (Raine et al., 1997) or characteristics of the individual and community level (Wikström & Loeber, 2000). More differentiated findings on this are to be anticipated from latent class analysis, cluster analysis, and configuration frequency analysis - procedures that are particularly appropriate for analyzing patterns of developmental data (see von Eye, 1990; Stemmler et al., 1998).

However, even without complex multivariate analyses, it is obvious that the assessment and prevention of antisocial development needs to pay attention to individual patterns of characteristics. This is particularly true when comorbidities are present (see Rutter, 1997). Comorbidities can lead to complex interactions in developmental mechanisms. For example, although it has been shown repeatedly that a shy and withdrawn temperament may, in general, have a protective function against juvenile delinquency (Lösel & Bender, in press), risk seems to be particularly high when children are shy and aggressive (e.g., Ensminger et al., 1983; Farrington et al., 1988). Similar personality patterns are also relevant for overcontrolled hostility in adulthood (e.g., Megargee, 1966; White & Heilbrun, 1995).

Whereas risk scales usually form sum scores of more or less equally weighted single factors, cluster and pattern analyses may lead to a theoretically and practically more differentiated view. For example, we have carried out an intensive study of "hardcore" soccer hooligans (Lösel et al., 2001). It revealed that most of them exhibited a accumulation of risk factors that is typical for violent young men: family problems such as a broken home, poor parenting, or alcohol abuse; deficits in school achievement, truancy, and other early behavior problems; repeated unemployment and downward professional careers; as well as enhanced scores on aggression-prone personality dimensions and personality disorders, alcohol and/or drug abuse, and further delinquency outside the context of soccer hooliganism. However, an inspection of the individual patterns of risk revealed that almost one-third had no serious problems in their family background. All such cases showed clear personality problems, and, at times, the Screening Version of the Psychopathy Checklist (Hart et al., 1995) even indicated a psychopathic personality disorder. A similar personality pattern was found in those hooligans exhibiting no serious problems with their careers.

Although the sum scores of risks may be similar, intervention for psychopathic or otherwise personality disordered individuals requires a different focus to that for those exhibiting mostly socialization deficits or achievement problems (see Blackburn, 2000; Lösel, 1998). Differential indications are also necessary for young persons in whom most delinquency risks are comparable with others, but they are additionally mentally retarded. Although various types of offender personality have been investigated (see Blackburn, 1993), most statistical predictions of serious juvenile antisocial behavior do not contain person-oriented pattern analyses. This indicates the need for more structured diagnoses that, while anchored in actuarial prediction, go beyond their purely statistical application. Such approaches are not to be equated with traditional clinical prediction, which has been found to exhibit low empirical validity (see Grove & Meehl, 1996).

5. Protective Factors and Mechanisms

Most research on antisocial behavior in young people concentrates on risks or deficits along with measures to reduce them. However, because - as mentioned above - only a portion of the children and adolescents at strong risk become seriously delinquent and others desist from deviant pathways, it is also necessary to ask about protective factors and processes (e.g., Lösel & Bliesener, 1990; Rutter, 1985; Werner & Smith, 1982). In conjunction with the concept of resilience, this issue has received increased attention since the 1980s (Cicchetti & Garmezy, 1993; Luthar et al., 2000). It is also very important for assessment and prevention practice, because, depending on the available personal or social resources, risks may have to be evaluated differently and other interventions may be indicated. However, only very few assessment instruments or prevention programs (e.g.,

Bartel & Forth, 2000; Dunst & Trivette, 1997) explicitly refer to protective factors. Furthermore, compared with research on risks, there are, only a few longitudinal studies that have focused particularly on protective factors against antisocial development (Lösel & Bender, in press). This research has to face a number of general problems with the resilience concept (e.g., Lösel & Bender, in press; Luthar et al., 2000). These involve, for example, the ambiguity in definitions and central terminology; the heterogeneity in risks experienced and competence achieved by "resilient" individuals; the instability of the phenomena of resilience; and the usefulness of the construct for developmental theory and prevention. Insofar, there is a risk of basing assessment and prevention on not yet sufficiently validated concepts of protective factors.

The various hypothetical models of protective mechanisms and study designs cannot be addressed here (e.g., Gest et al., 1993; Luthar et al., 2000).The most frequent approach is to compare two groups with a comparably high risk of delinquency: one of which has developed negatively; the other, positively or "resiliently." Personal or social characteristics that contribute to these different developments are viewed as protective factors (Rutter, 1985). However, these characteristics often have a negative pole that is simultaneously a risk factor for delinquent development (Farrington, 1994; Stouthamer-Loeber et al., 1993). For example, low intelligence has proved to be a delinquency risk; high intelligence, in contrast, seems to have a protective function (Kandel et al., 1988). Hence, one cannot talk about protective factors in general, but always only in relation to certain risks and outcomes: The more risk variables are integrated into a study, the less variance can be explained by additional protective factors.

Hence, it is particularly meaningful to assume a protective function only when a *high* risk of deviant development is buffered. For example, Werner and Smith (1992) studied children of a birth cohort who grew up into socially competent individuals despite four or more developmental risks in early childhood. They compared these "resilients" with those young people who exhibited major behavior problems in line with their risk exposure. Furthermore, they examined which factors helped a young person who had already become delinquent to make positive changes in development (desistance). Lösel and Bliesener (1994) compared high risk groups of young persons in care who exhibited relatively healthy psychosocial development with those in whom serious behavior problems had emerged. Approximately two-thirds of the sample maintained a stable resilience or deviance in adolescence; approximately one-third changed either positively or negatively. Other studies addressed, for example, children or adolescents who were abused or neglected in their families (Cicchetti & Rogosch, 1997), had criminal or mentally disturbed parents (Kandel et al., 1988), exhibited cumulative personal or familial risks (Stattin et al., 1997), or grew up in a high-risk milieu (Cowen et al., 1997).

Despite major differences in the risks, age ranges, social contexts, assessment methods, and criteria for resilience, there have been several consistent findings. In particular, the following characteristics seem to possess a protective function against the development or consolidation of delinquent behavior (see Lösel & Bender, in press): (a) genetic dispositions such as heightened autonomic arousal; (b) an "easy" or inhibited temperament; (c) a flexible adaptation of ego boundaries instead of rigid over- or undercontrol (ego resiliency); (d) above-average intelligence and realistic planning; (e) a secure bond to the mother or another reference person within or outside the family; (f) emotional warmth, supervision, and consistency in upbringing; (g) parents or other adults who serve as models of resilience under adverse circumstances; (h) an active and nonavoidant coping style; (i) social support from nondelinquent adults, friends, or, later on, partners; (j) success at school and a commitment to school values and norms; (k) social relations to nondelinquent peer groups or a degree of social isolation; (l) experiences of self-efficacy in nondelinquent activities and a positive but not unrealistically inflated self-image; (m) nonaggressive cognitive schemata, beliefs, and social information processing;

(n) experiences of structure and meaning in one's own life (e.g., sense of coherence); and (o) a socially integrated, nondeprived neighborhood, particularly for older children with a slighter risk of delinquency (Wikström & Loeber, 2000).

Obviously, many of these personal and social resources form the positive poles of the risks mentioned in Figure 1. Their protective effect is probably due to breaking the chain reactions of delinquent development described above. As with risk factors, it seems that an accumulation of several factors is particularly supportive of protective effects as well (Lösel et al., 1992; Stattin et al., 1997). The underlying processes and mechanisms are only beginning to be understood. What we do know, however, is that it is often the specific constellation that is decisive. An own study, for example, revealed that being not a clique member had a protective function for adolescents who exhibited externalizing behavior problems before (see Figure 2). The effect on youngsters with no prior aggression and delinquency was, in contrast, inverted. In these cases, social isolation seems to have a risk effect. Similar opposite effects as a function of prior deviance were found for satisfaction with social support (Bender & Lösel, 1997).

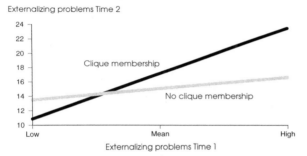

Figure 2: Interaction between clique membership and externalizing problems; Time 1 in the regression on externalizing problems. Time 2 for boys (from Lösel and Bender, 1997)

A positive self-image also seems to have a protective function for only some groups and some problems. For aggressive persons, it increases the risk of further deviance (Baumeister et al., 1996). Therefore, one should not make sweeping statements about risk and protective factors, but always ask "risk of what" and "protection against what?" Some characteristics that help high-risk youth to avoid a delinquent development or desist from a pattern of deviant behavior could represent a risk for internalizing disorders (see Lösel & Bender, in press). For example, Werner and Smith (1992) found that many resilients had no steady partnerships or reported health complaints. Such problems could well be the price they pay for a socially well-adjusted and successful development despite adverse circumstances.

Comparative research on the interaction of risk and protective mechanisms for various behavior disturbances in childhood and adolescence is only just beginning. It is therefore urgently necessary to also integrate characteristics that may have a "natural" protective function (without being general protective factors) when performing studies on the risk assessment and early prevention of antisocial behavior. Such studies of differential protective mechanisms may lead to a better causal understanding of developmental processes and improvement of prevention programs. We also still know far too little about how the developmental processes on the individual and microsocial level relate to macrosocial influences (see Rutter, 1995).

6. Risk Communication

Risk/need assessment cannot be reduced to technical issues regarding the respective instruments (e.g., reliability, validity, or cost utility). Measures of risk assessment are most meaningful when they are an integrated part of effective strategies of risk management. However, in practice, the availability of effective measures of risk assessment and management is not enough. Their potential efficacy is often not attained because of inadequate risk communication (e.g., Carson, 1998). This can be seen in a pilot study on the risk assessment of dangerous carers that we have performed as part of a European cooperation project (Lösel et al., 2000).

The study was designed to identify risk cases in which children have suffered serious abuse or neglect from their parents or other caregivers. An effective assessment and management of such cases is not just a direct protection of the children. It also contributes to the prevention of antisocial behavior, because abuse and neglect are risk factors in the family (Farrington & Loeber, 2001). In many of the cases in which children suffer from severe abuse and neglect, the family has already come to the attention of the social welfare services (Hagell, 1998). However, communication problems of professionals and between agencies impedes an effective intervention. Therefore, one goal of the project was to improve the assessment and management of dangerous caring through risk communication. This was performed with BridgeALERT, a practice-oriented instrument tapping the most important risk factors for abuse and neglect (see Jeyarajah Dent, 1998; Hagell, 1998). The 34 items cover three domains (characteristics of the family context, the carers, and the children). Each social service dealing with the family should report whether the risk factor in question definitely is present, seems to be present (but one is not certain), definitely is not present, or whether no statement can be made due to lack of information. Furthermore, the sources of information on which the response is based should also be reported.

Our German part of the pilot project used the BridgeALERT in 10 social services. Fifty-five cases with 88 children and 107 carers were assessed. In almost two-thirds of these cases, contact with the family had lasted more than 3 years or was already terminated. Nonetheless, social workers were unable to answer approximately 25% of the questions because of a lack of information. For approximately 14%, they were still uncertain. As Figure 3 shows, lack of information was particularly strong in the domain of carers.

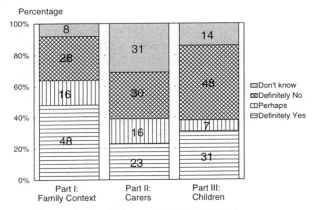

Figure 3: Response frequencies for the items of BridgeALERT risk assessment of dangerous care
(from Lösel, Bender and Holzberger, 2000)

For example, information was often missing on whether parents came from an abusing or disrupted family, whether they had behavioral problems, alcohol or drug abuse, mental health problems, a history of offenses and sentences, or noncompliance with treatment. The information deficits are even more striking when it is considered that the only data source for 70% of responses was personal experience or the core family. On the other hand, certainty about the answers was found to be related to the availability of various sources of information (e.g., teachers, day care, physicians, relatives, neighbors, or local authorities).

In part, information deficits were due to communication and cooperation problems arising from German specificities in the structure of social services and legal regulations. However, similar findings were also obtained in the pilot studies performed in England and Greece. To some extent, they are a consequence of more general problems of communication standards, institutional networks, and processes of case ownership in social work.

7. Prevention versus Treatment?

There are a number of good reasons for increasing early prevention of antisocial behavior in children and youth (e.g., Kazdin, 1993; Tremblay & Craig, 1995). Alongside such arguments, it is often also claimed that early prevention is necessary because later treatment of offenders is so unsuccessful (e.g., Gottfredson & Hirschi, 1990; Tremblay & Craig, 1995).

Naturally, scientists will always emphasize the merits of their own approach in the struggle to gain funding for projects. However, in material terms, the latter argument is too close to the "nothing works" ideology. In recent years, this has given way to a more constructive and differentiated view. A number of meta-analyses of controlled evaluation studies have shown that the treatment of juvenile or adult offenders does, in general, have positive effects (see Andrews et al., 1990; Lipsey, 1992; Lipsey & Wilson, 1998; Lösel, 1995a; McGuire, 2001). Although the mean effect size is rather small at approximately $r =$ .10, the large number of studies involved makes it highly significant. In addition, it is shown relatively consistently that certain types of programs have larger effects; others, smaller effects; whereas some have either no or even negative effects. Above-average effect sizes are found particularly for theoretically well-founded, cognitive-behavioral, and multimodal approaches that are tailored to fit the specific risk, the criminogenic needs, and the responsivity of the offenders (see Andrews et al., 1990; Lösel, 1995b). In controlled studies, such programs achieve effect sizes ranging from approximately .15 to .30. Figure 4 (left-hand side) summarizes the partial findings from 11 meta-analyses based on such programs (Lösel & Beelmann, 1998). The mean effect size weighted according to the number of studies was .22. This is still a small effect according to Cohen (1988). With an assumed recidivism rate of 50% in the untreated control group, however, an effect of .22 means that the treated group exhibits 22 percentage points or 44% less failure. From a methodological perspective, it is hardly realistic to anticipate larger treatment effects (Lösel, 1996, 2001), and in terms of cost-benefit, such effects may well be worthwhile (Welsh & 2001).

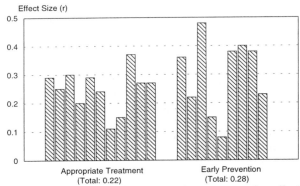

Figure 4: Mean effect sizes in various meta-analyses on appropriate offender treatment
and early prevention of anti-social behaviour
(from Lösel and Beelmann, 1998)

Although the argument that later treatment of antisocial behavior will fail is unfounded, it is still plausible for primary prevention or early intervention to have better effects because behavioral problems are not yet consolidated (e.g., Gottfredson & Hirschi, 1990). However, at present, this hypothesis also lacks clear confirmation. Although there is an increasing body of relatively controlled studies on universal, selective, or indicated prevention of antisocial behavior in children and youth (e.g., Beelmann, 2001; Catalano et al., 1998; Gottfredson, 2001; Wasserman & Miller, 1998), a general conclusion on the efficacy is difficult. Some studies report substantial long-term effects extending into adulthood (e.g., Schweinhart et al., 1993). However, it is questionable how far such results can be generalized. Other long-term evaluations give less ground for optimism, or have even reported negative effects for delinquency prevention (e.g., McCord, 1978). More typical are studies with much shorter follow-up intervals. Most of these use a broad spectrum of success measures on which findings may vary greatly. For criteria relating to antisocial behavior in daily life, recent studies on universal or selective prevention have produced effects of approximately r = .10 plus/minus .10 (e.g., Conduct Problems Research Group, 1999; Sanders et al., 2000). This means that they are of the same magnitude as the treatment effects reported above. As in offender treatment, it is not only outcomes that vary considerably but also types of program, research backgrounds, and contexts of implementation (e.g., early interventions in the family, preschool programs, school-based programs, special education, child guidance clinics, and community-oriented programs). Furthermore, it is doubted whether results from research-oriented model projects can be generalized to the everyday practice of prevention and intervention (Gottfredson, 2001; Weisz et al., 1995).

With such reservations in mind, Figure 4 (right-hand side) summarizes the mean effect sizes of various meta-analyses on early prevention and intervention for antisocial and other externalizing problems. The overall mean is .28, and thus has a similar magnitude to appropriate programs of offender treatment. This is plausible, insofar as most early prevention/intervention studies refer to behavioral, cognitive-behavioral, and multimodal programs. It should also be noted that this includes not only universal or selective approaches, but also programs of indicated prevention or child psychotherapy. The mean effects are generally more heterogeneous than for treatment, which is probably due to larger differences in the populations, contexts, and outcome measures. The use of methodologically stricter inclusion criteria seems to lead to smaller effects (e.g., Beelmann et al., 1994; Tremblay et al., 1999). The evaluations of early intervention programs also frequently have shorter follow-up intervals than those in studies on offender treatment.

Effects drop for follow-up intervals of more than 2 years (Beelmann et al., 1994). Furthermore, early intervention studies often include outcome measures that correspond closely to the contents of training or are "softer" than the recidivism data prevailing in offender treatment evaluations.

A sound comparison of different types of program would require more replication studies. Currently, social-cognitive skills training for children, cognitive-behavioral parent trainings, multisystemic interventions, and comprehensive programs that combine different approaches and contexts reveal the most promising outcomes (e.g., Beelmann et al., 1994; Farrington & Welsh, 1999; Kazdin, 1997; Tremblay et al., 1999). In all, it would seem that the basic characteristics of successful early intervention/prevention and later treatment programs are similar. However, there is a substantial lack of empirical evaluations of prevention programs, particularly based on studies with high internal and external validity and long-term behavioral outcomes (e.g., Lösel & Bliesener, 1999). We also know little about those characteristics of individuals, program implementation, settings, staff competence, and other moderators that lead to differential effects (e.g., Beelmann, 2001; Gottfredson, 2001). Here as well, conditions are similar to those found in offender treatment (Lösel, 2001). Instead of polarizations between early prevention and offender treatment, we need to develop systematic intervention packages that are suited to different phases of development and levels of deviance.

8. Further Issues

The issues I have discussed above are only a selection of the many that are relevant in early risk/need assessment and the prevention of seriously antisocial developments. A few other topics should be mentioned only briefly (for more details: LeBlanc, 1998; Tonry, 1987):

8.1. Ethical Issues

Although the early identification of youngsters who are particularly in need of help is carried out in the best interest of children, parents, and society, there is always a risk of negative effects. These may be a direct consequence of inappropriate programs or (perhaps more frequently) an indirect side effect of risk assessment and unsuccessful program participation. The latter result from labeling and stigmatization processes that can function as self-fulfilling prophecies. Although spectacular findings on self-fulfilling prophecies such as the Pygmalion effect (Rosenthal and Jacobson, 1968) have been insufficiently replicated (e.g., Elashoff & Snow, 1971), selective perceptions, attributions, and reactions of teachers or other significant others may contribute to deviant behavior (e.g., Carlen et a.l., 1992; Jussim, 1991). However, one should also not overlook the fact that labeling of "problematic" children takes place in everyday practice without a systematic assessment. In these cases, selective attributions may even be more dangerous because of the lack any correction through different data sources. Radical labeling theorists must also regard the ethical problems of a nonintervention strategy: Even though it knows better, should society do nothing to interrupt developmental pathways that lead to serious long-term problems for the youngsters themselves, the families, and many others?

There is no general solution for these dilemmas. Local program planners and field workers can only try to reduce them as far as possible. According to Tonry (1987), specific types of predictors should not be included in risk assessment (e.g., gender, SES, self-reported antisocial behavior). Universal prevention programs available to all children or families may be another solution (e.g., Catalano et al., 1998). However, they may be too expensive and/or not intensive enough for cases with higher risk. Recent findings in our research group suggest better program effects for the latter youngsters (Beelmann, 2001; Lösel et al., 2001). This is plausible, because universal prevention contains so many low-risk cases who reduce potential differences between program and control groups. Therefore, different modules or packages of programs based on multiple-gating assessments may be more adequate than universal programs alone.

8.2. Legal issues

Ethical issues of developmental delinquency prevention are closely related to legal questions. Data protection laws may forbid to assess specific areas of risk (see above). Other variables need to be excluded for reasons of discrimination (e.g., gender, ethnicity, religion, political attitudes). The legal problems of program implementation seem to be more difficult to solve. For example, it is a widespread experience that many families with the greatest needs for prevention programs are not willing to take advantage of them. For example, they do not accept program offers, show up irregularly, or drop out early. In most democratic countries, parents are not obliged to take part in any preventive programs. In Germany, for example, a substantial number of parents do not even complete all stages of the cost-free pediatric assessment program for young children. The state and its social services can intervene only without parental consent when very serious acts against the welfare of the child become obvious (e.g., physical abuse, sexual abuse, serious neglect). As mentioned above, even in those cases, parent's rights, regulations of confidence, and data protection laws may hinder effective early intervention. Although in some countries, social law contains lower thresholds for state interventions, the efficacy of forced parent trainings and the like is still unconfirmed.

The dilemma between basic rights of parents and developmental risks for children cannot be solved in a universal manner but must be tackled according to national conditions. School-based programs, for example, are easier to implement as part of the regular curriculum (Gottfredson, 2001). However, particularly for the highest risk groups, combinations with parental programs may be more promising than child trainings alone (Beelmann, 2001; Kazdin et al., 1992). Therefore, programs need thorough monitoring to reach the target groups, to reduce barriers, to offer adequate incentives, to promote positive group processes, and to create a climate of cooperation. In specific cases, such as young mothers who are living on welfare, smooth procedures of forced choice and supervision are also indicated. The development of a broader service network on the community level seems to be helpful in coping with problems of program implementation (e.g., Hawkins et al., 1992: "Community that cares").

8.3. Gender Issues

The majority of research on the antisocial behavior of young people addresses boys. This is due to the fact that young males are responsible for much more criminality and particularly for violent offenses. However, gender differences in prevalence are much less striking in areas such as shoplifting, drug offending, or relational forms of aggression (Crick & Grotpeter, 1995). Females are also particularly relevant for the transfer of risks to the next generation. Insofar, female antisocial behavior is a substantial part of the problem. More specific research on the aggression and delinquency of girls is needed because there seem to be developmental differences to males. For example, antisocial development in high-risk girls seems to be delayed (Silverthorn & Frick, 1999). Girls' delinquency also is related more closely to early puberty and heterosexual peers (Caspi et al., 1993; Stattin & Magnusson, 1996; Silbereisen & Kracke, 1993). Furthermore, internalizing problems and family factors seem to be more important for girls' deviant behavior (Lösel & Bliesener, 1998; McCord & Ensminger, 1997. Similar to risk factors, there are gender differences in protective mechanisms that are related, for example, to heterosexual partnerships (Quinton et al., 1993). Although these and other findings do not suggest a basically different approach to early developmental assessment and prevention, they require systematic testing of the generalizability of research based on male samples.

8.4. Cultural Issues

Similar to the gender topic, developmental risks and prevention strategies may vary between cultures or ethnic groups. On the international level, this issue is related to legal differences, cultural norms, and traditions. Insofar it is questionable whether the mainly Anglo-American findings in developmental criminology can be generalized to other nations and cultures. Although international cooperation has increased substantially, there are only very few longitudinal studies that systematically compare the development of criminals in different countries (e.g., Farrington & Wikström, 1994). Whereas we can expect more similarities between North America and Western Europe, the question of intercultural generalizability is more important with respect to Asia, South America, or Africa. The respective family structures and socialization, traditions of shame- and guilt orientation, the economic and political situation, and other framing characteristics may lead to differences in the risks and needs of young people. Karstedt (2000), for example, demonstrated that in contrast to some Western hypotheses, collectivism in societies seems to go along with more serious criminality than individualism. Cultural factors must also be taken into account when planning prevention programs. For example, North-American concepts based on expressive modes of self-presentation may be less adequate for European or, in particular, Asian societies that emphasize psychological intimacy and saving face. Similar to international cultural differences, they should also be considered on the national level (e.g., ethnic or religious minorities, aborigines, migrants). For example, whereas harsh discipline is more normative in Turkish families in Germany, broken homes are less prevalent (Bliesener & Lösel, in press). Cultural differences are not only relevant on the scientific level but even more important for the practice of program implementation. The respective issues range from language problems to the necessity of selecting only female staff for home visits to Muslim mothers.

9. Conclusions

Longitudinal findings on early starting and relatively persistent antisocial children and youth strongly suggest the value of systematic risk/need assessment and related early prevention/intervention measures for this group. However, although this is a very promising path of policy, much more research is needed to develop well-replicated strategies of risk assessment, risk communication, and risk management. The present chapter addresses a selection of issues that should be taken into account to ensure successful development in this field:

There is a very large number of risks for antisocial behavior that mostly have small predictive efficacy. Models and instruments are needed that both integrate and differentiate empirically well-founded predictions of these factors according to principles of equifinality and multifinality in development. Although many predictions may attain validity coefficients that are highly relevant for practice, there remains a substantial amount of unavoidable errors. These should be regarded not just as a technical problem of instruments but also as a basic aspect of change in human development that is not yet addressed in models of prediction. However, predictive accuracy is only one aspect in risk management. The efficacy of the respective prevention/intervention programs, their cost-utility, potential negative side effects, and other issues are also relevant for choosing more or less universal, selective, or indicated approaches of prevention. Relatively valid assessments of youngsters must be based on multiple-setting-multiple-informant approaches. They should include not only risks/needs in the sense of deficits but also protective factors and mechanisms. However, the negative or positive function is not a fixed quality of a variable but can vary in relation to different behavioral outcomes and combinations with other factors. With respect to such mechanisms, assessment should not only be based on variable-oriented developmental approaches but also include person-oriented strategies (e.g., pattern and configurational analyses). In any case, effective assessment and management need adequate structures of communication and cooperation between various institutions or practitioners. If this is not ensured, there is a danger that adequate help may even not be provided for children with very serious risks/needs. An integrative framework is also suggested for the various developmental phases of intervention. Although there is still a lack of more controlled evaluations, universal and selective prevention programs seem to have similar effects as indicated prevention or later offender treatment and rehabilitation programs. Principles of successful programming and evaluation problems also reveal more similarity than differences. At all levels of risk and intervention, research and practice need to tackle further issues such as ethical questions, legal regulations, gender differences, and cultural influences.

This chapter reveals both substantial knowledge and many unsolved problems in the field of developmental risk/need assessment and prevention/intervention. My impression is that the NATO ARW at Krakau strengthened the first part of this conclusion. However, one should not weigh both parts against each other, because progress in research always solves some problems while increasing our sensitivity for others.

References

Achenbach, T.M., McConaughy, S.H., & Howell, C.T. (1987). Child/adolescent behavioral and emotional problems: Implications of cross-informant correlations for situational specifity. *Psychological Bulletin, 101,* 213-232.

Andrews, D.A., Zinger, I., Hoge, R.D., Bonta, J., Gendreau, P., & Cullen, F.T. (1990). Does correctional treatment work? A clinically relevant and psychologically informed meta-analysis. *Criminology, 28,* 369-404.

Augimeri, L.K., Koegl, C.J., Webster, C.D., & Levene, K.S. (2001). *Early assessment risk list for boys (EARL-20B), version 2.* Toronto: Earlscourt Child and Family Centre.

Bandura, A. (1986). *Social foundations of thought and action.* Englewood Cliffs, NJ: Prentice Hall.

Bartel, P., & Forth, A. (2000, March). The development and use of the structural assessment of violence risk in youth (SAVRY). *Paper presented at the Biennial Conference of the American Psychology-Law Society,* New Orleans.

Baumeister, R.F., Smart, L., & Boden, J.M. (1996). Relation of threatened egotism to violence and aggression: the dark side of high self-esteem. *Psychological Bulletin, 103,* 5-33.

Beelmann, A. (2001). *Prävention dissozialer Entwicklungen: Psychologische Grundlagen und Evaluation früher kind- und familienbezogener Interventionsmaßnahmen.* Habilitationsschrift. Universität Erlangen-Nürnberg: Philosophische Fakultät I.

Beelmann, A., Pfingsten, U., & Lösel, F. (1994). Effects of training social competence in children: A meta-analysis of recent evaluation studies. *Journal of Clinical Child Psychology, 23,* 260-271.

Bender, D., & Lösel, F. (1997). Protective and risk effects of peer relations and social support on antisocial behaviour in adolescents from multi-problem milieus. *Journal of Adolescence, 20,* 661-678.

Bergman, L.R., & Magnusson, D. (1997). A person-oriented approach in research on developmental psychopathology. *Development and Psychopathology, 9,* 291-319.

Blackburn, R. (1993). *The psychology of criminal conduct: Theory, research, and practice.* Chichester: Wiley.

Blackburn, R. (2000). Treatment or incapacitation? Implications of research on personality disorders for the management of dangerous offenders. *Legal and Criminological Psychology, 5,* 1-21.

Bland, M. (1996). *An introduction to medical statistics.* Oxford: Oxford University Press.

Bliesener, T., & Lösel, F. (in press). *Aggression und Gewalt unter Jugendlichen.* Neuwied: Luchterhand.

Block, J. (1971). *Lives through time.* Berkeley, CA: Bancroft Books.

Blumstein, A., Farrington, D,P., & Moitra, S.D. (1985). Delinquency careers: Innocents, desisters, and persisters. In M. Tonry & N. Morris (Eds.), *Crime and justice* (Vol. 6, pp. 187-219). Chicago: University of Chicago Press.

Boyce, W.T., Frank, E., Jensen, P.S., Kessler, R.C., Nelson, C.A., Steinberg, L., & The MacArthur Foundation Research Network on Psychopathology and Development (1998). Social context in developmental psychopathology: Recommendations for future research from the MacArthur Network on Psychopathology and Development. *Development and Psychopathology, 10,* 143-164.

Bronfenbrenner, U., & Ceci, .J. (1994). Nature-nurture reconceptualized in developmental perspective: A bioecological model. *Psychological Review, 101,* 568-586.

Campbell, S.B. (1995). Behavior problems in preschool children: A review of recent research. *Journal of*

Child Psychology and Psychiatry, 36, 113-149.

Carlen, P., Gleeson, D., & Wardhaugh, J. (1992). *Truancy: The politics of compulsory schooling*. Buckingham: Open University Press.

Carson, D. (1998). Reducing the riskiness of risk assessment. In R. Reyarajah Dent (Ed.), *Dangerous care* (pp. 90-102). London: The Bridge Child Care Development Service.

Caspi, A., Lynam, D., Moffitt, T.E., & Silva, P.A. (1993). Unraveling girls' delinquency: Biological, dispositional, and contextual contributions to adolescent misbehavior. *Developmental Psychology, 29*, 19-30.

Catalano, R.F., Arthur, M.W., Hawkins, J.D., Berglund, L., & Olson, J.J. (1998). Comprehensive community- and school-based interventions to prevent antisocial behavior. In R. Loeber & D.P. Farrington (Eds.), *Serious & violent juvenile offenders: Risk factors and successful interventions* (248-283). Thousand Oaks: Sage.

Cicchetti, D., & Garmezy, N. (1993). Editorial. Prospects and promises in the study of resilience. *Development and Psychopathology, 5*, 497-502.

Cicchetti, D., & Rogosch, F.A (1996). Equifinality and multifinality in developmental psychopathology. *Development and Psychopathology, 8*, 597-600.

Cicchetti, D., & Rogosch, F.A. (1997). The role of self-organization in the promotion of resilience in maltreated children. *Development and Psychopathology, 9*, 797-815.

Cohen, J. (1988). *Statistical power analysis for the behavioral sciences* (2nd ed). New York: Academic Press.

Conduct Problems Prevention Research Group (1999a). Initial impact of the Fast Track Prevention Trial for Conduct Problems: I. The high-risk sample. *Journal of Consulting and Clinical Psychology, 67*, 631-647.

Conduct Problems Prevention Research group (1999b). Initial impact of the Fast track Prevention trial for conduct problems: II. Classroom effects. *Journal of Consulting and Clinical Psychology, 67*, 648-657.

Cowen E.L., Wyman, P.A., Work, W.C., Kim, J.Y., Fagen, D.B., & Magnus, K.B. (1997). Follow-up study of young stress-affected and stress-resilient urban children. *Development and Psychopathology, 9*, 565-677.

Crick, N.R., & Dodge, K.A. (1994). A review and reformulation of social information-processing mechanisms in children's social adjustment. *Psychological Bulletin, 115*, 74-101.

Crick, N.R., & Grotpeter, J.K. (1995). Relational aggression, gender, and social psychological adjustment. *Child Development, 66*, 710-722.

Douglas, K.S., Cox, D.N., & Webster, C.D. (1999). Violence risk assessment: Science and practice. *Legal and Criminological Psychology, 4*, 149-184.

Dunst, C.J., & Trivette, C.M. (1997). Early intervention with young at-risk children and their families. In R.T. Ammerman & M. Hersen (Eds.), *Handbook of prevention and treatment with children and adolescents* (pp. 157-180). New York: Wiley.

Elashoff, J.D., & Snow, E. (1971). A case study in statistical inference: Reconsideration of the Rosenthal-Jacobson data on teacher expectancy. Stanford: Stanford University Press.

Ensminger, M.E., Kellam, S.G., & Rubin, B.R. (1983). School and family origins of delinquency: Comparison by sex. In K.T. Van Dusen & S.A. Mednick (Eds.), *Prospective studies of crime and delinquency* (pp. 73-97). Boston: Kluwer-Nijhoff.

Farrington, D.P. (1987). Early precursors of frequent offending. In J.Q. Wilson & G.C. Loury (Eds.), *From children to citizens* (Vol. 3, pp. 27-50). New York: Springer.

Farrington, D.P. (1990). Age, period, cohort, and offending. In D.M. Gottfredson & R.V. Clarke (Eds.), *Policy and theory in criminal justice: Contributions in honour of Leslie T. Wilkins* (pp. 51-75). Aldershot: Avebury.

Farrington, D.P. (1992). Psychological contributions to the explanation, prevention, and treatment of offending. In F. Lösel, D. Bender, & T. Bliesener (Eds.), *Psychology and law: International perspectives* (pp. 35-51). Berlin, New York: De Gruyter.

Farrington, D.P. (1994). *Protective factors in the development of juvenile delinquency and adult crime*. Invited lecture at the 6th Scientific Meeting of the Society for Research in Child and Adolescent Psychopathology. London, June 1994.

Farrington, D.P. (1997). Early prediction of violent and nonviolent youthful offending. *European Journal on Criminal Policy and Research, 5*, 51-66.

Farrington, D.P., Gallagher, B., Morley, L., St. Ledger, R.J., & West, D.J. (1988). Are there any successful men from criminogenic backgrounds? *Psychiatry, 51*, 116-130.

Farrington, D.P., & Loeber, R. (2000). Some benefits of dichotomization in psychiatric and criminological research. *Criminal Behaviour and Mental Health, 10*, 100-122.

Farrington, D.P., & Loeber, R. (2001). Summary of key conclusions. In R. Loeber & D.P. Farrington (Eds.), *Child delinquents* (pp. 359-384). Thousand Oaks, CA: Sage

Farrington, D.P., Loeber, R., Elliott, D.S., Hawkins, J.D., Kandel, D.B., Klein, M.W., McCord, J., Rowe,

D.C., & Tremblay, R.E. (1990). Advancing knowledge about the onset of delinquency and crime. In B.B. Lahey & A.E. Kazdin (Eds.), *Advances in clinical and child psychology* (Vol. 13, pp. 283-342). New York: Plenum.

Farrington, D.P., & Tarling, R. (Eds.). (1985). *Prediction in criminology*. Albany, NY: State University of New York Press.

Farrington, D.P., & Welsh, B.C. (1999). Delinquency prevention using family-based interventions. *Children and Society, 13*, 287-303.

Farrington, D.P., & Wikström, P.-O. (1994). Criminal careers in London and Stockholm: A cross-national comparative study. In E.G.M. Weitekamp & H.-J. Kerner (Eds.), *Cross-national research on human development and criminal behavior* (pp. 65-89). Dordrecht, NL: Kluwer.

Frick, P.J. (1998). *Conduct disorders and severe antisocial behavior*. New York: Plenum Press.

Gendreau, P., Little, T., & Goggin, C. (1995). A meta-analysis of the predictors of adult offender recidivism: What works? *Criminology, 34*, 575-607.

Gest, S.D., Neeman, J., Hubbard, J.J., Masten, A.S., & Tellegen, A. (1993). Parenting quality, adversity, and conduct problems in adolescence: Testing process-oriented models of resilience. *Development and Psychopathology, 5*, 663-682.

Gigerenzer, G., & Selten, R. (Eds.)(2001). *Bounded rationality: The adaptive toolbox*. Cambridge, MA: MIT Press.

Glueck, S., & Glueck, E. (1950). *Unraveling juvenile delinquency*. Cambridge, MA: Harvard University Press.

Gottfredson, D.C. (2001). *Schools and delinquency*. Cambridge, UK: Cambridge University Press.

Gottfredson, D.M., & Tonry, M. (Eds.)(1987). Prediction and classification: Criminal justice decision making. *Crime and Justice: An Annual Review of Research* (Vol. 9). Chicago: University of Chicago Press.

Gottfredson, M., & Hirschi, T.M. (1990). A general theory of crime. Stanford, CA: Stanford University Press.

Grove, W.M., & Meehl, P.E. (1996). Comparative efficiency of informal (subjective, impressionistic) and formal (mechanical, algorithmic) prediction procedures: The clinical-statistical controversy. *Psychology, Public Policy and Law, 2*, 293-323.

Haapasalo, J., & Tremblay, R.E. (1994). Physically aggressive boys from ages 6 to 12: Family background, parenting behavior, and prediction of delinquency. *Journal of Consulting and Clinical Psychology, 62*, 1044-1052.

Hagell, A. (1998). *Dangerous care*. London: Policy Studies Institute.

Haney, C., & Zimbardo, P. (1998). The past and the future of U.S. prison policy. Twenty-five years after the Stanford Prison Experiment. *American Psychologist, 53*, 709-727.

Hart, S.D., Cox, D.N., & Hare, R.D. (1995). *The Hare Psychopathy Checklist: Screening Version (PVL:SV)*. Toronto: Multi-Health Systems.

Hawkins, J.D., Herrenkohl, T., Farrington, D.P., Brewer, D., Catalano, R.F., & Harachi, T.W. (1998). A review of predictors of youth violence. In R. Loeber & D.P. Farrington (Eds.), *Serious & violent juvenile offenders* (pp. 106-146). Thousand Oaks, CA: Sage.

Huesmann, L.R. (1997). Observational learning of violent behavior: Social and biosocial processes. In A. Raine, P.A. Brennan, D.P. Farrington, & S.A. Mednick (Eds.), *Biosocial bases of violence* (pp. 69-88). New York: Plenum Press.

Jeyarajah Dent, R. (1998). BridgeALERT: Key information for identifying children in danger. In R. Jeyarajah Dent (Ed.), *Dangerous care* (pp. 63-78). London: The Bridge Child Care Development Service.

Jussim, L. (1991). Social perception and social reality: A reflection-construction model. *Psychological Review, 98*, 54-73.

Kandel, E., Mednick, S.A., Kirkegaard-Sorensen, L., Hutchings, B., Knop, J., Rosenberg, R., & Schulsinger, F. (1988). IQ as a protective factor for subjects at high risk for antisocial behavior. *Journal of Consulting and Clinical Psychology, 56*, 224-226.

Karstedt, S. (2001). Die moralische Stärke schwacher Bindungen. Individualismus und Gewalt im Kulturvergleich. *Monatsschrift für Kriminologie und Strafrechtsreform, 84*, 226-243.

Kazdin, A.E. (1993). Treatment of conduct disorder: Progress and directions in psychotherapy research. *Development and Psychopathology, 5*, 277-310.

Kazdin, A.E. (1997). Parent management training: Evidence, outcomes, and issues. *Journal of the American Academy of Child and Adolescent Psychiatry, 36*, 1349-1356.

Kazdin, A.E., Siegel, T.C., & Bass, D. (1992). Cognitive problem-solving skills training and parent management training in the treatment of antisocial behavior in children. Journal

Lahey, B.B., Loeber, R., Hart, E., & Frick, P. (1995). Four-year longitudinal study of conduct disorder in boys: Patterns and predictors of persistence. *Journal of Abnormal Psychology, 104*, 83-93.

Lavigne, J.V., Arend, R., Rosenbaum, D., Binns, H.J., Christoffel, K.K., & Gibbons, R.D. (1998). Psychiatric disorders with onset in the preschool years: I. Stability of diagnosis. *Journal of the American Academy of Child and Adolescent Psychiatry, 37*, 1246-1254.

LeBlanc, M. (1998). Screening of serious and violent juvenile offenders: Identification, classification, and prediction. In R. Loeber & D.P. Farrington (Eds.), *Serious & violent juvenile offenders: Risk factors and successful interventions* (pp. 167-193). Thousand Oaks, CA: Sage.

LeBlanc, M., & Loeber, R. (1993). Precursors, causes and the development of offending. In D.F. Hale & A. Angold (Eds.), *Precursors, causes and psychopathology* (pp. 233-264). New York: Wiley.

Lipsey, M.W. (1992). The effect of treatment on juvenile delinquents: Results from meta-analysis. In F. Lösel, D. Bender & T. Bliesener (Eds.), *Psychology and law: International perspectives* (pp. 131-143). Berlin, New York: de Gruyter.

Lipsey, M.W., & Derzon, J.H. (1998). Predictors of violent or serious delinquency in adolescence and early adulthood: A synthesis of longitudinal research. In R. Loeber & D.P. Farrington (Eds.), *Serious & violent juvenile offenders* (pp. 86-105). Thousand Oaks, CA: Sage.

Lipsey, M.W., & Wilson, D.B. (1998). Effective intervention for serious juvenile offenders: A synthesis of research. In R. Loeber & D.P. Farrington (Eds.), *Serious & violent juvenile offenders* (pp. 313-345). Thousand Oaks, CA: Sage.

Loeber, R., & Dishion, T.J. (1983). Early predictors of male delinquency: A review. *Psychological Bulletin, 94,* 68-99.

Loeber, R., & Farrington, D.P. (1994). Problems and solutions in longitudinal and experimental treatment studies of child psychopathology and delinquency. *Journal of Consulting and Clinical Psychology, 62,* 887-900.

Loeber, R., & Farrington, D. (Eds.) (1998). *Serious and violent juvenile offenders: Risk factors and successful interventions.* Thousand Oaks, CA: Sage.

Loeber, R., & Farrington, D.P. (Eds.) (2001). *Child delinquents: Development, intervention, and service needs.* Thousand Oaks, CA: Sage.

Loeber, R., Farrington, D.P., & Waschbusch, D.A. (1998). Serious and violent juvenile offenders. In R. Loeber & D.P. Farrington (Eds.), *Serious & violent juvenile offenders* (pp. 13-29). Thousand Oaks, CA: Sage.

Loeber, R., & Stouthamer-Loeber, M. (1998). Development of juvenile aggression and violence: Some common misconceptions and controversies. *American Psychologist, 53,* 242-259.

Lösel, F. (1995a). Increasing consensus in the evaluation of offender rehabilitation? *Psychology, Crime and Law, 2,* 19-39.

Lösel, F. (1995b). The efficacy of correctional treatment: A review and synthesis of meta-evaluations. In J. McGuire (Ed.), *What works: reducing reoffending* (pp. 79-111). Chichester: Wiley.

Lösel, F. (1996). Changing patterns in the use of prisons: An evidence-based perspective. *European Journal on Criminal Policy and Research, 4 (3),* 108-127.

Lösel, F. (1998). Treatment and management of psychopaths. In D.J. Cooke, A.E. Forth, & R.B. Hare (Eds.), *Psychopathy: Theory, research, and implications for society* (pp. 303-354). Dordrecht, Netherlands: Kluwer.

Lösel, F. (2001). Evaluating the effectiveness of correctional programs: Bridging the gap between research and practice. In G.A. Bernfeld, D.P. Farrington, & A.W. Leschied (Eds.), *Offender rehabilitation in practice* (pp. 67-92). Chichester, UK: Wiley.

Lösel, F., & Beelmann, A. (1998). Is early delinquency prevention more effective than later offender treatment? *Paper presented at the 8th European Conference on Psychology and Law,* Cracow, PL.

Lösel, F., Beelmann, A., Jaursch, S., Koglin, & Stemmler, M. (2001). *Förderung von Erziehungskompetenzen und sozialen Fertigkeiten in Familien: Eine kombinierte Präventions- und Entwicklungsstudie zu Störungen des Sozialverhaltens (Zwischenbericht).* Universität Erlangen-Nürnberg: Institut für Psychologie.

Lösel, F., & Bender, D. (in press). Resilience and protective factors. In D.P. Farrington & J. Coid (Eds.), *Prevention of adult antisocial behavior.* Cambridge: Cambridge University Press.

Lösel, F., Bender, D., & Holzberger, D. (2000). *Risk assessment of dangerous carers in Germany: Results of the pilot study on BridgeALERT-G.* Report for the Daphne Program of the European Union. Erlangen-Nürnberg: Institute of Psychology.

Lösel, F., & Bliesener, T. (1990). Resilience in adolescence: A study on the generalizability of protective factors. In K. Hurrelmann & F. Lösel (Eds.), *Health hazards in adolescerce* (pp. 299-320). Berlin: de Gruyter.

Lösel, F., & Bliesener, T. (1994). Some high-risk adolescents do not develop conduct problems: A study of protective factors. *International Journal of Behavioral Development, 17,* 753-777.

Lösel, F., & Bliesener, T. (1998). Zum Einfluß des Familienklimas und der Gleichaltrigengruppe auf den Zusammenhang zwischen Substanzengebrauch und antisozialem Verhalten von Jugendlichen. *Kindheit und Entwicklung, 7,* 208-220.

Lösel, F., & Bliesener, T. (1999). Germany. In P.K. Smith, Y. Morita, J. Junger-Tas, D. Olweus, R. Catalano, & P. Slee (Eds.), *The nature of school bullying: A cross-national perspective* (pp. 224-249). London: Routledge.

Lösel, F., Bliesener, T., & Köferl, P. (1991). Erlebens- und Verhaltensprobleme bei Jugendlichen: Deutsche

Adaption und kulturvergleichende Überprüfung der Youth Self-Report Form der Child Behavior Checklist. *Zeitschrift für Klinische Psychologie, 20*, 22-51.

Lösel, F., Bliesener, T., Fischer, T., & Pabst, M. (2001). *Hooliganismus in Deutschland*. Berlin: Bundesministerium des Innern.

Lösel, F., Kolip, P., & Bender, D. (1992). Stress-Resistenz im Multiproblem-Milieu: Sind seelisch widerstandsfähige Jugendliche Superkids"? *Zeitschrift für Klinische Psychologie, 21*, 48-63.

Luthar, S.S., Cicchetti, D, & Becker, B. (2000). The construct of resilience: A critical evaluation and guidelines for future work. *Child Development, 71*, 543-562.

Lynam, D.R. (1996). Early identification of chronic offenders: Who is the fledgling psychopath? *Psychological Bulletin, 120*, 209-234.

Magnusson, D., & Bergman, L.R. (1988). Individual and variable-based approaches to longitudinal research on early risk factors. In M. Rutter (Ed.), *Studies of psychosocial risk: The power of longitudinal data* (pp. 45-61). Cambridge: Cambridge University Press.

Magnusson, D., & Cairns, R.B. (1996). Developmental science: Toward a unified framework. In R.B. Cairns, G.H. Elder Jr., E.J. Costello, & A. McGuire (Eds.), *Developmental science* (pp. 7-30). Cambridge: Cambridge University Press.

McCord, J. (1978). A thirty-year follow-up of treatment effects. *American Psychologist, 33*, 284-289.

McCord, J., & Ensminger, M.E. (1997). Multiple risks and comorbidity in an African-American population. *Criminal Behaviour and Mental Health, 7*, 339-352.

McCord, W., McCord, J., & Zola, I.K. (1959). *Origins of crime*. New York: Columbia University Press.

McGuire, J. (2001). What works in correctional intervention? Evidence and practical implications. In G.A. Bernfeld, D.P. Farrington & A.W. Leschied (Eds.), *Offender rehabilitation in practice* (pp. 25-43). Chichester, UK: Wiley.

Megargee, E.I. (1966). Undercontrolled and overcontrolled personality types in extreme antisocial aggression. *Psychological Monographs, 80*, Whole No. 611.

Moffitt, T.E. (1993a). Adolescence-limited and life-course-persistent antisocial behavior: A developmental taxonomy. *Psychological Review, 100*, 674-701.

Moffitt, T.E., Caspi, A., Dickson, N., Silva, P., & Stanton, W. (1996). Childhood-onset versus adolescent-onset antisocial conduct problems in males: natural history from ages 3 to 18 years. *Development and Psychopathology, 8*, 399-424.

Nagin, D., & Tremblay, R.E. (1999). Trajectories of boys' physical aggression, opposition, and hyperactivity on the path to physically violent and nonviolent juvenile delinquency. *Child Development, 70*, 1181-1196.

Olweus, D. (1979). Stability of aggressive reaction patterns in males: A review. *Psychological Bulletin, 86*, 852-875.

Patterson, G.R., Forgatch, M.S., Yoerger, K.L., & Stoolmiller, M. (1998). Variables that initiate and maintain an early-onset trajectory for juvenile offending. *Development and Psychopathology, 10*. 531-547.

Plomin, R. (1994). *Genetics and experience*. Newbury Park, CA: Sage.

Quinton, D., Pickles, A., Maughan, B., & Rutter, M. (1993). Partners, peers and pathways: Assortive pairing, and continuities in conduct disorder. *Development and Psychopathology, 5*, 763-783.

Raine, A. (1993). *The psychopathology of crime*. San Diego: Academic Press.

Raine, A. (1997). Antisocial behavior and psychophysiology: A biosocial perspective and a prefrontal dysfunction hypothesis. In D.M. Stoff, J. Breiling, & J.D. Maser (Eds.), *Handbook of antisocial behavior* (pp. 289-304). New York: Wiley.

Raine, A., Farrington, D.P., Brennan, P., & Mednick, S.A. (Eds.) (1997). *Biosocial bases of violence*. New York: Plenum Press.

Robins, L.N. (1978). Sturdy childhood predictors of adult antisocial behavior: Replications from longitudinal studies. *Psychological Medicine, 8*, 611-622.

Rosenthal, R., & Jacobson, L. (1968). *Pygmalion in the classroom*. New York: Holt, Rinehart & Winston.

Rutter, M. (1985). Resilience in the face of adversity. Protective factors and resistance to psychiatric disorder. *British Journal of Psychiatry, 147*, 598-611.

Rutter, M. (1995). Causal concepts and their testing. In M. Rutter & D.J. Smith (Eds.), *Psychosocial disorders in young people: Time trends and their causes* (pp. 7-34). Chichester: Wiley.

Rutter, M. (1997). Comorbidity: concepts, claims and choices. *Criminal Behaviour and Mental Health, 7*, 265-285.

Sameroff, A.J., & Fiese, B.H. (2000). Transactional regulation: The developmental ecology of early intervention. In J.P. Shonkoff & S.J. Meisels (Eds.), *Handbook of early child intervention* (pp. 135-159). Cambridge: Cambridge University Press.

Sanders, M.R., Markie-Dadds, C., Tully, L.A., & Bor, W. (2000). The Triple P-Positive Parenting Program: A comparison of enhanced, standard, and self-directed behavioral family interventions for parents of children with early onset conduct problems. *Journal of Consulting and Clinical Psychology, 68*, 624-640.

Schweinhart, L.L., Barnes, H.V., & Weikhart, D.P. (1993). Significant benefits: The *High/Scope Perry Preschool Study through age 27*. Ypsilanti, MI: High/Scope Press.

Silbereisen, R.K., & Kracke, B. (1993). Variation in maturational timing and adjustment in adolescence. In S. Jackson & H. Rodriguez-Tomé (Eds.), *Adolescence and its social worlds* (pp. 67-94). Hove: Erlbaum.

Silverthorn, P., & Frick, P. (1999). Developmental pathways to antisocial behavior: The delayed-onset pathway in girls. *Development and Psychopathology, 11*, 101-126.

Stattin, H., Romelsjö, A., & Stenbacka, M. (1997). Personal resources as modifiers of the risk for future criminality. *British Journal of Criminology, 37*, 198-223.

Stattin, H., & Magnusson, D. (1996). Antisocial development: A holistic approach. *Development and Psychopathology, 8*, 617-645.

Stemmler, M., Lösel, F., & Erzigkeit, H. (1998). Analysis of longitudinal data: Introduction to special issue. *Methods of Psychological Research, 3*, 1-6.

Stouthamer-Loeber, M., Loeber, R., Farrington, D.P., Zhang, Q., van Kammen, W., & Maguin, E. (1993). The double edge of protective and risk factors for delinquency: Interrelations and developmental patterns. *Development and Psychopathology, 5*, 683-701.

Susman, E.J., & Ponirakis, A. (1997). Hormones-context interactions and antisocial behavior in youth. In A. Raine, P.A. Brennan, D.P. Farrington, & S.A. Mednick (Eds.), *Biosocial bases of violence* (pp. 251-269). New York: Plenum.

Thornberry, T.P. (1998). Membership in youth gangs and involvement in serious and violent offending. In R. Loeber & D.P. Farrington (Eds.), *Serious & violent juvenile offenders: Risk factors and successful interventions* (pp. 147-166). Thousand Oaks, CA: Sage.

Tonry, M. (1987). Prediction and classification: Legal and ethical issues. In D.M. Gottfredson & M. Tonry (Eds.), *Prediction and classification: Criminal justice decision making. Crime and Justice: An Annual Review of Research* (Vol. 9, pp. 367-423). Chicago: Chicago University Press.

Tremblay, R.R. (2000). The development of aggressive behavior during childhood: What have we learned in the past century? *International Journal of Behavioral Development, 24*, 129-141.

Tremblay, R.E., & Craig, W.M. (1995). Developmental crime prevention. In M. Tonry & D. Farrington (Eds.), *Building a safer society: Strategic approaches to crime prevention. Crime and Justice: An Annual Review of Research* (Vol. 19, pp. 151-236). Chicago: University of Chicago Press.

Tremblay, R.E., LeMarquand, D., & Vitaro, F. (1999). The prevention of oppositional defiant disorder and conduct disorder. In H.C.Quay & A.E. Hogan (Eds.), *Handbook of disruptive behavior disorders* (pp. 525-555). New York: Kluwer Academic Publisher/Plenum Press.

von Eye, A. (1990).*Statistical methods in longitudinal research*, 2 vols. San Diego: Academic Press.

Wasserman, G.A., & Miller, L.S. (1998). The prevention of serious and violent juvenile offending. In R. Loeber & D.P. Farrington (Eds.), *Serious & violent juvenile offenders* (pp. 197-247). Thousand Oaks, CA: Sage.

Weisz, J.R., Donenberg, G.R., Weiss, B., Han, S.S. (1995). Bridging the gap between laboratory and clinic in child and adolescent psychotherapy. *Journal of Consulting and Clinical Psychology, 63*, 688-701.

Welsh, B.C., & Farrington, D.P. (2001): Evaluating the economic efficiency of correctional intervention programs. In G.A. Bernfeld, D.P. Farrington, & A.W. Leschied (Eds.), *Offender rehabilitation in practice* (pp. 45-65). Chichester, UK: Wiley.

Werner, E.E., & Smith, R.S. (1982). *Vulnerable but invincible*. New York: McGraw-Hill. Press.

White, A.J., & Heilbrun, K. (1995). The classification of overcontrolled hostility: comparison of two diagnostic methods. *Criminal Behaviour and Mental Health, 5*, 106-123.

Wikström, P.-O, & Loeber, R. (2000). Do disadvantaged neighborhoods cause well-adjusted children to become adolescent delinquents? A study of male juvenile serious offending, individual risk and protective factors, and neighborhood context. *Criminology, 38*, 1109-1142.

Wolfgang, M.E., Figlio, R.M., & Sellin, T. (1972). *Delinquency in a birth cohort*. Chicago: Chicago University Press.

Zumkley, H. (1994). The stability of aggressive behavior: A meta-analysis. *German Journal of Psychology, 18*, 273-281.

Multi-Problem Violent Youth
R.R. Corrado et al. (Eds.)
IOS Press, 2002

58

Are Pre and Perinatal Factors Related to the Development of Criminal Offending?*

Sheilagh Hodgins, Lynn Kratzer and Thomas F. McNeil

Events occurring in the pre and perinatal periods affect the developing fetus just as events occurring during the rest of the life course impact on the individual. Little is known, however, about the influence of pre and perinatal events that modify brain development. For example, the full range of events which can negatively affect the fetus, whether or not these events change the brains of all persons in the same way, the times during the pre and perinatal periods when specific brain structures are modified, and whether or not the modifications to the fetal brain are permanent and resistant to change. Harm to the fetus may result from the genes it carries, from its mother's biological characteristics, from its mother's behaviors and emotions, from events occurring to its mother, or from interactions of these factors. Depending on the severity of the event and the time during development when it occurs, the location and extent of the fetal brain will vary (Nowakowski & Hayes, 1999). Severe events, for example the mother being beaten in the stomach or drinking large quantities of alcohol every day during the pregnancy, will cause generalized damage to the brain that can be visually documented with scans. Less severe events cause changes to the developing brain that are more subtle and that have been found to be associated with the development of different disorders and patterns of behavior later in life. For example, it has been shown that severe psychosocial stress on the mother during pregnancy increases the risk of schizophrenia in the offspring (Huttunen & Niskanen, 1978), as does exposure to the flu virus (Mednick, Huttunen, & Machon, 1994), or severe malnutrition (Susser et al., 1996). Maternal exposure to the flu virus, earthquakes, famine have each been associated with the development of major affective disorders (Brown, van Os, Driessens, Hoek, & Susser, 2000; Machon, Mednick, & Huttunen, 1997; Watson, Mednick, Huttunen, & Wang, 1999). Events occurring during the pre and perinatal periods may alter the developing fetal brain so as to directly lead to a disorder or pattern of behavior or they may interact with or add to other factors that determine these adult outcomes. If there is an association between prenatal and perinatal events and criminality, it is likely to vary by sub-type of offender. The offenders most likely to have been affected by such factors are those that have shown an early onset of behavior problems that are stable pattern across the lifespan.

This chapter reviews what is known about pre and perinatal factors and criminal offending. The chapter is divided into four sections. Section I briefly reviews methodological features of investigations of the associations between pre and perinatal events and criminality. These methodological issues are important for assessing the results of studies in this field. Section II, the most voluminous, reviews the investigations which have examined the associations between events noted in obstetrical files and criminal offending in adulthood, and presents two recent studies. Section III reviews recent investigations of the link between maternal smoking during pregnancy and criminality among male offspring in adulthood. Section IV reviews investigations of exposure to lead in utero and aggressive behavior.

1. Methodological Features of Investigations of the Role of Pre and Perinatal Factors and Offending

1.1 Definitions of Pre and Perinatal Events

The first important methodological feature to consider in reviewing this literature is the definition of pre and perinatal factors. As was noted, little is known about fetal brain development and the factors that facilitate or interrupt it. Consequently, at this stage of our understanding, any factor which disturbs the development of the foetus should be examined. However, this is not really feasible nor has it been done. Definitions of pre and perinatal events vary greatly from one investigation to another. Most investigations have narrowly defined pre and perinatal factors as obstetrical complications (OCs). These are defined as the broad class of deviations from an expected, normal course of events and offspring development during pregnancy, labor-delivery and the early neonatal period (McNeil, 1988).

Timing. Some studies have included only complications that occurred during pregnancy, whereas others include complications that have occurred at birth and during the neonatal period. The time during development when an insult resulting in a complication occurs is critical. While the type and severity of the insult may determine whether or not there is damage, the timing of the insult determines which brain structure is affected. As different parts of the brain are actively growing at different times during development, insults which interrupt development modify the parts of the brain that are active at the time the insult occurs (Mednick & Huttunen, 2000).

Source of information. A third and related methodological feature of these investigations which must be considered when interpreting the results is the source of information about the complications, and the way in which this information is extracted and coded. All of the published studies on pre and perinatal factors and antisocial or criminal behavior reviewed in Section II have obtained information from medical files. These clinical records of pregnancies and births are advantageous because of the objective nature of the information they include, but they have been assembled by staff whose primary responsible is not uncovering the association between fetal brain development and criminality, but rather keeping the mother and baby alive and healthy. Consequently, the quantity and quality of information in such files which is relevant to fetal brain development varies from one study to another and it may be insufficiently precise to measure an association with a pattern of behavior which becomes apparent in adulthood. Medical files do not include information that physicians have not considered life threatening, for example, maternal smoking and moderate drinking during pregnancy. This information may be critical to understanding the role of pre and perinatal factors in the development of offending, but it can only be obtained from the mothers themselves. It is obviously preferable to obtain information about the pregnancy from mothers prospectively in order to limit memory problems and/or distortion.

Coding of pre and perinatal information. The way in which the information extracted from medical files is coded has been shown to change the results (McNeil, Cantor-Graae, & Sjöström, 1994). This is not surprising given how little is known about brain development during the fetal period and about the longterm consequences for the brain of biological and psychosocial events which occur during this period. This means, however, that the importance of various bits of information is often unknown and leads to practical problems, for example, of not only knowing what exactly should be included as a "complication", but also of not knowing the relation between the numbers of complications and the severity of these complications and fetal development. Further, coding maternal behaviors which may pose a risk to the fetus is problematic because little is known about the associations between different degrees of severity and chronicity of these behaviors and harm to the fetus. For example, it is known that consumption by the mother of large quantities of drugs and alcohol during the pregnancy causes damage to the fetus that is evident at birth. Whether or not lower doses, or less frequent consumption of such substances has consequences for the fetus which are not detected at birth but which relate to behavior patterns which emerge later in life is not known.

Experimental design. Another important methodological feature to consider when assessing the extant literature is the experimental design. Two designs have been used to study the relation between perinatal factors and offending. Cross-sectional studies have compared the prevalence of perinatal complications experienced by offenders and nonoffenders. The principal weakness of such studies is the limited generalizability of the results due to the use of biased samples. A more powerful design for verifying the association between pre and perinatal complications and offending is a longitudinal, prospective study of an unselected birth cohort.

The sample. A sample may include sub-groups of subjects who are at differential risk to commit crimes and/or at differential risk to experience pre and perinatal events. For example, males are more likely to offend than females, persons who develop major mental disorders and those who are intellectually handicapped are more likely to offend than are non-disordered persons and more likely to experience OCs. If there is an association between OCs and offending in one of these sub-groups and not in another, this can only be detected by distinguishing the groups and conducting separate statistical analyses. Thus, in order to understand the relations between OCs and offending, it is essential to examine samples that are homogeneous with respect to risk of offending and risk of OCs.

The patterns of criminality. Finally, in attempting to understand the long-term consequences of events that occurred during the fetal period it is necessary to be specific about the behavior pattern being examined. The definitions of criminal, violent, aggressive, impulsive vary considerably across investigations. Further, caution must be exercised when using convictions for a crime, which are socially defined and determined, to index behavior. The extent to which an official conviction reflects a behavior and the extent to which multiple convictions reflect a stable pattern of behavior varies from one investigation to another, for different sub-groups within a population, and for different types of behavior. The most likely candidates to have been affected by pre and/or perinatal events which have altered fetal brain development in such a way as to increase the risk of offending are those who present stable patterns of illegal behaviors from an early age, the early-starters or life course persistent offenders.

2. A Review of Studies of the Associations Between Pre and Perinatal Events Noted in Obstetrical Files and Criminal Offending

2.1 Cross-Sectional Studies

Shanok and Lewis (1981) examined differences in OCs among groups of delinquent girls who varied in aggressive behavior. Three groups of girls, randomly selected and matched for age and race were examined: 1) incarcerated delinquents; 2) never incarcerated delinquents; and 3) nondelinquents. No differences were found in the prevalence of OCs between the delinquent nonincarcerated girls and the nondelinquent girls. Differences in the prevalence of obstetric complications, however, were found between the two groups of offenders: Six of the nine delinquent incarcerated girls as compared to two of the 19 delinquent nonincarcerated girls had experienced perinatal difficulties. The majority of the delinquent incarcerated girls had been institutionalized because their aggressive behavior made them unmanageable at home or in other residential settings, whereas most of the delinquent nonincarcerated girls had committed only property offenses. Although these results suggest a relationship between OCs and violent offending but not offending generally in young females, they must be viewed with extreme caution due to the small number of subjects and the lack of information on the definition and coding of OCs. In a previous investigation by the same authors (Lewis & Shanok, 1971), similar results were obtained indicating no association between OCs and offending in general.

Longitudinal Prospective Cohort Studies. Consider first the investigations that have not found an association between OCs and later criminality. Denno (1990) studied the Philadelphia cohort (487 males and 500 females, African-Americans) of the Collaborative Perinatal Project. She found no evidence for a relationship between pregnancy and delivery complications and police contacts and arrests from ages 7 to 22. The subjects in this cohort had, on average, a slightly lower socioeconomic status than the general U.S. population and included a disproportionate number of young mothers. Denno attributed the lack of association between OCs and criminal behavior to the infrequent occurrence of certain pregnancy and birth complications and to the homogeneity of the sample.

Farrington (1997) has prospectively followed a cohort of boys from a poor neighborhood in London. He reported that "pregnancy and delivery complications did not significantly predict official or self-reported violence" in adulthood (p. 20). More detailed information about this finding, however, cannot be found.

The Danish Perinatal Project is a prospective longitudinal investigation of more than 9,000 consecutive births. Three studies that have examined samples from this cohort found no relation between pregnancy and/or birth complications and criminality (Kandel & Mednick, 1991; Raine, Brennan, & Mednick, 1994, 1997). Kandel and Mednick (1991) sampled three groups of males and females from this cohort. In the first group ($n=72$), subjects had a parent who was schizophrenic and in the second group ($n=72$), subjects had either a mother with a character disorder or a father with psychopathy. The third group of subjects ($n=72$) consisted of children of parents with no psychiatric history. The three groups were matched on a number of variables (i.e., social class, sex of child, race, multiple birth status, pregnancy number, sex of ill parent, mother's age, mother's height, and father's age). The authors did not clearly report whether a random sample was drawn for each of the first two groups or whether they sampled all parents with the given psychiatric condition. In Kandel and Mednick's study, pregnancy complications were not related to violent offending up to age 22. Raine and colleagues, in two separate studies, (1994, 1997) selected all the male subjects ($n=4,269$), with the exception of those who were born at less than 20 weeks. In this all male sample, birth complications were not associated with violent offending between the ages of 20 and 22 years of age and 33 and 35 years of age.

Consider now the investigations that claim to have found a relationship between OCs and offending. Litt (1972) examined a Danish birth cohort from the 1930s. He reported that although OCs were not related to criminality in general, documented up to age 36, they were associated with impulsive offenses. In order to identify individuals with impulsive behavior, all subjects with a criminal record were rated on a "poor inhibitory control" (PIC) scale. "The PIC scale was developed in order to provide a set of indicators which would identify those criminals whose anti-social actions might be related more specifically to pbcs [pregnancy and birth complications], such that neurological damage could reasonably be inferred from the impulsive nature of their behavior" (p. 61). The PIC scale was used to create five groups of offenders: 1) not impulsive; 2) repetition (more than 3 crimes of exactly the same type or more than 5 convictions of any type; 3) nonviolent sex; 4) violent sex, minor violence; and 5) major violence. The severity of pregnancy and delivery complications were rated on a 6-point scale. The sum of the severity ratings and the severity score for the most severe OC were compared across the groups. The repetition group and the violent sex/minor violence group had the highest mean ratings for the most severe delivery complication. Based on these results, Litt concluded that there was evidence of an association between OCs and impulsive criminality. This conclusion should be viewed cautiously for three reasons. One, as Litt acknowledged, in order to assess the degree of impulsivity associated with a criminal act, a more detailed scale that takes account of the circumstances and motivation is needed. Additionally, no details were provided on how the subjects were assigned to the PIC categories, whether or not these categories were mutually exclusive, and why certain categories of offenders, specifically the groups with more severe OCs, were considered to have engaged in more impulsive crimes than the other categories of offenders. Two, the overall "F" comparing the PIC groups on mean ratings for most severe delivery complication was significant at the probability value of .05, but given the large number of tests conducted in the study, a more conservative alpha value should have been used to reduce the possibility of chance findings. Furthermore, Litt used t-tests to compare the individual mean scores of the five PIC groups and again no adjustment of alpha values was made to take account for chance findings as a result of multiple tests. The third reason for cautious interpretation of the results was provided by Litt himself, "the large number of negative insignificant results seems to indicate that, in general, pbcs [pregnancy and birth complications] considered in isolation are not a major etiological factor in criminal behavior" (p. 278).

In a study by Denno (1982), the results are also interpreted as supporting a relationship between OCs and a particular type of offending. That is, female offenders with a history of multiple offenses as compared to one-time female offenders had experienced more pregnancy and birth complications. This conclusion, like Litt's (1972) is also questionable because of statistical issues. Denno conducted an analysis of variance in order to compare four groups of subjects, nonoffenders, one-time offenders, offenders with two to four convictions, and offenders with five or more convictions. Even though the overall "F" was not significant, simple effects tests (i.e., Duncan Multiple Range) were used to compare the means for each group. The text suggests that the Duncan's Multiple Range test resulted in significant differences in the mean number of birth complications between certain groups, but the tables (tables 5.6 and 5.7) in her study indicate that there were no significant differences in group means.

Two studies that have examined samples drawn from the Danish Perinatal Project reported associations between OCs and criminal, aggressive, and impulsive behaviors (Baker & Mednick, 1984; Kandel & Mednick, 1991). As reported previously, Kandel and Mednick's sample was composed of males and females who had a parent with a mental disorder (schizophrenic, character-disorder, psychopathy) and a comparison group with parents with no psychiatric history. They found that birth complications, but not pregnancy complications, were associated with violent crimes committed up to age 22. Birth complications, however, only accounted for 1.6% of the variance in violent crime, which was more accurately predicted by an interaction between birth complications and parental mental disorder. Furthermore, the main effect for birth complications was only significant after controlling for socio-economic status, parental psychiatric diagnosis, mother's age, sex of the child and number of offenses. Baker and Mednick (1984) sampled males and females from this same cohort who were high and low in medical risk during the neonatal period and at the age of one year. Subjects assigned to the high-risk group (n=43) scored high in complications either during the neonatal period and/or at one year. Subjects assigned to the low risk group (n=32) had not experienced complications during either of these two periods. Infant medical risk was found to be significantly, but weakly associated with teacher's ratings of aggressive behavior at age 18. Similar to the study above, an interaction effect between perinatal factors and family factors (infant medical risk by family intactness) better accounted for aggressive behavior than perinatal factors alone. Furthermore, whereas infant medical risk alone was only weakly associated with teacher's reports of aggressive behavior, it was strongly associated with teacher's ratings of impulsivity at age 18. Specifically, approximately 9% of cohort members with low medical risk were rated as impulsive as compared to 50% of those with high medical risk.

Thus, all of the evidence that an interaction between OCs and psychosocial adversity contributes to the development of offending comes from studies of samples from a single investigation, the Danish Perinatal Project. As already presented, Kandel and Mednick (1991) demonstrated an interaction between birth complications and parental mental disorders and violent offending. Eighty percent of violent offenders as compared with 47% of nonoffenders scored in the high range of birth complications, however, the majority of violent offenders with birth complications had parents with a mental disorder (schizophrenia, character disorder, psychopathy). Furthermore, no significant differences were observed in birth complications between property offenders and nonoffenders. Although Kandel and Mednick (1991) included both males and females in their sample, analyses were not conducted separately for each sex.

Raine and colleagues (1996) examined a random sample of 397 males from the Danish Perinatal Project and found that a biosocial or interaction model better accounted for total crime, thievery and violent crime, up to age 22, than OCs alone. Convictions for crime were compared across three groups of subjects: 1) biosocial neurological problems in the first week of life, slow motor development at age one, early maternal rejection, family conflict, family instability and a parent with a criminal record; 2) OCs pregnancy complications, birth complications, prematurity and slow motor development; and 3) poverty relatively lower social, economic, education, employment, and living status. A greater proportion of males in the biosocial group than in the other two groups were convicted of crimes, and a greater proportion of males in the biosocial group than in the obstetric complications group were convicted for violent crimes.

In another study, Raine and colleagues (1994) selected all male subjects (with the exception of deliveries under 20 weeks) from the entire Danish (n=4269) cohort. They demonstrated an interaction between birth complications and early maternal rejection, but not between birth complications and socioeconomic risk factors in predicting violent crime up to age 18. Early maternal rejection was defined as public institutional care of the infant, attempts to abort the foetus, and/or an unwanted pregnancy. Raine and colleagues (1997) extended this finding of an association between the interaction of birth complications by maternal rejection and violent crime by following the subjects to age 34. The interaction was found to be predictive of the most serious forms of violent crime (robbery, rape, and murder). Notably, the interaction only predicted violent crime among males whose first violent offence was committed before age 18. Institutionalization during the first year of life was the form of maternal rejection most often associated with violent offending in interaction with birth complications.

Summary. The results of studies that have examined the relationship between OCs and adult criminality and violence are contradictory. Prospective investigations of the prevalence of OCs experienced by members of a cohort who offended compared to those with no criminal record, found no differences (Denno, 1990; Farrington, 1997) or provided very weak or questionable evidence of differences (Denno, 1982; Kandel & Mednick, 1991; Litt, 1972). Similarly, retrospective studies have found no differences in the prevalence of OCs experienced by female adolescent offenders and female non-offenders (Lewis & Shanok, 1971; Shanok & Lewis, 1981). In contrast, studies of samples from a single birth cohort provide evidence that birth complications and difficulty in the neonatal period are related to impulsivity, and in combination with some type of family adversity (parental mental disorder, maternal rejection) to aggressive behavior and violent criminality (Baker & Mednick, 1984; Kandel & Mednick, 1991; Raine et al., 1994; 1997; Raine,

Brennan, Mednick, & Mednick, 1996). Further, this evidence suggests that the association between a combination of OCs and adversity and violent crime may only apply to males who presented a stable pattern of conduct problems from childhood onwards (Raine et al., 1996).

Almost all of the evidence to support the conclusion that OCs play some role in the development of some kinds of offending among certain sub-types of offenders comes from studies of samples of subjects drawn from one cohort, the Danish Perinatal Project. Given the paucity and lack of precision of the available information on the role of perinatal factors in the development of offending and its potential importance, two studies were undertaken. Our goal was to clarify the dependent variable, offending generally, nonviolent offending, violent offending, the type and timing of the associated OCs, the characteristics of individuals for whom these factors are associated, and the circumstances and nature of the psychosocial adversity which may potentiate the consequences of OCs.

2.2 Two Recent Investigations

We examined a birth cohort followed from early pregnancy to age 30. OCs were rated using a standardized and validated scale and psychosocial adversity was more narrowly defined as low socio-economic status of the family of origin and inadequate parenting. Further, to achieve our goal, we reduced the heterogeneity of the cohort by excluding intellectually handicapped subjects who are at increased risk both for offending and for OCs, by conducting analyses separately for groups at differential risk for offending and for OCs, that is, males, females, and early-start persistent offenders, and by investigating cohort members with major mental disorders separately from the large majority of subjects who did not develop mental illnesses by age 30.

2.3 Participants

The cohort is composed of all 15,117 persons born in Stockholm in 1953 and residing there in 1963 (Janson, 1984). Of them, 94% were still alive and living in Sweden at age 30. Excluded from the present analyses are those individuals who were institutionalized before beginning school, and those who were mentally retarded.

Participants with at least one conviction for a criminal offence by age 30 were classified as offenders and those with at least one violent offence as violent offenders.

2.4 Measures

Criminal convictions were documented from the records of the Swedish National Police in 1983. Violent offences were defined as crimes involving the use or threat of physical violence (assault, rape, robbery, unlawful threat, and molestation).

Information on any abnormality of the mother or the fetus was extracted from the files of mid wives, obstetricians, and hospitals in the early 1970s. This information was coded using the McNeil-Sjöström Scale for Obstetric Complications (McNeil & Sjöstrom, 1995) by one of the authors (TM). The severity of each complication is rated on a 6-point scale reflecting the ordinal degrees of inferred potential harm to the baby: severity level 1, not harmful or relevant (e.g., maternal heartburn, maternal fatigue); severity level 2, not likely harmful or relevant (e.g., maternal nose bleed, maternal headache, maternal ischias (pain due to compression of the spinal cord and specifically the ischias nerve), severity level 3, potentially but not clearly harmful or relevant (e.g., maternal febrile cystitis, maternal sinus infection, induction of labor); severity level 4, potentially clearly harmful or relevant (e.g., mild preeclampsia, breech delivery); severity level 5, potentially clearly greatly harmful/relevant (e.g., severe preeclampsia, fetal asphyxia); and severity level 6, very great harm to or deviation in offspring (e.g., eclampsia, severe neonatal distress, offspring hypoxic-ischemic cerebral injury). The scale is a reliable and valid research instrument for measuring somatic complications and conditions occurring during pregnancy (PC), labor-delivery (LDC), and the neonatal period (NNC) (McNeil, Cantor-Graae, & Weinberger, 2000). For each period and for each subject, two scores were calculated: (1) the number of different OCs above a severity level of three; and (2) the sum of the severity scores for OCs with a severity score above three. The McNeil-Sjöström Scale has been used with considerable empirical success in identifying the complications associated with schizophrenia (McNeil, Cantor-Graae, & Ismail, 2000) and is more sensitive to OCs than other scales (McNeil, Cantor-Graae, & Sjöström, 1994).

Socio-economic status (SES) of subjects' family of origin was indexed using Swedish norms (Janson, 1984). Parents' occupations, at the time of the subject's birth, were used to assign individuals a score ranging from 5 (unskilled workers) to 1 (upper or upper middle socio-economic status). Inadequate parenting was documented from the reports of the Child Welfare Committee, which at this time in Sweden had a broad mandate to ensure children's well being. Each subject's file was initially divided into three sections: from birth to six years, from age seven to age 12, and from age 13 to 18. Scores were then assigned for each of the three age periods. Decisions by the Child Welfare Committee to intervene because of inadequate or inappropriate parenting were assigned a score of 1 if the subject was left with his/her parents and a 2 if the subject was removed from the family home. These scores were added to those assigned to placements. If the subject was placed in a foster home a score of 1 to 6 was assigned depending on the length of the placement. If the subject was placed in an institution a score of 2 to 12 was assigned depending on the length of the placement (Kratzer & Hodgins, 1997).

2.5 Analyses

Differences in OCs, SES, and parenting were compared between: 1) offenders and non-offenders, 2) violent offenders and non-offenders, and 3) early-starters and non-offenders. When reporting t tests, numbers in parentheses refer to degrees of freedom or approximate degrees of freedom corrected for heteroscedasticty. When reporting chi squares, numbers in parentheses refer to the degrees of freedom and the number of subjects in the analysis. In some analyses, the dependent variable was dichotomized (non-offending/offending, non-offending/violent offending, non-offending/early-start offending), and for many of the analyses the independent variables were dichotomized (PCs: 0 versus 1 or more; LDCs 0 versus 1 or more; NNCs 0 versus 1 or more; SES: high classes 1-3 versus low-classes 4 and 5; inadequate parenting: no intervention versus intervention). It has recently been shown that dichotomization of variables facilitated the study of risk factors for delinquency, encouraged a focus of individuals, and most importantly, showed no signs of producing misleading conclusions (Farrington & Loeber, 2000).

2.6 Study 1

This study (Hodgins, Kratzer, & McNeil in press) examined the cohort members who did not have an intellectual handicap and who did not develop a major mental disorder by age 30. (For further details, see Hodgins, 1992). Included are data from 7,101 males and 6,751 females. Early-start offenders were defined as those who were convicted for an offence both before and after the age of 18, and who committed at least one crime during each of three or more age periods (before age 15, 15-17,18-20, and 21-30 years) (for more details, see Kratzer & Hodgins, 1999).

Results Males. The mean number and mean severity ratings of PCs, LDCs, and NNCs did not differ for offenders as compared to non-offenders, and for violent offenders as compared to non-offenders.[1] Early-starters as compared to non-offenders had fewer LDCs (ES: $M = 0.62$, $SD = 0.73$; non-offenders: $M = 0.73$, $SD = 0.79$), t (436.27) = -2.57, $p < .01$, and also had a lower mean severity rating for LDCs [ES: $M = 2.27$, $SD = 2.70$; non-offenders: $M = 2.70$, $SD = 3.03$), t (442.42) = -2.83, $p < .005$]. There were no differences between any of the groups in the proportions of subjects who had experienced PCs, LDCs, and NNCs.

As compared to non-offenders (SES: $M = 2.94$, $SD = 1.39$; FP: $M = -0.32$, $SD = 0.47$), men who committed an offence had been raised in families of lower SES [offenders: $M = 3.35$, $SD = 1.34$, t (6852) = 11.66, $p < .001$; violent offenders: $M = 3.58$, $SD = 1.30$, t (5103) = 9.85, $p < .001$]; were early-starters [$M = 3.67$, $SD = 1.22$, t (522.22) = 11.62, $p < .001$], and had experienced more severely inadequate parenting [offenders: $M = 0.06$, $SD = 0.55$), t 3827.55) = 6.69, $p < .001$; violent offenders: $M = 0.14$, $SD = 0.70$, t (4966) = 7.35, $p < .001$; early-starters: $M = 0.15$, $SD = 0.68$, t (446.59) = 5.27, $p < .001$].

Table 1: Percentages of male nonoffenders, offenders, violent offenders, and early-start offenders with each characteristic

	Non offenders ($n = 4756$)	Offenders ($n = 2345$)	Violent offenders[1] ($n = 525$)	Early-start offenders[1] ($n = 441$)
Obstetrical complications				
PCs	889 (18.7%)	438 (18.7%)	98 (18.7%)	84 (19.0%)
LDCs[a]	2117 (55.4%)	999 (53.0%)	223 (54.0%)	177 (49.7%)
NNCs[b]	704 (18.4%)	378 (20.1%)	89 (21.5%)	83 (23.3%)
Psychosocial adversity				
Poor parenting	760 (16.0%)	593 (25.3%)	167 (31.8%)	143 (32.4%)
Low SES[c]	1991 (43.2%)	1268 (56.5%)	316 (63.6%)	285 (68.0%)
Obstetrical complications and psychosocial adversity				
PCs + poor parenting	108 (2.3%)	110 (4.7%)	34 (6.5%)	25 (5.7%)
PCs + Low SES[d]	419 (8.8%)	250 (10.7%)	56 (10.7%)	56 (12.7%)
LDCs + poor parenting[e]	245 (5.5%)	208 (9.6%)	60 (12.5%)	44 (10.9%)
LDCs + Low SES[f]	930 (21.6%)	552 (27.0%)	138 (31.2%)	110 (29.6%)
NNCs + poor parenting[g]	93 (2.1%)	102 (4.7%)	32 (6.7%)	26 (6.5%)
NNCs + Low SES[h]	340 (7.9%)	218 (10.6%)	56 (12.6%)	60 (16.1%)

1. As noted in the Method Section, violent and early-start offenders are sub-groups among the offenders.

Note: Some subjects have missing data. See below for total number of subjects in each cell. Percentages are calculated using the number of subjects with complete data for that item.

3824 nonoffenders, 1886 offenders, 413 violent offenders, 356 early-start offenders

3823 nonoffenders, 1885 offenders, 413 violent offenders, 356 early-start offenders

4608 nonoffenders, 2246 offenders, 497 violent offenders, 419 early-start offenders

4748 nonoffenders, 2341 offenders, 522 violent offenders, 440 early-start offenders

4444 nonoffenders, 2157 offenders, 480 violent offenders, 402 early-start offenders

4303 nonoffenders, 2048 offenders, 442 violent offenders, 371 early-start offenders

4444 nonoffenders, 2157 offenders, 480 violent offenders, 402 early-start offenders

h. 4310 nonoffenders, 2058 offenders, 446 violent offenders, 373 early-start offenders

As presented in Table 1, a larger proportion of offenders (56.5%), X^2 (1,6854) = 106.27, $p < .001$, violent offenders (63.6%), X^2 (1, 5105) = 75.18, $p < .001$, and early-starters (68.0%), X^2 (1, 5057) = 95.43, $p < .001$) as compared to non-offenders (43.2%) had been raised in families with low SES. Additionally, a larger proportion of offenders (25.3%), X^2 (1,7107) = 88.23, $p < .001$, violent offenders (31.8%, X^2 (1,5281) = 81.86, $p < .001$), and early-start offenders (32.4%), X^2 (1,5197) = 76.04, $p < .001$) as compared to the non-offenders (16.0%) experienced inadequate parenting.

Table 2: Odds Ratios (and 95% confidence intervals) for offending and for violent offending

	Males		Females	
	Offending	Violent offending	Offending	Violent offending
Inadequate parenting	1.39 (1.28-1.50) ($n = 4857$)	2.02 (1.67-2.44) ($n = 4294$)	2.09 (1.70-2.56) ($n = 5614$)	1.77 (0.98-3.18) ($n = 5283$)
Pregnancy complications	0.96 (0.87-1.06) ($n = 5748$)	0.90 (0.70-1.17) ($n = 4354$)	0.87 (0.64-1.17) ($n = 5543$)	0.71 (0.30-1.67) ($n = 5278$)
Inadequate parenting and pregnancy complications	1.64 (1.43-1.89) ($n = 5774$)	2.86 (2.09-3.91) ($n = 3651$)	1.79 (1.16-2.75) ($n = 4794$)	1.81 (0.57-5.79) ($n = 4553$)

Logistic regressions[2] indicated that the interaction between PCs and parenting was significant in predicting offending, Wald X^2 (1) = 5.73, $p < .02$, and violent offending, Wald X^2 (1) = 5.34, $p < .02$. Table 2 presents the results of comparisons of men who had experienced only inadequate parenting (19.1%), men who had experienced only PCs (18.7%), men who had experienced both poor parenting and PCs (3.1%), and men who had experienced neither poor parenting nor PCs (59.1%). As can be observed, poor parenting increased the risk of both offending generally and violent offending, slightly less than did the combination of poor parenting and PCs.

There were significant differences in the mean number of crimes (Kruskal-Wallis, *df* = 3, 126.08, *p* < .001) and mean number of violent crimes (Kruskal-Wallis, *df* = 3, 65.34, *p* < .001) of the four aforementioned groups. As these one-way non parametric analyses of variance comparing men who experienced both PCs and poor parenting, only PCs, only poor parenting, and neither for both total number of crimes and total number of violent crimes were statistically significant, Mann Whitney U tests were used to compare group means. The males who had experienced both PCs and poor parenting committed on average more crimes (*M* = 8.22, *SD* = 27.27) and more violent crimes (*M* = 0.47, *SD* = 1.70) than the males with no PCs and adequate parenting (crimes *M* = 2.70, SD = 14.10, *p* < .001; violent crimes *M* = 0.18, *SD* = 1.18, *p* < .001), and than those who had experienced only PCs (crimes *M* = 2.26, *SD* = 8.97, *p* < .001; violent crimes *M* = 0.14, *SD* = 0.80, *p* < .001). Those who had experienced both PCs and poor parenting had committed, on average, more crimes than those who had experienced only poor parenting (crimes *M* = 5.73, *SD* = 20.03, *p* = .012) and similar numbers of violent crimes (*M* = 0.33, *SD* = 1.43, ns).

The types of complications experienced by the males with PCs and poor parenting were examined. The most frequent complications were preeclampsia-related conditions affecting 82% of the males with PCs and poor parenting. The prevalence rates of the six most frequent complications (toxemia plus other complications, toxemia alone, anesthesia, RH, twin, other) were compared for the offenders and the non-offenders. No differences were found. Neither poor parenting nor PCs nor both predicted early-start offending as tested in a logistic regression.

Results Females. The mean number and mean severity ratings of PCs, LDCs, and NNCs did not differ for offenders as compared to non-offenders, violent offenders as compared to non-offenders, and early starter offenders as compared to non-offenders.

As compared to non-offenders (SES: *M* = 3.04, *SD* = 1.38), all groups of female offenders had been raised in families of lower SES (offenders: *M* = 3.41, *SD* = 1.36, *t*(6538) = 5.39, *p* < .001; violent offenders: *M* = 3.63, *SD* = 1.33, *t* (6166) = 3.32, *p* < .001; early-starters: *M* = 3.82, *SD* = 1.28, *t*(6132) = 2.97, *p* < .003). Additionally, offenders (*M* = 0.17, *SD* = 0.99) as compared to non-offenders (*M* = -0.03, *SD* = 0.37) had experienced poor parenting (*t* (425.33) = 4.22, *p* < .001).

Table 3: Percentages of female nonoffenders, offenders, violent offenders, and early-start offenders with each characteristic

	Non offenders (*n* = 6297)	Offenders (*n* = 454)	Violent offenders[1] (*n* = 65)	Early-start offenders[1] (*n* = 30)
Obstetrical complications				
PCs	1070 (17.0%)	67 (14.8%)	9 (13.8%)	7 (23.3%)
LDCs[a]	2650 (52.2%)	169 (46.8%)	29 (53.7%)	12 (46.2%)
NNCs[b]	911 (18.0%)	43 (11.9%)	7 (13.0%)	3 (11.5%)
Psychosocial adversity				
Poor parenting	1066 (16.9%)	142 (31.3%)	18 (27.7%)	15 (50.0%)
Low SES[c]	2864 (46.9%)	256 (59.0%)	38 (61.3%)	19 (67.9%)
Obstetrical complications and psychosocial adversity				
PCs + poor parenting	174 (2.8%)	20 (4.4%)	3 (4.6%)	3 (10.0%)
PCs + Low SES[d]	525 (8.3%)	44 (9.7%)	6 (9.2%)	4 (13.3%)
LDCs + poor parenting[e]	366 (6.2%)	47 (11.5%)	6 (9.7%)	6 (21.4%)
LDCs + Low SES[f]	1266 (22.4%)	94 (24.3%)	15 (27.8%)	8 (29.6%)
NNCs + poor parenting[g]	140 (2.4%)	14 (3.4%)	2 (3.2%)	2 (7.1%)
NNCs + Low SES[h]	457 (8.1%)	28 (7.2%)	4 (7.3%)	3 (11.1%)

1. As noted in the Method Section, violent and early-start offenders are sub-groups among the offenders.

a. 5076 nonoffenders, 361 offenders, 54 violent offenders, 26 early-start offenders

b. 5074 nonoffenders, 361 offenders, 54 violent offenders, 26 early-start offenders

c. 6106 nonoffenders, 434 offenders, 62 violent offenders, 28 early-start offenders

d. 6290 nonoffenders, 454 offenders, 65 violent offenders, 30 early-start offenders

e. 5880 nonoffenders, 408 offenders, 62 violent offenders, 28 early-start offenders

f. 5660 nonoffenders, 387 offenders, 54 violent offenders, 27 early-start offenders

g. 5880 nonoffenders, 408 offenders, 62 violent offenders, 28 early-start offenders

h. 5674 nonoffenders, 389 offenders, 55 violent offenders, 27 early-start offenders

As can be noted in Table 3, there was only one significant difference between any of the groups in comparisons of the proportions of the different subject groups with OCs. More of the non-offenders (18.0%) than the offenders had neonatal complications, χ^2 (1, 5435) = 8.50, $p < .004$). A larger proportion of offenders (59.0%, χ^2 (1, 6751) = 6.46, $p <$

.011) and early-starters (67.9% χ^2 (1, 6327) = 23.05, $p < .001$) as compared to non-offenders (46.9%) had been raised in families of low SES. Additionally, a larger proportion of offenders (31.3% X^2 (1.6751) = 59.34, $p < .001$) and early-starters (50.0% χ^2 (1,6327) = 23.05, $p < .001$) as compared to the non-offenders (16.9%) had experienced poor parenting.

The logistic regressions indicated no significant interactions between PCs, LDCs, NNCs and SES or poor parenting in association with offending, violent offending and early-start offending. This may well be due to the small number of female offenders, and the even smaller numbers in the various comparisons. Table 2 presents comparisons of the risks of offending generally and of violent offending among the females who had experienced only inadequate parenting (18.1%), among those who had experienced only pregnancy complications (17.0%), among those who had experienced both poor parenting and pregnancy complications (4.0%), as compared to those who had experienced neither poor parenting nor pregnancy complications (60.9%). As among the males, poor parenting increased the risk of offending and of violent offending only slightly less than did the combination of poor parenting and pregnancy complications. Among the early-start females, only 37% had not experienced either PCs or inadequate parenting, 50% had experienced inadequate parenting, 10% of them also PCs, and another 13% had experienced PCs but not inadequate parenting.

Conclusions. Among both men and women no relation was identified between pregnancy, birth, and neonatal complications, occurring in the absence of inadequate parenting, and offending and violent offending. Early-start offenders were characterized by fewer and less severe labor and delivery complications than non-offenders. The associations between both low SES and inadequate parenting and offending, violent offending, and early-start offending were found, as in many previous investigations, to be powerful (Loeber, Farrington, & Stouthamer-Loeber, 1986; Patterson & Capaldi, 1991). Pregnancy complications combined with poor parenting in the early years of life slightly increased the risk of offending and more than doubled the risk of violent offending. The combination of PCs and poor parenting affected only 3% of the males and 4% of the females, and it increased the risk of crime and of violent crime only slightly more than did inadequate parenting alone. This is important because inadequate parenting was much more common affecting another 16% of the males and 18% of the females (who did not experience PCs). To illustrate the significance of this finding for preventing crime, consider the following numbers. Of all the men born in Stockholm in 1953, 1,135 experienced inadequate parenting and did not experience pregnancy complications. Of these 1,135, 483 (42.6%) were convicted of an offence. By contrast, 218 of the male cohort members experienced both inadequate parenting and PCs, and 110 (50.5%) of them were convicted of criminal offences. In other words, four times ($n=483$) more male offenders experienced inadequate parenting than inadequate parenting combined with pregnancy complications ($n=110$). However, while few males had experienced both PCs and inadequate parenting, of those who did half became offenders and 16% of them violent offenders and they committed many offences.

The finding that inadequate parenting in combination with OCs increased the risk of offending concurs generally with studies of samples from the Danish Perinatal Project. However, results differ in three important ways. In the present study, (1) the complications associated with offending occurred during the pregnancy and not at birth or in the neonatal period; (2) complications were associated with offending generally and not only with violent offending; and (3) the association between OCs and inadequate parenting and offending was not observed for early-start offenders. One possible explanation for these differences in the findings is the exclusion of the mentally retarded and mentally ill cohort members from the sample examined in the present investigation. Such persons are at high risk for offending (Hodgins, 1992), at even higher risk for violent offending (Hodgins,

1992, 1998) and for homicide (Hodgins, 1994) than the general population, and for OCs particularly at birth and in the neonatal period (McNeil, Cantor-Graae, & Ismail, 2000). If samples inadvertently included disproportionate numbers of mentally retarded and/or mentally ill subjects, an association between a combination of OCs and family adversity and offending which applies only to them, may have been interpreted as characteristic of male offenders generally. This would be especially true in a country like Denmark where the violent crime rate is relatively low and the proportions of mentally retarded and mentally ill subjects among the offenders are relatively high.

A disproportionately high number of males who became persistent offenders and who had begun to offend at a young age, had experienced fewer than average, labor-delivery complications. Three possible explanations of this finding warrant further study. One, based on twin and adoptions studies (Eley, Lichenstein, & Stevenson, 1999; Lahey, Waldman, & McBurnett, 1999; Lyons et al., 1995), it would be expected that some elevated proportion of the mothers of early-start offenders would themselves present a history of antisocial behavior which is associated with low anxiety, fear, and arousal. These maternal characteristics could be associated with a reduction of labor and delivery complications. A second possible explanation relates to recent findings on body size. In the present investigation and in the longitudinal investigation of a New Zealand cohort it has been found that this type of early-start male offender is heavier than the average at birth (Kratzer & Hodgins, 1996). Further, large body size at age 3 has been found to be associated with aggressive behavior at age 11 (Raine, Reynolds, Venables, Mednick, & Farrington, 1998), high body mass index at various ages has been found to be associated with aggressive behavior (Ravaja, Keltikangas-Järvinen, 1995), and increased weight during the first twelve months of life has been associated with violent offending in adulthood (Räsänen, Hakko, Järvelin, & Tiihonen, 1999). Boys who are larger than their peers during early childhood may learn to be aggressive as a result of persistent provocation, or alternately, the various measures of body size used in these different investigations may be tapping a metabolic syndrome which is related to brain functioning and to impulsivity or reduced behavioral disinhibition.

2.7 Study 2

Sample. The sample Included only those subjects who developed a major mental disorder (schizophrenia, major affective disorders, paranoid states, and other non-alcohol and non-drug related psychoses) and were admitted to a psychiatric ward. Studies of the schizophrenia diagnoses in this register (Kristjansson, Allebeck, & Wistedt, 1987) and in other Swedish registers (Wetteberg & Farmer, 1990) have suggested that these diagnoses are valid and closely resemble DSM-III diagnoses. This report is based on data from 82 males and 79 females. Subjects with at least one registration for a criminal offence by age 30 were classified as offenders, those with at least one registration for a violent offence as violent offenders, and early-starters were defined as those who were convicted of their first crime before the age of 18.

Table 4: Percentages of mentally ill male nonoffenders, offenders, violent offenders, and early-start offenders with each characteristic

	Non offenders ($n = 41$)	Offenders ($n = 41$)	Violent offenders[1] ($n = 20$)	Early-start offenders[1] ($n = 26$)
Obstetrical complications				
PCs	3 (7.3%)	9 (22.0%)	3 (15.0%)	6 (23.1%)
LDCs[a]	21 (67.7%)	23 (63.9%)	14 (77.8%)	15 (31.8%)
NNCs[b]	1 (3.2%)	12 (33.3%)	4 (22.2%)	8 (36.4%)
Psychosocial adversity				
Poor parenting	8 (19.5%)	12 (29.3%)	3 (15.0%)	8 (30.8%)
Low SES[c]	12 (36.4%)	21 (55.3%)	10 (55.6%)	13 (56.5%)
Obstetrical complications and psychosocial adversity				
PCs + poor parenting	1 (2.4%)	2 (4.9%)	1 (5.0%)	1 (3.8%)
PCs + Low SES[d]	1 (16.7%)	5 (12.2%)	1 (5.0%)	2 (8.7%)
LDCs + poor parenting[e]	5 (12.5%)	6 (15.4%)	3 (15.0%)	4 (18.2%)
LDCs + Low SES[f]	7 (18.4%)	12 (31.6%)	7 (38.9%)	8 (36.4%)
NNCs + poor parenting [g]	1 (2.5%)	7 (17.9%)	3 (15.0%)	5 (22.7%)
NNCs + Low SES[h]	1 (2.6%)	8 (21.1%)	4 (22.2%)	5 (22.7%)

1. As noted in the Method Section, violent and early-start offenders are sub-groups among the offenders.

Note: Some subjects have missing data. See below for total number of subjects in each cell. Percentages are calculated using the number of subjects with complete data for that item.

31 nonoffenders, 36 offenders, 18 violent offenders, 22 early-start offenders

31 nonoffenders, 36 offenders, 18 violent offenders, 22 early-start offenders

39 nonoffenders, 38 offenders, 18 violent offenders, 23 early-start offenders

39 nonoffenders, 38 offenders, 18 violent offenders, 23 early-start offenders

31 nonoffenders, 36 offenders, 18 violent offenders, 22 early-start offenders

31 nonoffenders, 36 offenders, 18 violent offenders, 22 early-start offenders

31 nonoffenders, 36 offenders, 18 violent offenders, 22 early-start offenders

31 nonoffenders, 36 offenders, 18 violent offenders, 22 early-start offenders

Results for men with major mental disorders. The mean number and mean severity ratings of PCs, LDCs, and NNCs experienced by offenders, violent offenders, and early-start offenders were compared to those experienced by non offenders. Among the males, PCs, and LDCs did not differ for offenders as compared to non-offenders, violent offenders as compared to non-offenders and early-starter offenders as compared to non-offenders. Male offenders and early starters had experienced, on average, more NNCs (offenders: $M = 0.39$, $SD = 0.60$, $t(42.16) = 3.40$, $p < .001$; ES: $M = 0.41$, $SD = 0.59$, $t(23.78) = 2.90$, $p < .008$) than non-offenders ($M = 0.32$, $SD = 0.18$), and these complications were more severe (offenders: $M = 1.56$, $SD = 2.48$, $t(45.26) = 3.14$, $p < .003$; ES: $M = 1.54$, $SD = 2.20$, $t(26.01) = 2.79$, $p < .01$) than those experienced by the non-offenders ($M = 0.16$, $SD = 0.90$). Eleven of the 12 offenders who experienced NNCs, had an abnormal length of gestation, 4 too short, 7 too long. (The one non offender who experienced NNCs had a gestational age less than 245 days and weighed less than 2,000 grams.)

As can be observed in Table 4, while similar proportions of offenders, violent offenders and early-start offenders had experienced PCs and LDCs, one-third of the offenders had experienced neonatal complications as compared to only one of the non-offenders, $\chi^2 (1,67) = 9.65$, p < .002.

Table 5: Relative risk ratios for offending, violent offending, and early-start offending among men who developed mental illnesses

	Offending	Violent Offending	Early-Start Offending
NNCs	2.08	2.51	2.79
	(1.48-2.91)	(1.35-4.65)	(1.71-4.56)
NNCs x Inadequate parenting	1.88	2.04	2.50
	(1.25-2.82)	(1.01-4.11)	(1/39-4.48)
NNCs x Low SES	2.51	2.80	3.61
	(1.48-4.25)	(1.35-5.82)	(1.64-7.94)

Among men who developed a major mental disorder, having had a NNC increased the risk of offending by a factor of 2.1 (95% CI 1.48-2.91), of violent offending by a factor of 2.5 (95% CI 1.36-4.65), and of early-start offending by a factor of 2.8 (95% CI 1.71-4.56). Among these men who developed major mental disorders, those who had experienced NNCs committed, on average, more offences ($M = 18.00$, $SD = 23.96$), than those who had not experienced NNCs ($M = 6.26$, $SD = 13.94$; MW z = 3.0, $p < .003$). Of these 12 offenders who had experienced NNCs, eight were convicted of their first criminal offence before the age of 18.

As compared to non-offenders ($M = 2.49$, $SD = 1.41$), offenders ($M = 3.32$, $SD = 1.45$, $t (75) = 2.54$, p < .013) and early starters ($M = 3.48$, $SD = 1.41$, $t (60) = 2.67$, p < .01) had been raised in families with lower socio-economic status. There were no differences between any of the groups of subjects as to the mean scores for parenting, or the proportions of subjects who had been raised in families of low SES and by parents who provided inadequate or inappropriate care.

Most of the male subjects who experienced NNCs, also experienced inadequate parenting or low SES of the family of origin. Of the 12 offenders who had experienced NNCs, five came from low SES families and had experienced inadequate parenting, three came from low SES families, and two experienced inadequate parenting. Thus only two of the 12 had neither low SES or inadequate parenting. The one non offender who had experienced NNCs, also came from low SES family and had experienced inadequate parenting. However, as can be noted in Table 3, these factors did not increase the risks of offending, violent offending, and early-start offending, over and above the risk conferred by NNCs alone.

Table 6: Percentages of mentally ill female nonoffenders, offenders, violent offenders, and early-start offenders with each characteristic

	Non offenders ($n = 64$)	Offenders ($n = 15$)	Violent offenders[1] ($n = 5$)	Early-start offenders[1] ($n = 6$)
Obstetrical complications				
PCs	8 (12.5%)	4 (26.7%)	2 (40.0%)	2 (33.3%)
LDCs[a]	25 (52.1%)	6 (60.0%)	2 (66.7%)	0
NNCs[b]	7 (14.6%)	1 (10.0%)	0	1 (33.3%)
Psychosocial adversity				
Poor parenting	16 (25.0%)	3 (20.0%)	1 (20.0%)	3 (50.0%)
Low SES[c]	36 (57.1%)	4 (28.6%)	2 (50.0%)	3 (50.0%)
Obstetrical complications and psychosocial adversity				
PCs + poor parenting	1 (1.6%)	0	0	2 (33.3%)
PCs + Low SES[d]	6 (9.4%)	1 (6.7%)	1 (20.0%)	1 (16.7%)
LDCs + poor parenting [e]	6 (9.8%)	0	0	0
LDCs + Low SES[f]	15 (27.3)	1 (8.3%)	0	0
NNCs + poor parenting [g]	2 (3.3%)	0	0	0
NNCs + Low SES[h]	4 (7.3%)	0	0	0

1. As noted in the Method Section, violent and early-start offenders are sub-groups among the offenders.

48 nonoffenders, 10 offenders, 3 violent offenders, 3 early-start offenders

48 nonoffenders, 10 offenders, 3 violent offenders, 3 early-start offenders

63 nonoffenders, 14 offenders, 4 violent offenders, 6 early-start offenders

63 nonoffenders, 14 offenders, 4 violent offenders, 6 early-start offenders

48 nonoffenders, 10 offenders, 3 violent offenders, 3 early-start offenders

f. 48 nonoffenders, 10 offenders, 3 violent offenders, 3 early-start offenders

g. 48 nonoffenders, 10 offenders, 3 violent offenders, 3 early-start offenders

h. 48 nonoffenders, 10 offenders, 3 violent offenders, 3 early-start offenders

Results for women with major mental disorders. There were no differences between offenders and non-offenders in the mean number or mean severity ratings of pregnancy, labor and delivery, and neonatal complications, and in the mean scores for parenting. Non-offenders ($M = 3.28$, $SD = 1.3$), came from families of lower SES than offenders ($M = 2.21$, $SD = 1.58$, $t(75) = -2.68$, $p < .009$).

Conclusions. Among men who developed major mental disorders, having experienced complications in the neo-natal period increased the risk of offending two-fold, the risk of violent offending 2.5 times, and the risk of early-start offending 3 times. Most of these men had been raised in families with low socio-economic status, and slightly more than half had experienced inadequate parenting. These family factors, however, did not confer additional risk for offending, violent offending, or early-start offending. Most of these mentally ill offenders displayed a stable pattern of offending with an early onset in mid adolescence. Notably, one non offender had experienced all three events - neonatal complications, low SES, and inadequate parenting - that were associated with offending.

The etiological factors associated with offending among men who developed major mental disorders, appear to be specific to men with major mental disorders. These factors did not characterize the female offenders who developed major mental disorders, the non disordered male offenders as shown in Study 1.

Strengths and weaknesses of the two studies. The two studies reported above are characterized by a number of strengths that increase confidence in the validity of the results and in their generalizability. This was a large, unselected birth cohort born and raised in a society that provided good health care and social services to all of its citizens (Hodgins & Janson, 2001). Information from the obstetrical records was extracted by persons blind to the objectives of the present study and coded using a standardized and validated rating scale. Information on criminality was complete. Subjects were followed from pregnancy to age 30 with almost no attrition. Finally, the specificity of risk factors for offending for men and women without mental retardation and with and without mental illness were examined.

Like all investigations, however, this one has weaknesses. Five are of importance for interpreting the results. One, given the multiple comparisons, the findings should be interpreted cautiously until they are replicated. Two, no information was available on behaviors of the mothers during the pregnancy that have been recently found to increase the risk of violent criminality in the offspring. These investigations are reviewed in the next section.

Three, official criminal records were used to index behavior. This would lessen the strength of all associations except those related to serious violence such as murder that would almost always lead to criminal charges. Four, the follow-up period was not long enough to allow accurate classification of all persons who would develop major mental disorders. Consequently, some persons who developed these disorders after the age of 30 would have been classified as non-disordered. Five, the small number of female offenders may have prevented the detection of associations between Ocs and offending.

2.8 Conclusion – Are OCs Noted in Medical Files Related to Offending?

Obstetrical complications play different etiological roles among different types of offenders. Among male offenders who are not mentally retarded and who do not develop mental illness and who are not classified as stable early-start offenders, OCs play a very minor role, with social factors such as the socio-economic status of the family and inadequate parenting being much more important. Pregnancy complications in combination with inadequate parenting were found to affect a small group of these offenders and to be associated with serious, persistent offending. Further, among men who are neither mentally retarded or mentally ill, those who begin offending early and who become the most serious offenders have been found to have experienced less difficult births than non offenders. Finally, among men who develop major mental disorders in adulthood, complications in the neonatal period, but not social factors, significantly increase the risk of offending. Presently, the mechanisms by which OCs occurring at specific times do increase the risk of offending in specific types of offenders are unknown. The findings reviewed earlier from the Danish Perinatal Project suggest that these OCs index damage to the fetal brain which is associated with impulsivity, which over the course of development leads to illegal behaviors. Little has been learned about the role of OCs in the development of offending among women. Whether or not this is due to methodological characteristics of the investigations conducted to date or is due to the fact that OCs play no role is unknown.

3. Maternal Smoking During Pregnancy

Evidence has accumulated in recent years indicating that maternal smoking during pregnancy is associated with conduct problems in childhood and adolescence, and with early onset, persistent and violent offending in adulthood among male offspring. One study examined boys referred for behavior problems and asked mothers about smoking during the pregnancy (Wakschlag et al., 1997). Another examined the offspring of parents with and without affective disorders and also collected information on smoking during pregnancy, retrospectively (Weissman, Warner, Wickramaratne, & Kandel, 1999). Three investigations have examined population cohorts and gathered information about the pregnancy prospectively, thereby increasing the confidence that the association between maternal smoking and antisocial behavior is real. In New Zealand, a birth cohort of 1,265 children were followed from pregnancy to age 18 when they underwent diagnostic interviews (Fergusson, Woodwar, & Horwood, 1998). In Finland, all the 5,636 males born in two Northern provinces in 1966 were followed to age 27 when official criminal records were obtained (Räsänen et al., 1999). In Denmark, a birth cohort composed of 4,169 males were followed to age 34 when official criminal records were obtained (Brennan, Grekin, & Mednick, 1999). The results of these investigations are remarkably similar in demonstrating an independent association between maternal smoking and antisocial and criminal behavior. The available evidence suggests the association is either restricted to males or stronger in males than females. The quantity of cigarettes necessary to observe the effect is still unclear. Maternal smoking during pregnancy is associated specifically with early onset, persistent behavior problems that develop into violent offending and is not related to other types of disorders or symptoms.

The association between maternal smoking during pregnancy and persistent antisocial behavior across the lifespan could be due to other factors characterizing women who smoke while pregnant. To test whether or not this is true, the investigations described above examined the association between maternal smoking and antisocial and criminal behavior among male offspring, controlling for a number of other factors known to be associated with the development of antisocial and criminal behavior. These factors included: parental antisocial personality, other mental disorders, substance abuse and criminality, maternal alcohol consumption during the pregnancy and other obstetrical complications, mother's age, mother's education, socio-economic status of the family of origin, poor supervision, harsh parenting, and other parenting practices. None of these factors accounted for the association between maternal smoking and persistent antisocial behavior.

However, women who smoke while pregnant are more likely than women who do not smoke to be characterized by these other factors which increase the risk of antisocial behavior in their offspring. For example, in the New Zealand investigation, the more the mother had smoked during pregnancy the lower her level of education, the younger she was at the time of the pregnancy, the lower her socio-economic status, the more likely the pregnancy was unplanned, the more likely she was to have consumed alcohol and drugs during the pregnancy, the lower her scores on ratings of emotional responsiveness to the child, the more likely she was to abuse the child, to separate from the child's father, and the more likely it was for either her or the father or both to have a history of criminality and substance abuse. Thus, while maternal smoking during pregnancy remains directly associated with persistent antisocial behavior among male offspring, it is also associated with several other factors known to contribute to the development of this behavior pattern. The results of the Finnish investigation illustrate this conclusion. Maternal smoking during pregnancy was found to increase the risk of persistent criminality 2.4 times (95%CI 1.6-3.7) after adjustment for socio-demographic characteristics of the family of origin, obstetrical complications, developmental lags, and family factors. Maternal smoking during pregnancy when combined with other factors increased the risk of persistent offending to an even greater extent: plus a mother younger than 20, 10.8 times (95%CI 4.3-26.9); plus single parent family, 11.9 times (95%CI 5.7-24.6); plus a delay in walking or speaking, 5.5 times (95%CI 2.6-11.7); plus an unwanted pregnancy, 9.1 times (95%CI 4.5-18.3); plus a single parent family, and an unwanted pregnancy, and developmental lags, 14.2 times (95%CI 5.3-38.5).

4. Exposure to Environmental Toxins in Utero

It is suspected that certain environmental toxins negatively affect the fetal brain, but despite findings clearly supporting this notion, little research on the topic has been conducted. Lead accumulates in the bones and teeth. Levels of lead measured in childhood are associated with the development of aggressive and delinquent behavior (Needleman, Riess, Tobin, Biesecker, & Greenhouse, 1996). Higher levels of both lead and cadium have been found to distinguish violent from non-violent offenders (Pihl & Ervin, 1990).

5. Conclusion

There is evidence suggesting that events which occur during the pre and perinatal periods are associated with the development of patterns of antisocial behavior which emerge in childhood and are stable across the life-span among men. Presently, it appears that the most important of these factors is maternal smoking during pregnancy. Investigations of birth cohorts using prospectively collected information have found that the association is independent, even after controlling for a multitude of other factors known to be associated with the development of antisocial and criminal behavior. These same investigations demonstrate however, that mothers who smoke during the pregnancy are characterized by a number of other factors which also increase the risk of antisocial and criminal behavior in their offspring and that they mate with men who are also characterized by factors known to increase the risk of antisocial and criminal behaviors in their offspring. Thus, maternal smoking during pregnancy is associated with early onset, persistent antisocial behavior in male offspring and with maternal and paternal characteristics that increase the risk of these behaviors in male offspring.

Obstetrical complications noted in medical files do not appear to be associated with offending generally. Complications during the pregnancy in interaction with inadequate parenting is associated with an increased risk of offending and violent offending among males, but occurs infrequently. There is evidence to hypothesize that environmental toxins, specifically lead, may contribute to the development of antisocial behavior. Exposure to such toxins during pregnancy merits study.

Pre and perinatal events associated with offending appear to differ by sub-type of offenders. Maternal smoking during pregnancy specifically increases the risk of early onset and persistent offending which includes violence, among males. The interaction of pregnancy complications and inadequate parenting specifically increases the risk of offending, and more so of violent offending, among males who are neither mentally retarded nor develop mental illnesses in adulthood. Complications in the neo-natal period are strongly associated with offending among men who develop major mental disorders.

Knowledge about fetal development is limited. As seen, knowledge about associations between events occurring in the pre and perinatal periods and behavior patterns emerging years later is even further limited. Investigations addressing this issue are difficult to conduct, as they need to be prospective and longitudinal and include large numbers of subjects. Yet, given the vulnerability of the fetal brain to injury and the possibility of intervening to prevent the development of persistent antisocial and criminal behavior, this is an area of study worth pursuing.

Notes

[*] This investigation was completed with funds from the Social Sciences and Humanities Research Council of Canada to S. Hodgins.
[1] In the first step of the analyses, t tests were used to compare the mean number of PCs, LDCs, NNCs; the mean severity level of OCs at each reproductive period; and the mean scores for SES and parenting. These analyses measure group differences on each of the mentioned variables and presume that the groups of subjects are relatively homogeneous with respect to the variable being measured. In order to verify the extent to which group differences applied to all subjects within the group, chi square tests were conducted to compare the prevalence of each measure within each group of subjects. An alpha value of $p < .01$ was used to adjust for the large number of tests that were conducted. Since the number of crimes committed by subjects varied widely, non-parametric statistics were used to compare mean numbers of crimes and of violent crimes.
[2] In the second step of the analyses, logistic regressions were conducted to examine interactions between OCs and psychosocial adversity. Separate logistic regressions were carried out for three different dependent variables: offending/non-offending, violent offending/non-offending, and early starter offending/non-offending. The predictor variables entered into the logistic regressions were the following: OCs (PCs, or LDCs, or NNCs), SES, parenting, OCs (PCs or LDCs or NNCs) x SES, and OCs (PCs or LDCs or NNCs) x parenting. Thus, separate logistic regressions were conducted for OCs that occurred at three different periods of development. One possible outcome of having entered two interactions simultaneously into models is that, if the two interactions were strongly correlated, then the effect of one interaction may have cancelled the effect of the other in the model. Thus, we conducted a second set of logistic regressions in which a single interaction (e.g., pregnancy complications x family problems) along with an OC complication, SES and family problems were entered into the model. This second set of analyses was conducted only with interactions that had a significant probability of less than 0.05 in the first set of analyses. We adopted this liberal criterion in conducting this second set of analyses because there were so few significant interactions in the first set. It is important to note that the two significant interactions in the first set of the analyses were still significant in the second set and no additional interactions were significant in the second set.

References

Baker, R. L., & Mednick, B. R. (1984). Influences on perinatal outcomes. In S. A. Mednick (Ed.), *Influences on human development: A longitudinal perspective.* Boston: Nijohoff.

Brown, A. S., van Os, J., Driessens, C., Hoek, H. W., & Susser, E. S. (2000). Further evidence of relation between prenatal famine and major affective disorder. *American Journal of Psychiatry, 157,*190-195.

Denno, D. J. (1982). Sex differences in cognition and crime: Early developmental, biological, and sociological correlates. *Dissertation Abstracts International, 43*(11), 0349B. (University Microfilms No. AAC83-07303)

Denno D. (1990). *Biology and violence: From birth to adulthood.* Cambridge: Cambridge University Press.

Eley, T. C., Lichenstein, P., & Stevenson, J. (1999). Sex differences in the etiology of aggressive and nonaggressive antisocial behavior: Results from two twin studies. *Child Development, 70,* 155-168.

Farrington, D. P. (1997). Predictors, causes and correlates of male youth violence. In M. Tonry & M. H. Moore (Eds.), *Youth violence crime and justice* (Vol. 24). Chicago: University of Chicago Press.

Farrington, D. P., & Loeber, R. (2000). Some benefits of dichotomization in psychiatric and criminological research. *Criminal Behavior and Mental Health, 10,* 100-122.

Fergusson, D. M., Woodward, L. J., & Horwood, L. J. (1998). Maternal smoking during pregnancy and psychiatric adjustment in late adolescence. *Archives of General Psychiatry, 55,* 721-727.

Hodgins, S. (1992). Mental disorder, intellectual deficiency and crime: Evidence from a birth cohort. *Archives of General Psychiatry, 49,* 476-483.

Hodgins, S. (1994). Schizophrenia and violence: Are new mental health policies needed? *Journal of Forensic Psychiatry, 5,* 473-477.

Hodgins, S. (1998). Epidemiologic investigations of the associations between major mental disorders and crime: Methodological limitations and validity of the conclusions. *Social Psychiatry & Psychiatric Epidemiology, 33,* S29-S37.

Hodgins, S., & Janson, C.-G. (2001). *Criminality and violence among the mentally disordered: The Stockholm Metropolitan Project.* Cambridge: Cambridge University Press.

Hodgins, S., Kratzer, L. & McNeil, T. F. (in press). Obstetrical complications, parenting practices, and risk of criminal behavior. *Archives of General Psychiatry.*

Huttunen, M. O., & Niskanen, P. (1978). Prenatal loss of father and psychiatric disorders. *Archives of General Psychiatry, 35,* 429-431.

Janson, C.-G. (1984). *A longitudinal study of a Stockholm cohort* (Research Report No. 21). Stockholm: University of Stockholm, Department of Sociology.

Kandel, E., & Mednick, S. A. (1991). Perinatal complications predict violent offending. *Criminology, 29,* 519-529.

Kratzer, L. & Hodgins, S. (1996). *Patterns of crime and characteristics of female as compared to male offenders.* Paper presented at the Life History Research Society Meeting, London.

Kratzer, L., & Hodgins, S. (1997). Adult outcomes of child conduct problems: A cohort study. *Journal of Abnormal Child Psychology, 25,* 65-81.

Kratzer, L., & Hodgins, S. (1999). A typology of offenders: A test of Moffitt's theory among males and females from childhood to age 30. *Criminal Behavior and Mental Health, 9,* 57-73.

Kristjansson, E., Allebeck, P., & Wistedt, B. (1987). Validity of the diagnosis schizophrenia in a psychiatric inpatient register. *Nordisk Psykiatrisk Tidsskrift, 41,* 229-234.

Lahey, B. B., Waldman, I. D., & McBurnett, K. (1999). Annotation: The development of antisocial behavior—An integrative causal model. *Journal Child Psychology and Psychiatry, 1999, 40,* 669-682.

Lewis, D. O., & Shanok, S. (1971). Medical histories of delinquent and nondelinquent children: An epidemiological study. *American Journal of Psychiatry, 134,* 1020-1025.

Litt, S. (1972). Perinatal complications and criminality. *Dissertation Abstracts International, 33*(05), 0622B. (University Microfilms No. AAC72-27880)

Loeber, R., Farrington, D. P., & Stouthamer-Loeber, M. N. (1986). Family factors as correlates and predictors of juvenile conduct problems and delinquency. In M. Tonry, & N. Morris (Eds.), *Crime and justice: An annual review of research.* Chicago: University of Chicago Press, 1986.

Lyons, M. J., True, W. J., Eisen, S. A., Goldberg, J., Meyer, J. M., Faraone, S. V., Eaves, L. J., & Tsuang, M. T. (1995). Differential heritability of adult and juvenile antisocial traits. *Archives of General Psychiatry, 52,* 906-915.

Machon, R. A., Mednick, S. A., Huttunen, M. O. (1997). Adult major affective disorder after prenatal exposure to an influenza epidemic. *Archives of General Psychiatry, 54,* 322-328.

McNeil, T. F. (1988). Obstetric factors and perinatal injuries. In M. T. Tsuang & J. C. Simpson (Eds.), *Handbook of schizophrenia, Vol. 3: Nosology, epidemiology and genetics.* New York: Elsevier.

McNeil, T. F., Cantor-Graae, E., & Ismail, B. (2000). Obstetric complications and congenital malformation. *Brain Research Revue, 31,* 166-178.

McNeil, T. F., Cantor-Graae, E., & Sjöström, K. (1994). Obstetric complications as antecedents of schizophrenia: Empirical effects of using different obstetric complication scales. *Journal of Psychiatric Research, 28,* 519-530.

McNeil, T. F., Cantor-Graae, E., & Weinberger, D. R. (2000). Obstetric complications and brain structure size differences in monozygotic twin pairs discordant for schizophrenia. *American Journal of Psychiatry, 157,* 203-212.

McNeil, T. F., & Sjöström, K. (1995). *McNeil-Sjöström Scale for Obstetric Complications.* Malmö, Sweden: Department of Psychiatry, Lund University.

Mednick, S. A., & Huttunen, M. (2000). Lessons from the wings of drosophila. In L. R. Bergman, R. B.

Cairns, L.-G. Nilsson, & L. Nystedt (Eds.), *Developmental science and the holistic approach* (pp. 203-208). Mahwah, NJ: Erlbaum.

Mednick, S. A., Huttunen, M. O., & Machon, R. A. (1994). Prenatal influenza infections and adult schizophrenia. *Schizophrenia Bulletin, 20*, 263-267.

Nowakowski, R. S., & Hayes, N. L. (1999). CNS development: An overview. *Development & Psychopathology, 11*, 395-417

Patterson, G. R., & Capaldi, D. M. (1991). Antisocial parents: Unskilled and vulnerable. In P. A. Cowan & E. M. Hetherington (Eds.), *Family transitions. Advances in family research series.* Hillsdale NJ: Erlbaum.

Raine, A., Brennan, P., Mednick, B., & Mednick, S. (1996). High rates of violence, crime, academic problems, and behavioral problems in males with both early neuromotor deficits and unstable family environments. *Archives of General Psychiatry, 53*, 544-549.

Raine, A., Brennan, P., & Mednick, S. A. (1994). Birth complications combined with early maternal rejection at age 1 year predispose to violence crime at age 18 years. *Archives of General Psychiatry, 51*, 984-988.

Raine, A., Brennan, P., & Mednick, S. A. (1997). Interaction between birth complications and early maternal rejection in predisposing individuals to adult violence: Specificity to serious, early-onset violence. *American Journal of Psychiatry, 154*, 1265-1271.

Raine, A., Reynolds, C., Venables, P. H., Mednick, S. A., & Farrington, D. P. (1998). Fearlessness, stimulation-seeking, and large body size at age 3 years as early predispositions to childhood aggression at age 11 years. *Archives of General Psychiatry, 55*, 745-751.

Räsänen, P., Hakko, H., Isohanni, M., Hodgins, S., Järvelin, M.-R., & Tiihonen, J. (1999). Maternal smoking during pregnancy and risk of criminal behavior among adult male offspring in the Northern Finland 1966 birth cohort. *American Journal of Psychiatry, 156*, 857-862.

Räsänen, P., Hakko, H., Järvelin, M. R., & Tiihonen, J. (1999). Is a large body size during childhood a risk factor for later aggression? *Archives of General Psychiatry, 56*, 283-284.

Ravaja, N., & Keltikangas-Järvinen, L. (1995). Temperament and metabolic syndrome precursors in children: A three-year follow-up. *Preventive Medicine, 24*, 518-527.

Shanok, S., & Lewis, D. O. (1981). Medical histories of female delinquents. *Archives of General Psychiatry, 38*, 211-213.

Susser, E., Neugebauer, R., Hoek, H. W., Brown, A. S., Lin, S., Labovitz, D., & Gorman, J. M. (1996). Schizophrenia after prenatal famine: Further evidence. *Archives of General Psychiatry, 53*, 25-31.

Wakschlag, L. S., Lahey, B. B., Loeber, R., Green, S. M., Gordon, R. A., & Leventhal, B. L. Maternal smoking during pregnancy and the risk of conduct disorder in boys. *Archives of General Psychiatry, 54*, 670-676.

Watson, J. B., Mednick, S. A., Huttunen M., & Wang X. (1999). Prenatal teratogens and the development of adult mental illness [Review]. *Development & Psychopathology, 11*, 457-466.

Weissman, M. M., Warner, V., Wickramaratne, P. J., & Kandel, D. (1999). Maternal smoking during pregnancy and psychopathology in offspring followed to adulthood. *Journal of the American Academy of Child and Adolescent Psychiatry, 38*, 892-899.

Wetteberg, L., & Farmer, A. E. (1990). Clinical polydiagnostic studies in a large Swedish pedigree with schizophrenia. *European Archives Psychiatry Clinical Neuroscience, 240*, 188-190.

Multi-Problem Violent Youth
R.R. Corrado et al. (Eds.)
IOS Press, 2002

Behavioral Aggression in Pre-Schoolers: Identification, Assessment and Understanding

Pratibha N. Reebye

Dillon, hardly 3 ½ years old, holds the front admirably. In contrast, his parents, Beth and Andy, look exhausted. Dillon has been suspended from his preschool for repeatedly hitting his peers and is awaiting a decision regarding his expulsion. He punches other children, destroys his classmates' toys, and viciously upsets his classmates' play. His behavior is escalating. Dillon is seen as an attention seeking, self-centered child, who cannot share playthings or attention from caregivers. Dillon's demanding behavior annoyed his peers. They saw him as a pest and a bully and did not want to include him in their play. Both adults and children were upset by Dillon's behavior. Dillon always states that he does not want to fight, all he wants to do is to play.

Mary is 4 years old and, unlike Dillon, is an ideal pupil in her preschool and daycare. Teachers and classmates alike love her. But at home, Mary's rages are getting out of control. Her mother, a single parent, worries about the landlord's reaction and eviction seems certain because of Mary's screaming episodes. Mary gets very angry with her mother and hits her; she hurts family pets, and writes all over the apartment walls and furniture — something her mom interprets as an act of overt aggression. Mary recognizes that her actions are wrong. In therapy, Mary sometimes talked about being angry with her mother, stating that she definitely wanted to teach her mother a lesson. In sharp contrast, Mary's mother considered Mary's raging behavior to be "out of control" and thought that Mary did not know what she was doing.

Johnny has a sharp, vigilant look. At the age of 4 ½ years, he pulled the trigger of a loaded gun and nearly killed his baby brother. Johnny is difficult to understand. He is not able to adapt to any group setting. Currently, he is in a therapeutic foster home requiring constant supervision. At times, he is very connected to the group home staff; at other times, he has attacked and bitten them. Johnny never mentions his parents. Johnny had many play therapy sessions, but not once did he talk about his intentions or plans to hurt his stepbrother. On direct questioning, he would shrug his shoulders stating he did not know.

The behavioral difficulties of these three preschoolers fall along a broad spectrum of aggressive acts. The traditional notion of infant innocence and the assumption that aggression is rare in young children have been undermined by the findings of recent research. In a longitudinal survey of 16,038 Canadian children, the developmental trajectories of children aged 4 to 11 years were examined (Tremblay et al., 1999). The results indicated that four-year-old boys and girls exhibited the highest levels of aggression. The peak age of physical aggression was the third year after birth. One important implication of this finding is that the best window for preventive interventions may lie not in adolescence but much earlier in life. This chapter addresses issues related to the definition, assessment, and causes of aggression in children aged 5 or younger, sometimes referred to as toddlers or pre-schoolers.

1. Definition of Aggression

Contemporary definitions of aggression are based on experiences with adults and older children. Many misconceptions and myths exist about what toddlers are or are not capable of doing. Also, views vary widely regarding what is normal behavior for toddlers and are influenced by parental, societal, and cultural expectations. Although studies of aggressive preschoolers are few relative to those found in the literature with aggressive adolescents and adults, an important shift is occurring with developmental psychopathology. We are now studying carefully longitudinal pathways to adult antisocial behavior, tracing causal connections dating back to early childhood.

In definitions of adult aggression, intent to harm is a central concept. In many cases, the definition is restricted to include behaviors intended to cause physical injury in another person. This definition can be problematic when applied to aggression in preschoolers. Preschoolers may lack the ability to cause physical injury, or at least serious physical injury. When expanded to include any behavior intended to cause injury or fear of injury in another, the definition seems more appropriate. Also, as evident from the three examples in the introduction, intent often must be inferred by adults or peers and may not be recognized or admitted by the troubled children themselves. Indeed, intent in preschoolers sometimes must be understood as occurring at the subconscious or preconscious level. For example, although Johnny refuses to discuss why he pulled the trigger on the gun, it may be reasonable to infer the act resulted of repressed hostility toward his baby brother; and Dillon's aggression toward his peers may be a paradoxical and dysfunctional attempt to acquire a connection with them by frightening them.

Consistent with previous views (e.g., Grusec & Lytton, 1988; Maccoby, 1980), the following definition of aggression will be used in this chapter: *A child's behavior is aggressive when its goal, intent, or motivation is directed at harming or frightening another person.*

2. Assessment Issues

2.1 Is Aggression Abnormal?

Statistical Deviation. One way to conceptualize aggressive at behavior is in terms of statistical deviation from the typical behavior of same-aged children. The problem with this is that the set point of normative mean behavior will change according to family, parental, and socio-cultural values. Also, behavior deviation from the mean in either direction, positive or negative, would then be seen as unacceptable.

Qualitative Deviation from Expected Norms. Gelfand, Jenson, and Drew (1997) consider behavior as deviant when *the type, goal and intensity* of that behavior is taken into account. Take for example, the typical *rough and tumble play* (RTP) of preschoolers, which consists of playful fighting and chasing. Usually these play behaviors are seen as prosocial behaviors and can be seen as a healthy progression from RTP to assertive play. RTP is seen as distinctive from other dominance oriented or aggressive behaviors, having a differential salience of interpersonal cues and having a gender related function (DiPietro, 1981). RTP in young children, by Gelfand et al.'s definition, would be unacceptable only when it becomes harmful to others, even if unintentional. The intensified or exaggerated forms of this play can then be construed as aggressive behavior.

2.2 Age-Related Changes

Age Specificity. At a given age, some behaviors have specific qualitative interpretations. The temper tantrums of "terrible twos" right up to age three are seen as benign. Even if a younger child is thrashing around, kicking and sometimes destroying objects, the tantrums are not likely to be seen as serious or aggressive at that stage.

Some authors ascribe toddler aggressive acts as belonging to a specific developmental stage and having a specific behavioral set. Fagot and Hagan (1991) distinguished between younger and older toddler aggressive acts. Young toddlers' acts are more aggressive but brief, while older toddlers have fewer aggressive exchanges, which last longer. The most common acts are to grab or take objects, then to hit followed by a verbal assault.

In a population sample of Canadian children, age of onset and frequency of aggressive behavior was studied. The onset of physical aggression was traced to 80% of this population by age 17 months. However, most children had learnt to inhibit their aggressive impulses by the time of school entry. The authors have suggested that there is a sensitive period when children learn to inhibit aggressive tendencies and it offers the best time for prevention efforts (Tremblay et al., 1999).

Aggression in infancy poses its own challenges. Aggressive infants evoke moralistic judgments and uneasy feelings in most of us. We propose that intentionality to damage, which can cause fearful responses, needs to be interpreted within an infant-caregiver context. A detailed study of infant developmental factors is beyond the scope of this chapter. However, four important concepts must be cited in accepting this proposed viewpoint. First, infants are not passive and definite bi-directionality is observed in the infant's responses to his caregivers. Secondly, infant expression of the feeling states is amodal (Stern, 1985). Thirdly, preverbal communication is quite sophisticated in human infants who do have the ability to express many simple and blended emotions (Izard, Huebner, Rasser, McGuinness, & Doherty, 1982). Fourthly, most literature on the subject accepts that ages 0 through 3 define the infancy period.

Symptom Specificity. From early infancy to the late preschool period, symptoms of aggression change. The progression of control over physical actions (i.e. reaching objects by age 5 months to mobility by first year) to speech and language development and understanding the use of objects in their functional and non functional capacity are responsible for these changes. For example, using a crayon as a writing object and then using it to poke someone with the intention of causing harm is due to increased cognitive awareness. Four dimensions of aggressive behavior that have been investigated are anger, hostility, verbal aggression and physical aggression. in preschoolers, however, impulsiveness emerges as an important determinant as well. Unfortunately, impulsivity is not a well-operationalized construct. Taken in the context of aggressive children, the core problem is disinhibition. Problems with disinhibition start around 3-4 years of age, in the target group under discussion. This capacity for inhibitory control constitutes one of the key executive functions critical in self-regulation and thus behavioral controls.

The notion of symptom specificity is challenged with findings that prevalence of each symptom of conduct and oppositional defiant disorders has been reported in a study of 79 clinic referred children. Starting fights was the most commonly endorsed symptom stealing with confrontation and forcing sexual activities were the least endorsed symptoms of conduct disorder (Keenan & Wakschlag, 2000).

2.3 Aggression and Gender

Generally, there is reluctance in accepting female aggression, especially in little girls. In clinical psychiatric samples, males predominate. However, in North America, female violence has risen substantially over the last decade. Statistics Canada has reported a 127%

increase in charges for violent crimes among females over this period (Savioe, 2000). In the preschool period as well, boys are seen as more aggressive, noncompliant and hyperactive than girls. The gender appropriate acceptable behavior in aggressiveness emerges after early toddler years (Sanson & Prior, 1993).

2.4 Stability of Aggression

The concept of stability is central in understanding if and how young aggressive children become aggressive youth and adults. Longitudinal follow-up studies of aggressive pre-school children offer us important information about the stability of aggressive psychopathology. Several studies have looked at the persistence of aggressive symptoms throughout the lifespan and have determined links between early disruptive behavior, later conduct disorder and adult antisocial personality disorders have been demonstrated (Campbell, 1991; Moffitt, 1990).

Moderate to strong continuity of aggressive behavior was shown in a cohort of low-income boys. There was a persistence of externalizing factors from age 2 to age 6 for 62% of the original sample (Shaw, Gilliom, & Giovannelli, 2000). Thus, the prevailing myth about children growing out of their problems does not seem to be supported by the latest research findings. Child's locus of control and ability to achieve self-regulation and inhibitory control could deflect the progression of aggressive, violent behavior. Studies of temperament precisely explore that view. Isolated aspects of temperament such as a high emotionality are seen as leading to a state of vulnerability (Gjone & Stevenson, 1997

The hypothesis that young child's interpersonal and social relationships can predispose him/her to unmodulated aggression is coming on the forefront. For any child, the first social relationships, which have meaning in a personal sense, are the primary and secondary attachments for the infant. It is commonly accepted that secure attachment provides the intersubjective platform for building confidence in a child's ability to problem solve. This capacity would allow young children to be able to seek alternative solutions to problems, rather than choosing to act out aggressively. An important link has been found between children with insecure avoidant histories and their tendencies to victimize their peers (Troy & Sroufe, 1987).

Disorganized attachment has direct and indirect consequences on aggression stability. To begin with, mothers who reported higher partner violence were likely to have infants with disorganized attachment (Steiner, Zeanah, Stuber, Ash, & Angell, 1994). Secondly, a parent's unresolved trauma impedes optimal parental care of infants, offering insufficient parental protection and guidance regarding self-regulation. The Minnesota Mother-Infant Project showed that early attachments were predictive of aggression in preschool and elementary school aged children. The more significant finding was that the importance of less than optimal transactional patterns were important, even in securely attached children. This sample consisted of an at risk population determined by the poverty of the subjects.

The conclusion of Loeber and Hay (1994) about the influence of peer groups on sustaining aggressiveness in groups of children is one of the landmark findings in this literature. Their non-deviant peers did not only reject children who manifested early conduct problems but they attributed aggressive meanings to normal behavior of deviant children. Targeted children also saw aggressive motives in other's actions with a gradual development in deviant peer groups becoming bullies themselves).

3. Causal Factors

3.1 Emotion and Personality

Surprisingly little attention is given in the literature about the composition of the different emotions underlying aggressive acts in young children. Dillon's matter of fact affect even when caught acting aggressively, Mary's overwhelming anger, and Johnny's confused lost feeling brings forward an important lesson. The extent of a child's feelings, whether angry or sad or frustrated, is not always predictable or consistent with the act of aggression. Very young infants may smile and laugh and yet also surreptitiously commit aggressive acts. For example, pulling the tail of the family pet may be a game for the infant and not construed as an aggressive act. Such overt acts are important in understanding why the notion of intent to harm has to be determined according to the developmental stage of the infants. If an older child, let us say, Dillon, engaged in this action as compared to a young infant of 11-12 months, his behavior could be interpreted as an aggressive act.

Anger is one of the universally recognized emotions underlying aggressive acts. The experience of anger as well as the ability to deal with the interpersonal anger around them, are important issues for infants and toddlers (Cummings, 1987).

Rage, shame and lack of empathy also have occupied a central role in understanding aggressive behavior of infants and preschoolers (Kaufman, 1996). Rage can be seen as an inflation or magnification of anger affect. Ablation of orbitofrontal cortex after early infancy releases the sympathetic and somatomotor responses of rage behavior in animals resulting in aggression (Kolb, 1984).

Empathy or lack of empathy is an important consideration in assessing relational or interpersonal aggressiveness. Between 10 and 14 months the earliest forms of an empathic reaction can be seen. By 18 months an infant can establish prosocial altruistic behavior in the form of offering comfort to others. Orbitofrontal cortex again has an important part to play in regulating empathic and moral behavior (Schore, 1994).

Impatience and the need to win as a personality and emotional determinant of aggressive behavior are described as well. These preschoolers may represent a subset of young children (Vega-Lahr & Field, 1986). Preschoolers coping with inter-parental hostility perceived more negative affect than those children coming from low conflict families. This hostility was exhibited by the preschoolers in their heart rate reactivity and skin conductance response (El-Sheikh, 1994).

Arsenio and Lover (1997) addressed the connections between the emotional displays of preschoolers and their aggressive and non-aggressive peer disputes. The proportion of baseline anger, aggression-related happiness, and aggression related anger differed in the children's initiation of aggressive disputes in their study. Preschoolers who experienced loneliness, dissatisfaction, and withdrawal were seen as at risk of developing aggressive behavior or being victims of aggressive behavior as well (Ladd & Burgess, 1999).

The assimilation of moral feeling is a gradual process. Moral emotions such as fear and guilt also follow developmental progression. Young children will obey to avoid punishment. Perspective taking ability and intelligence are necessary; but they are not sufficient conditions for the development of high levels of moral thought (Maccoby, 1980).

3.2 Psychiatric or Medical Syndromes

Thus far, aggression has been considered in isolation as a form of normal or abnormal behavior, rather than as the symptom of specific mental disorders. But there is a clear link between between mental disorder and childhood aggression. For example, the fourth edition of the *Diagnostic and Statistical Manual of Mental Disorders* (American Psychiatric Association, 1994) documents and groups externalizing disorders such as the attention deficit /hyperactivity disorder, and disruptive behavior disorders. Oppositional defiant disorder, conduct disorder and disruptive behavior disorder not otherwise specified fall under the subheading of disruptive behaviors where propensity for aggressiveness is recognized (Speltz, McClelland, DeKlyen, & Jones, 1999). Similarly, post-traumatic stress disorder in infants, toddlers, and preschoolers can be associated with the onset of aggression toward peers and adults and animals. Behavioral aggression in preschoolers can present also as part of other established neuro-psychiatric disorder such as seizure disorders, brain tumors, and syndromes such as Lysch Nyhan syndrome. A new group of impulsively aggressive young children is rapidly becoming a social nightmare. This group comprises of children with prenatal exposure to alcohol and substances of abuse and HIV infection. In New York City, 80 of every 10,000 children born are addicted to chemicals (Doweiko, 1990).

3.3 Exposure to Violence

Direct violence describes an infant's perception of witnessing of violence and the actual experience of violence, whatever the degree. If exposed to indirect violence, infants themselves may not experience traumatic effects, but because of their dependence on caregivers and the importance of that relationship, any trauma or perceived trauma inflicted on the caregiver may also influence the infant. No other developmental stage is affected as much by indirect violence than this young population. The previously held view that infants and young children were not as seriously affected by experiencing or witnessing violence has been challenged by recent research. This young population can exhibit stress in the four key areas, emotional distress, immature and regressive behaviors, physical complaints and loss of skills, notably language skills. Infants and toddlers are often targets of direct or indirect violent events.

Media Violence. The exposure to violence through television, computer, and war games is under serious scrutiny because these are potentially preventable stressors. The results are mixed regarding the effects of TV viewing by preschoolers. The important findings suggest that when a program provokes aggressive fantasies, preschoolers seem to be most susceptible. The aggressive content can lose its impact if children view TV with a trusted adult who is able to guide and explain. Studies have suggested a causal direction from heavy TV viewing to aggressive behavior (Singer & Singer, 1981; Silvern & Williamson, 1987).

Family Violence. Children are most likely to be victims of violence within the walls of their own family home (Straus, 1974). Infants spend a maximum amount of time within their homes; therefore experiences of violence are mostly confined to homes. Family violence is a major potential stressor in a developing infant's life. Maltreatment and neglect often go hand in hand. In 1995 the validated figures of abused children in the United States was over 1,000, 000 (National Center on Child Abuse and Neglect, 1995).

Most serious physical abuse occurs in the first year of a child's life. Failure to thrive in this age category could be an important clue of abuse and neglect. These infants are described having a frozen watchfulness. Gaze avoidance, and avoidance of communication with the caregiver by non-vocalizations, are common in the first year. Maltreated infants show a different developmental progression than non-maltreated infants. They show emotional and social developmental lags, and aggressive behavior toward their peers. They use fewer words to describe their internal states (Beeghly & Cicchetti, 1994), and have disturbed attachment patterns with their caregivers (Carlson, Cicchetti, Barnett, & Braunwald, 1989).

In general, young children exposed to trauma engage in destructive behaviors more than their non-maltreated peers do. Physically abused children and youth (this study consisted of 415 children ranging from 4-17 years) engaged in more other –directed destructive behaviors. Sexually maltreated children engaged more in self directed destructive behaviors (Taussig & Litrownik, 1997). Even spanking as a disciplinary procedure has come under criticism as children who are spanked show more aggressive behavior towards their peers (Strassburg, Dodge, Pettit, & Bates, 1994). Preschoolers who witness their parent's death through violence exhibit a mixture of emotional and behavioral disturbances (Payton & Krocker, 1988). They are especially at risk for developing aggressive behavior. They often develop a conflicting sense of loyalty, and can even develop aggressive behavior as if to identify with the aggressor (Van Dalen & Glasserman, 1997).

Community Violence. Unfortunately violence at large is the reality of our modern world. Out of 856 gun -shot injuries seen in a Los Angeles emergency department over a 29 month period, there were 272 gang related incidents, of which there were 55 pediatric and 217 adult victims. Since 272 were gang-related incidents, one wonders about the pediatric casualties and whether they were protected from experiencing and witnessing violence (Song, Naude, Gilmore, & Bongard, 1996). Community violence is shown to affect violent behavior of preschoolers in a differential manner. Those who witness violence seem show internalized symptoms while those children who were victimized by violence exhibit externalizing behaviors (Shahnifar, Fox, & Leavitt, 2000).

Children who are victims of severe psychosocial stressors such as poverty, living in crowded living quarters and living in dangerous communities are identified as targets of the chronic violence around them. The problems of these children are seen in terms of not having opportunities to enjoy or master normal expected developmental tasks, such as engaging in play. The second set of problems for these children stems from a likely absence of adequate supervision from distressed, unwilling, or emotionally fragile parents. Children living in violent neighborhoods are aptly described as children living in urban war zones (Garbarino, Kostelny, & Dubrow, 1991). The sad result for children living in these dangerous communities is that they may look for security in antisocial solutions. Parental figures are not often seen as competent or trustworthy; incapable of offering protection. This lack of community support and strong family connection can sometimes create conditions leading to a reliance on gangs or antisocial groups for protection.

3.4 Attachment Problems

The attachment system between infant and caregiver is said to be essentially a regulatory system. Secure early attachment lays solid ground for learning affect regulation, behavioral regulation, behavioral synchrony and early representations (Weinfeld, Sroufe, Egeland, & Carlson, 1999). In the first three years of life there are two ways in which neural connections occur: experience expectant and experience dependent processes. In experience expectant process, the synapse overproduction that occurs is genetically driven (Siegel, 2001) and requires minimum species specific environmental stimulation. The circuits, which are important for emotional and social functioning, are activated for the first time during the early years. Specifically, orbitofrontal region of the brain may be particularly dependent on experience influenced environment. This largely experience dependent development of neural connections needs experience, thus the importance of attachment in early years.

When the optimal experience of secure infant-caregiver attachment does not take place, the insecurity can take many forms. The disorganized, disoriented subtype is of interest here. Predisposition to relational violence can result from early maternal neglect and later parental abuse as expressed in disorganized type of infant-caregiver attachment. This could then be a pathway ready to be explored further from disturbed infant-caregiver attachment to insufficient regulation of infant rages with consequences for unmodulated aggression (Schore, 2001). Youths with the history of disorganized attachments are seen as at risk of expressing hostility with their peers and later on, for interpersonal violence (Lyons-Ruth & Jacobwitz, 1999). The core dynamics of disorganized attachment is explained in terms of having a frightened, frightening or disoriented behavior with the child (Main & Hesse, 1990). This is an important landmark in understanding the intergenerational violence pattern. If the caregivers are frightened, or frightening (abusive), and if they have not resolved trauma, their caregiving pattern can be affected by this factor, carrying forward a dysfunctional parental style and inducing disturbances of attachment with their progeny.

4. Conclusion

The research described in this chapter supports the view that children formative, impressionable, and vulnerable in their preschool years are. During this critical stage of a child's life, the blueprints for behavioral activity, self-regulation, and problem solving are developed. It may be quite normal, even normative, for preschoolers to engage in mild aggression; but a wide range of factors may disturb normal development and increase risk for serious aggression in the short or long term. This suggests that comprehensive assessment and clinical intervention with aggressive children and youths must consider the developmental pathways down which they have traveled. It also offers a ray of hope: The same things that make children vulnerable to disturbance in their preschool years may also make them more responsive to prevention and intervention services they receive during that stage.

References

American Psychiatric Association. (1994). The diagnostic and statistical manual of mental disorders (4th ed.). Washington, DC: Author.
Arsenio W. F., & Lover A. (1997). Emotions, conflicts and aggression during preschoolers' freeplay. *British journal of Developmental Psychology, 15,* 531-542.
Beeghley, M., & Cicchetti, D. (1994). Child maltreatment, attachment and the self-system: Emergence of an internal state lexicon in toddler at high social risk. *Development and Psychopathology, 6,* 5-30.
Campbell, S. B. (1991). Longitudinal studies of active and aggressive preschoolers: Individual differences in early behavior and outcome. In D. Cicchetti & S. L. Toth (Eds.), *Internalizing and externalizing expressions of dysfunction* (pp. 57-90). Hillsdale, NJ: Erlbaum.
Carlson, V., Cicchetti, D., Barnett D., & Braunwald, K. (1989) Disorganized/disoriented attachment relationships in maltreated infants *Developmental Psychology, 25,* 525-531.
Cummings, E. M. (1987). Coping with background anger in early childhood. *Child Development, 58,* 796-84.
DiPietro, J. A. (1981). Rough and tumble play: A function of gender. *Developmental Psychology, 17,* 50-58.
Doweiko, H. E. (1990). *Concepts of chemieal dependency.* Pacific Grove, CA: Brooks/ Cole.
El-Sheikh, M. (1994). Childrens' emotional and physiological responses to interadult angry behavior: The role of history of interparental hostility. *Journal of Abnormal Child Psycholgy, 22,* 661-678.
Fagot, B. I., & Hagan, R. (1991). Observations of parent reaction to sex-stereotyped behaviors: Age and sex effects. *Child Development, 62,* 617-628.
Garbarino, J., Kostelny, K., & Dubrow, N. (1991). What children can tell us about living in danger. *American Psychologist, 46,* 376-383.
Gelfand, D., Jenson ,W., & Drew, C.(1997).*Understanding child behavior disorders* (3rd ed.). Fort Worth, TX: Harcourt Brace.
Gjone, H., & Stevenson, J. (1997). A longitudinal twin study of temperament and behavior problems: Common genetic or environmental influences? *Journal of the American Academy of Child and Adolescent Psychiatry, 36,* 1448-1456.
Grusec, J.E., & Lytton, H. (1988). *Social development.* New York: Springer–Verlag.
Izard, C.E., Huebner, R.R., Risser, D., McGuiness, G., & Dougherty, L. (1982). The young infant's ability to produce discrete emotional expressions. *Developmental Psychology, 16,* 132-140.
Kaufman, G. (1996). *The psychology of shame: Theory and treatment of shame-based syndromes.* New York:

Springer

Keenan K., & Wakschlag, L. S. (2000). More than the terrible twos: The nature and severity of behavior problems in clinic–referred preschool children. *Journal of Abnormal Child Psychology, 28,* 33-46.

Kolb, B. (1984). Functions of the frontal cortex in the rat: A comparative review. *Brain research Reviews, 8,* 65-98.

Loeber, R., & Hay, D. F. (1994). Developmental approaches to aggression and conduct problems. In M. Rutter & D. F. Hay (Eds.), *Development through life: A handbook for clinicians* (pp. 488-515). Oxford: Blackwell Scientific.

Ladd, G.W., & Burgess K.B. (1999). Charting the relationship trajectories of aggressive, withdrawn, and aggressive/withdrawn children during early grade school. *Child Development, 70,* 910-929.

Lyons-Ruth, K., & Jacobwitz, D. (1999). Attachment disorganization: Unresolved loss, relational violence, and lapses in behavioral and attentional strategies. In J. Cassidy & P. R. Shaver (Eds.), *Handbook of attachment: Theory, research, and clinical applications* (pp. 520-554). New York: Guilford.

Maccoby, E. E. (1980). *Psychological growth and the parent–child relationship.* NY: Harcourt Brace Jovanovich.

Main, M., & Hesse, E. (1990). Parents' unresolved traumatic experiences are related to infant disorganized status: Is frightened and /or frightening parental behavior the linking mechanism? In M. Greenberg, D. Cicchetti, & M. Cummings (Eds.), *Attachment in the preschool years* (pp. 161-182). Chicago: University of Chicago Press.

Moffitt, T. E. (1990). Juvenile delinquency and attention deficit disorder: Boy's developmental trajectories from age 3 to age 15. *Child Development, 61,* 893-910.

National center on Child Abuse and Neglect, U.S. Department of Health and Human Services. 1995). Child Maltreatment 1993: Reports from the states to the National Center on Child Abuse and Neglect. Washington, DC: U.S. Government Printing Office.

Payton J.B., & Krocker–Tuskan, M. (1988). Children's reactions to loss of parent through violence. *Journal of the American Academy of Child & Adolescent Psychiatry, 27,* 563-566.

Sanson, A., & Prior, M. (1993). Gender differences in aggression in childhood: Implications for a peaceful world. *Australian Psychologist, 28,* 86-92.

Savioe, J. (1999). *Youth violent crime* (Statistics Canada, Catalogue no. 85-002-XPE, Vol. 19, no. 3). Ottawa: Juristat: Canadian Centre for Justice Statistics.

Schore, A. N. (1994). *Affect regulation and the origin of the self.* Hillsdale, NJ: Erlbaum.

Schore, A. N. (2001). The effects of early relational trauma on right brain development, affect regulation, and infant mental health. *Infant Mental Health Journal, 22,* 201-269.

Shahnifar, A., Fox, N. A., & Leavitt, L. A. (2000). Preschool children's exposure to violence: Relation of behavior problems to parent and child reports. *American Journal of Orthopsychiatry, 70,* 115-125.

Shaw, D. S., Giliom, M., & Giovannelli, J. (2000). Aggressive behavior disorders. In C. H. Zeanah (Ed.), *Handbook of infant mental health* (2nd ed., pp. 397-411). New York: Guilford.

Siegel, D. J. (2001). Toward an interpersonal neurobiology of the developing mind: Attachment relationships, "mindsight," and neural integration. *Infant Mental Health Journal, 22,* 67-94.

Silvern, S. B., & Williamson, P. A. (1987). The effects of video game play on young children's aggression, fantasy, and prosocial behavior. *Journal of Applied Developmental Psychology, 8,* 453-462

Singer, J. L., Singer, D. G. (1981). *Television, imagination and aggression: A study of preschoolers.* Hillsdale, NJ: Erlbaum.

Song, D. H., Maude, G. P., Gilmore, D. A., & Bongard, F. (1996). Gang warfare: The medical repercussions. *Journal of Trauma-Injury infection and Critical care, 40,* 810-815.

Speltz, M.L., McClellan, J., DeKlyen, M., & Jones, K. (1999). Preschool boys with oppositional defiant disorder: Clinical presentation and diagnostic change. *Journal of American Academy of Child and .Adolescent.Psychiatry, 38,* 838-845.

Steiner, H., Zeanah, C.H., Stuber, M., Ash, P., & Angell, R. (1994). The hidden faces of trauma: An update on child psychiatric traumatology. Scientific Proceedings of the Annual meeting of the American Academy of Child and Adolescent Psychiatry.

Stern, D. (1985). *The interpersonal world of the infant.* New York: Basic Books.

Strassberg, Z., Dodge, K. A., Pettit, G. S., & Bates, J. E. (1994). Spanking in the home and children's subsequent aggression toward kindergarten peers. *Development and Psychopathology, 6,* 445-461.

Straus, M. (1974) Cultural and organizational influences on violence between family members. In R. Prince & D. Barried (Eds), *Configurations: Biological and cultural factors in sexuality and family life* (pp. 121-132). Washington, DC: Heath.

Taussig, H. N., & Litrownik, A. J. (1997). Child maltreatment. *Journal of the American Professional Society on the Abuse of Children, 12,* 172-182.

Tremblay, R.E., Japel, C., Perusse, D., McDuff, P., Boivin M., Zoccolillo, M., & Montplaisir, J. (1999). The search for the age of 'onset' of physical aggression: Rousseau and Bandura revisited. *Criminal Behaviour & Mental Health, 9,* 8-23.

Troy, M., & Sroufe, L. A.(1987) Victimization among preschoolers: The role of attachment relationship theory. *Journal of the American Academy of Child and Adolescent Psychiatry, 26,* 166-172.

Van Dalen, A., & Glasserman, M. (1997). My father, Frankenstein: A child's view of battering parents. *Journal of the American Academy of Child and Adolescent Psychiatry, 36,* 1005-1007.

Vega-Lahr, N., & Field, T.M. (1986). Type A behavior in preschool children. *Child Development, 57,* 1333-48.

Weinfeld, N. S., Sroufe, L. A., Egeland, B., & Carlson, E. A. (1999). The nature of individual differences in infant-caregiver attachment. In J. Cassidy & P. R. Shaver (Eds.), *Handbook of attachment: Theory, research, and clinical applications* (pp. 68-88) New York: Guilford.

Multi-Problem Violent Youth
R.R. Corrado et al. (Eds.)
IOS Press, 2002

Child Neuropsychiatric Disorders:
A Review of Associations with Delinquency and Substance Use

Niklas Långström

This literature review attempts to delineate major trends in the scientific literature concerning the relative importance of the following child neuropsychiatric disorders for the development or persistence of antisocial behaviors/delinquency and substance abuse in children and adolescents; *autism spectrum disorders,* specifically *Asperger syndrome, Attention Deficit/Hyperactivity Disorder (ADHD),* and *Tourette's syndrome* (American Psychiatric Association, 1994). These disorders were chosen first because clinical experience suggests that antisocial behavior or substance use could be over-represented in individuals diagnosed with these disorders. Second, these disorders are often identified and diagnosed already in childhood and could thus be expected to have been included in prospective studies of young individuals. Outcomes other than delinquency and substance use disorders (SUD) were not considered. The review is based on a computerized search of the databases MedLine, PsycInfo up to, and including the autumn of 2000 completed with a manual search of other references known by the author. It focuses on cross-sectional and prospective studies of children and adolescents (and occasionally adults) reported upon during the last 15 years. Population-based studies including male and female subjects as well as studies of individuals referred for evaluation or treatment were included. Since most clinical studies have focused on boys, it is important to note that reviewed studies from clinical settings concern males unless otherwise stated.

It is crucial to remember that the associations reviewed in this chapter could be due to different underlying mechanisms and should not automatically be conceived of as indicating direct *causal* relationships. First, associations of two observed phenomena – for example, a low level of a marker of the brain's serotoninergic neurotransmitter system measured in the blood with criminal offending – could initially just be a random finding that needs to be repeated across settings and samples to strengthen the evidence of a true association. After this test of generalizability, the *direction* of a possible causal relationship should be delineated. In a cross-sectional study, this is usually more complicated than in a prospective study. In the latter case, the temporal sequence of occurring events lends stronger support to conclusions concerning directionality. Further, an association could be caused by a shared (latent) underlying risk factor or process that, in our example, affects the serotoninergic system itself *and* the development of criminal offending in adulthood. That is, associated phenomena could be different time- or setting-specific expressions of a common single underlying process. In recent years, the latter model of conceptualization has been applied in studies of developmental trajectories or paths leading from specified constellations of risk- and protective factors to chains of specific future outcomes. In the words of Patterson et al., "covariation of symptoms may itself serve as a marker variable signaling the presence of a shared underlying process. A wide spectrum of symptoms with very different topographic features may be generated by a single process that unfolds over time" (Patterson, DeGarmo, & Knutson, 2000, p. 91). We are at present just beginning to get a deeper understanding of such underlying processes.

1. Conduct Disorder

There are several reasons to consider Conduct Disorder (American Psychiatric Association, 1994) as a risk factor for the development of antisocial behaviors, together with the neuropsychiatric disorders. The diagnosis of CD captures a repeated and persistent pattern of behaviors that violates the basic rights of other people or age-appropriate basic social norms and rules and interferes with social, academic or occupational functioning. The behaviors associated with CD include running away from home, truancy, property destruction, fraud, and aggressive behaviors towards humans and animals including recurrent fights, physical cruelty, robbery and sexual offending. CD has a multi-determined cause involving various biopsychosocial factors interacting in ways only partly elucidated to date, despite considerable achievements in recent years (for reviews see Brennan & Raine, 1997; Lahey, Waldman, & McBurnett, 1999; Loeber & Hay, 1997; Loeber, Burke, Lahey, Winters, & Zera, 2000). A variety of population-based studies indicate that the prognosis for children with an early start in persistent delinquency or an early diagnosis of CD is grim and includes persistent adolescent and adult antisocial behavior, antisocial personality disorder, substance abuse and other forms of psychosocial maladaption (Kratzer & Hodgins, 1997; Robins & Price, 1991; Stattin & Magnusson, 1991; see also Rutter, 1995; Loeber & Farrington, 1998; 2000 for reviews). Symptoms of CD carry the least favorable long-term prognosis of all diagnoses established in this age-period (Rutter, 1995). In addition, treatment effects are modest at best (see Loeber & Farrington, 1998 for reviews).

Although CD is a heterogeneous disorder, it has shown consistent convergent validity concerning co-morbidity with substance use, depressive, and anxiety disorders and quite robust predictive validity. Thus, early onset CD (i.e. before 10 years of age) is associated with poor long-term prognosis. Several large longitudinal studies indicate that most children who meet criteria for this diagnosis before the onset of puberty commit offences as adults or suffer from psychiatric disorders, typically antisocial personality disorder (ASPD) and substance use disorders (SUDs) (Kratzer & Hodgins, 1997; Rutter, 1995; AACAP, 1997). Oppositional Defiant Disorder (ODD, American Psychiatric Association, 1994) is often construed as a mild or prodromal form of CD that may precede the latter with several years in clinical studies. ODD is characterized by a persistent pattern of negativistic, defiant, disobedient and oppositional behavior. Population-based studies, however, suggest that CD and ODD tap somewhat separate dimensions of psychopathology (Loeber et al., 2000).

2. Attention Deficit/Hyperactivity Disorder (ADHD)

ADHD is one of the most common and well-researched diagnostic entities used in contemporary child and adolescent psychiatry, with a prevalence of 3-5% in the general population. The hallmarks of this disorder are persistent problems with inattention and/or hyperactivity/impulsivity obvious in at least two different settings associated with a clinically significant reduction in everyday functioning. The current conceptualization in the fourth edition of the *Diagnostic and Statistical Manual of Mental Disorders* or DSM-IV (American Psychiatric Association, 1994) differentiates between primarily inattentive, primarily hyperactive and combined ADHD subtypes. The tenth edition of the *International Statistical Classification of Diseases and Related Health Problems* or ICD-10 (World Health Organization, 1992) has adopted a more restrictive definition of the "hyperkinetic syndrome."

ADHD and CD/ODD co-occurrence. Studies based on non-referred as well as clinical populations have indicated that in 12 to 50% of all boys with ADHD, this disorder co-occurs with CD. Recently, this has also been found to hold for girls (Biederman et al., 2000). In older studies, CD and ADHD were not always carefully distinguished, leading to inflated correlations between ADHD and the long-term development of criminality and SUD (Lilienfield & Waldman, 1990; Rutter, 1995), since this association was confounded by high levels of concurrent aggression or CD. Most studies suggest that CD and ADHD are separate nosologic entities with somewhat different risk factors (see AACAP, 1997; Taylor, Chadwick, Hepinstall, & Danckaerts, 1996).

A body of research has identified the co-occurrence of inattention, hyperactivity, and impulsivity – that is, ADHD plus ODD or CD – as a diagnostic combination in children that is associated with serious problems in both concurrent functioning and long-term outcome. Particularly, it seems as if children diagnosed with the combined subtype of ADHD, that is with both symptoms of inattention and hyperactivity/impulsivity are at a heightened risk for associated ODD and CD as well as other co-morbid disorders both in clinic-referred (Faraone, Biederman, Weber, & Russell, 1998b) and non-referred samples (Willcutt, Pennington, Chhabildas, Friedman, & Alexander, 1999). Negative outcomes include social maladaption, long-term antisocial behavior, and SUDs (e.g., Moffitt, 1990; Satterfield, Swanson, Schell, & Lee, 1994; MacDonald & Achenbach, 1996, see also Rutter, 1995; Lynam, 1997 for discussions). Data from longitudinal, prospective studies indicate that ADHD increases the risk for development of CD, especially the early-onset form (Taylor et al., 1996; Loeber, Green, Keenan, & Lahey, 1995). However, although ADHD appears to be a risk factor particularly for early onset CD, CD does not in itself cause hyperactivity (Rutter, 1995). Also, non-referred elementary school children with co-occurring ADHD and ODD were shown to be significantly more impaired in terms of externalizing, internalizing and social problems than children with either "pure" ODD or ADHD (Carlson, Tamm, & Gaub, 1997).

3. ADHD and Concurrent Problems

In a cross-sectional study, Gray, Pithers, Busconi, and Houchens (1999) reported on 6 - 9 year-old boys and girls referred for assessment and treatment because of "developmentally unexpected sexual behaviors." Reasons for admissions included exploitative contact- as well as non-contact sexual behaviors (public exposure and public or compulsive masturbation). The authors found CD among 76% and ADHD in 48% of the children. At present, we know very little about the relation between child neuropsychiatric disorders and sexual problem behavior or aggression. However, Kafka and associates (Kafka & Prentky, 1998) have reported on overrepresentation of ADHD in outpatient adults that suffer from paraphilias and non-paraphilic hypersexual behaviors.

Several studies have found that the strength of the relationship between ADHD and CD is affected by psychosocial risk factors. Although a large body of research indicates, and a majority of experts in the field agree, that ADHD is a neuropsychological disorder mainly caused by biological factors, newer data indicate that the *persistence* of ADHD, psychiatric co-morbidity, and negative outcomes are affected by non-neuropsychological risk factors. Thus, in a cross-sectional study of Caucasian boys, Biederman et al. (1995a) found a dose-response relationship between the number of psychosocial risk factors in children (low socio-economic status (SES), pervasive parental conflicts, criminality and psychiatric disorders among parents, etc.) and concurrent child risk for ADHD as well as psychiatric and psychosocial problems (including CD). This relation remained significant even when controlling for family history of ADHD. The authors found a similar increasing risk for psychosocial and psychiatric problems among these children with increasing number of familial risk factors, independently of the presence of ADHD. In a follow-up study of the same boys (Biederman et al., 1995b), ADHD and poor social functioning in school and with peers increased with greater number of family problems at baseline (low SES, family conflicts and parental psychiatric disorder). ADHD was not necessary for the lowering of social functioning as a function of the number of family problems for each child. Thus, no direct support for ADHD contributing to individual child vulnerability for adverse psychosocial factors was identified. On the diagnostic level, family factors did not contribute to increased prevalence of CD, which probably indicates etiological differences between ADHD-related and non-ADHD related CD (see also Lynam, 1997; Frick, Barry, & Bodin, 2000a; Frick, Bodin, & Bary, 2000b). In another follow-up of the same sample of Caucasian boys, Biederman et al. (1996) found that family history of ADHD and psychosocial risk factors both increased risk for the persistence of the ADHD syndrome. Thus, genetic, other biological as well as psychosocial risk factors seem to interact to cause the phenotypic expression of ADHD.

In the Dunedin study of a large birth cohort, Moffitt (1990) found that boys with both ADHD and delinquency at age 13 had histories of earlier and more persistent antisocial behavior than those with only ADHD *or* delinquency. Moffitt (1993) later posited that within a relatively small group of individuals with an early onset of delinquency, neuropsychological difficulties interact with criminogenic environments and psychosocial risk factors in the development of more antisocial personalities. These criminally persistent individuals would become more violent, often remain delinquent throughout adolescence into adult age and tend to contribute to a large proportion of adult criminality. Consistent with Moffitt's theory (Moffitt, 1993), early starters were more criminally active, more diverse and more violent than late starters in a large Swedish birth cohort followed from childhood to age 30 (Kratzer & Hodgins, 1997). However, Moffitt's theory better described male than female offenders (see also Silverthorn & Frick, 1999). These authors also concluded that early-start offenders constitute a heterogeneous group and that further studies should try to identify subgroups among these early-start, life-course persistent offenders. Patterson et al. (Patterson, Forgatch, Yoerger,, & Stoolmiller, 1998) also presented longitudinal data supporting Moffitt's theory. Finally, results from a study based on a large Danish birth cohort (Raine, Brennan, & Mednick, 1994; 1997; Raine, Brennan, Mednick, & Mednick, 1996) are also in agreement with Moffitt's theory. Raine and colleagues found that an interaction between delivery complications (likely to have increased risks for neuropsychological deficits in subjects) or early neuromotor problems and adverse family factors specifically predicted early onset of serious violent behavior in adolescence and young adulthood beyond that of either factor alone. However, further studies are needed to test empirically.

4. Moffitt's Early- and Late-onset Typology

Lynam (1998) presented data that indicated that ADHD + concurrent CD, contrasted against ADHD only, CD only and normal control boys 12-13 years of age could be the childhood diagnostic correlate most closely resembling adolescent or adult psychopathy. Lynam used the 12-item Childhood Psychopathy Scale (CPS), modeled after the PCL-R, other measures of personality and self-reported delinquency, computer tests assessing response modulation and delay of gratification, and two tests of frontal lobe functioning. Apart from the CPS, all measures had previously been shown to differentiate between adult psychopaths and non-psychopaths. Similar to findings with adult psychopaths, the ADHD +

CD group differed significantly from the other groups on all tests. Lynam's results are in agreement with data suggesting that ADHD with concurrent antisocial behavior in children or parents could be regarded as a nosologically and clinically meaningful diagnostic construct separate from ADHD only (Biederman et al., 1995b; Faraone, Biederman, Mennin, Russell, & Tsuang, 1998a, see also Jensen, Martin, & Cantwell, 1997). This is also an approach followed in the current ICD-10 hyperkinetic conduct disorder construct (World Health Organization, 1992). Thus, CD + co-morbid ADHD, in contrast to either disorder on its own, taps into many of the biological and behavioral risks known to be associated with the adult psychopathy construct.

5. Predictive validity of ADHD

In the short term, problems with hyperactivity persist in many children diagnosed with ADHD. Longer follow-ups into early adulthood have shown a significant reduction of the persistence of the full ADHD syndrome even though established co-morbidity like SUD and ASPD does not seem to decrease considerably over time once established (Hill & Schoener, 1996; Klein & Manuzza, 1991; Manuzza, Klein, Bessler, Malloy, & LaPadula, 1993; Rasmussen & Gillberg, 2000). Thus, in an overview of prospective studies, Hill and Schoener (1996) found that the persistence of ADHD diagnoses established in childhood decreased exponentially with age. With a 4% prevalence in childhood, the adult prevalence of the full ADHD syndrome in an unselected non-referred population could be extrapolated to 0.8% at age 20 and 0.05% at age 40 (see also Toone & van der Linden, 1997). However, it is important to note that the analysis by Hill and Schoener only discussed the maintenance of full diagnostic status that is *syndromatic* persistence in contrast to *symptomatic* persistence. For example, Weiss et al. (1995) found that about one third of 25-year-old men diagnosed with ADHD at age 10 reported mild to severe problems with at least one of the core symptoms of ADHD at follow-up (i.e., hyperactivity, attention deficits and impulsivity). Further, Biederman, Mick and Faraone (2000) presented data indicating that the *definition* of remission strongly affected reported remission rates at age 19. Thus, although only 38% of boys met criteria for the full ADHD syndrome by age 19, 90% showed evidence of clinically significant impairment. Symptoms of hyperactivity and impulsivity declined at higher rates than inattention symptoms. Biederman, Mick, and Faraone (1998) also found that children with persisting ADHD have variable emotional, educational and social outcomes, despite the persistence of the syndrome itself. Psychiatric co-morbidity and impulsivity, as well as exposure to maternal psychopathology and larger family size predicted failure to normalize functioning among children with persistent ADHD. The authors suggested that the normalization of function and persistence of ADHD may be partially independent.

Several prospective studies of clinic-referred boys indicate that a childhood diagnosis of ADHD clearly increases the risk for adolescent antisocial behavior and substance abuse (Satterfield et al., 1994; Loeber et al., 1995), adult antisocial personality disorder (ASPD, Weiss, Hechtman, Milroy, & Perlman, 1985; Satterfield & Schell, 1997) and adult substance abuse (Weiss et al., 1985). In a series of papers, Manuzza, Klein and colleagues followed up a cohort of boys aged 6 to 12 who were clinically diagnosed with ADHD *without* concomitant CD. At an average age of 18 years (Gittelman, Manuzza, Schenker, & Bonagura, 1985; replicated in a parallel cohort: Manuzza et al., 1991) less than one third of those who had been diagnosed with ADHD as children were diagnosed with CD or ASPD (8% of controls), whereas some 10% (2% of controls) were diagnosed with ongoing SUD. Thus, childhood ADHD predicted ASPD and SUD at age 18. At an average age of 25 (Manuzza et al., 1993), 18% fulfilled criteria for ASPD and 16% suffered from concurrent

SUD (4% and 4%, respectively, among controls). At follow-up of the parallel cohort (Manuzza et al., 1998) at 24 years of age, 12% of the subjects were now diagnosed with ASPD and 12% with SUD (4% and 4%, respectively of controls). Specifically, abuse of alcohol did not differ between controls and ADHD subjects. Rey et al. (1995) found ASPD in 27% of young Australian adults who had been clinic-referred and diagnosed with ADHD *without* comorbid CD six years earlier. The corresponding point-prevalence of ASPD at follow-up for those diagnosed with CD only was 28%, whereas 50% of adolescents diagnosed with ADHD + CD at baseline fulfilled criteria for ASPD six years later.

In a Swedish population-based study (Rasmussen & Gillberg, 2000; Hellgren, Gillberg, Bågenholm, & Gillberg, 1994), ADHD with co-occurring developmental coordination disorder (DCD) was specifically studied. At a 10-year follow-up of children diagnosed with ADHD + DCD at age 7, the prevalence of ASPD was not different from that found in the control group (7% *vs.* 7%, Hellgren, et al., 1994). The corresponding figures for SUD were 14% and 2%, respectively. Due to the low sample power, this difference did not reach conventional levels of statistical significance. Within the group of individuals diagnosed with ADHD + DCD, the risk for persistent antisocial behavior, as well as ASPD and borderline personality disorder, was the largest among those that had developed antisocial behaviors by age 10. At a 15-year follow-up of the same sample (Rasmussen & Gillberg, 2000), ADHD (with or without DCD) at age 7 significantly increased the risk for ASPD (18% *vs.* 2%) and SUD (31% *vs.* 7%), as compared to the non-ADHD control group. The authors argued that DCD symptoms associated with ADHD strongly contributed to persistent problems at follow-up. Notably, in accordance with therapeutic practices in Sweden, none of the subjects had ever received stimulant medication.

In the Dunedin study, Caspi and co-workers (Caspi, Moffitt, Newman, & Silva, 1996) found that undercontrolled behavior (including children who were impulsive, restless and distractible) among 3-year olds of both sexes predicted a higher risk of being diagnosed with ASPD or having committed a violent crime at age 21. Similar results were obtained in a Swedish population-based study of 13-year old boys followed up for registered alcohol problems or violent offending to age 25 (af Klinteberg, Andersson, Magnusson, & Stattin, 1993)).

6. Predictive Validity of ADHD Plus ODD/CD

Criminal offending. In a population-based longitudinal study, Farrington et al. (1990) found independent effects of attention deficits and conduct disordered behavior on adolescent and adult criminal offending. In yet another large prospective study, Taylor et al. (1996) found that hyperactivity assessed at age 6 had a unique predictive effect on violent behavior in late adolescence also when controlling for baseline conduct problems. In addition, Elander, Simonoff, Pickles, Holmshaw, and Rutter (2000) reported that hyperactivity contributed independently to the prediction of criminal convictions and repeated convictions in young adult age among male child psychiatric patients. In a Swedish study (Dalteg & Levander, 1998) of a group of incarcerated juvenile law-breakers diagnosed with CD, ADHD diagnoses were assigned retrospectively based on file reviews. The authors reported an association between a diagnosis of ADHD and accumulated total number of offences rather than violent offending at a follow-up more than 15 years later. They concluded that a retrospective diagnosis of ADHD could be associated with extent of adult criminal offending over and above the effect posed by CD.

Other studies on clinical samples indicate that ADHD *without* the development of childhood CD does not increase risk for severe criminal offending in adulthood. No incremental predictive effect of ADHD over ODD/CD for adult criminality was found in clinical groups of children followed up in adulthood (Barkley, Fischer, Edelbrock, & Smallish, 1990; Satterfield & Schell, 1997), nor in large population-based prospective studies from New Zealand (Fergusson, Lynskey, & Horwood, 1997), Sweden (Alm, 1996) or the US (MacDonald & Achenbach, 1996). MacDonald and Achenbach did not find any differential effects of the two determinants between the sexes using gender-specific ADHD and CD criteria sets. They pointed out that the relative contributions of attention and conduct problems to later delinquent behavior probably vary with both type of outcome and sample with independent and additive effects being more likely in population-based samples than in clinical ADHD samples. The former suggestion is supported by data from a population-based study of 9-year olds with hyperactivity/impulsivity and control children, followed up after 17 years (Babinski, Hartsough, & Lambert, 1999). Babinski et al. found that hyperactivity/impulsivity and CD at baseline both contributed uniquely to the prediction of any *official arrest* and non-violent offences in adulthood among boys. However, only childhood CD contributed to official arrests for adult violent offending in multiple logistic regression models. Likewise, only CD contributed significantly to *self-reported* violent offending in separate logistic regression models, whereas only hyperactivity/impulsivity contributed to self-reported property offending. Hyperactivity/impulsivity and CD each contributed significantly to *self-reports* of 10 or more offences. The authors concluded that it was predominantly the symptoms of hyperactivity-impulsivity, but not inattention, that contributed to risk for offending over and above the risk associated with early CD alone. Low statistical power for analyses involving girls precluded the examination of gender differences. Finally, in a large Canadian population-based study of different developmental trajectories for 6-15 year old boys and resulting juvenile delinquency up to 17 years of age (Nagin & Tremblay, 1999), hyperactivity was not a powerful predictor of juvenile delinquency when controlling for opposition and aggression.

Substance Use Disorder (SUD). In a study of adults with a lifetime history of SUD, Wilens, Biederman and Mick (1998) found that co-morbid ADHD was associated with a longer duration of SUD and a significantly slower remission rate. Whether the familial co-occurrence of ADHD and SUD is due to common underlying mechanisms it not yet known (see Milberger, Faraone, Biederman, Chu, & Wilens, 1998). However, ongoing behavioral genetic studies with twin samples are likely to provide answers in a near future. An assumed risk for development of substance use in ADHD subjects treated with low-dose amphetamines has not received empirical support. Rather, subjects with ADHD medicated with amphetamine at baseline were at substantially (85%) and significantly reduced risk for a SUD at follow-up relative to untreated subjects with ADHD (Biederman, Wilens, Mick, Spencer, & Faraone, 1999). These results suggest that the association between ADHD and SUD is weakened instead of strengthened as a result of adequate pharmacological treatment. This also holds true for the association between ADHD and aggression or CD, according to results from the large NIMH-funded MTA study, where the relative effects of different treatment modalities for ADHD were studied (Jensen et al., 2001).

Chilcoat and Breslau (1999) found that levels of externalizing problems in a community-based sample of normal and low birthweight children diagnosed with ADHD at age 6 predicted risk for drug use at age 11. A large population based cross-sectional investigation of adolescent twins of both sexes did not find an incremental risk for substance use of ADHD over CD (Disney, Elkins, McGue, & Iacono, 1999). The Dunedin study (Fergusson, Lynskey, & Horwood, 1997b) also failed to establish a relation between ADHD and later substance use while controlling for CD. Results from the Pittsburgh Youth Study of boys ages 7 to 18 (Loeber, Stouthamer-Loeber, & White, 1999) showed that whereas persistent delinquency in childhood predicted persistent substance use at follow-ups, ADHD did not predict persistent substance use or delinquency when controlling for co-occurring delinquency.

Among subjects diagnosed with ADHD with or without co-morbid conditions, it may be important to explore other traits, skills and behaviors that affect or mediate outcome. For instance, low *social skills* in ADHD boys at baseline was a predictor of CD and SUD at a four-year follow-up also when controlling for baseline CD and levels of aggression and attention problems (Greene, Biederman, Faraone, Sienna, & Garcia-Jetton, 1997). In the MTA study, co-morbid *anxiety* in ADHD children clearly conferred partly positive effects on ADHD children of both sexes, regardless of the presence of ODD or CD (Jensen et al., 2001). That is, anxiety seemed to exert ameliorating effects on concurrent ODD/CD and improved outcome at follow-up. Similarly, in a large Swedish cross-sectional study of a community sample of male and female 16-year-olds (Andershed, Kerr, & Stattin, 2001), self-reported *callous/unemotional personality traits* were significantly more commonamong serious and violent juvenile offenders compared to other groups of problem youths. Importantly, this held true for both sexes in gender-specific analyses also after statistically controlling for factors correlated with both callous/unemotional traits and conduct problems, such as hyperactivity-impulsivity-inattention, thrill-seeking, irresponsibility, lying and manipulation.

7. Tourette's Syndrome

Simple motor or vocal tics (i.e., involuntary psychomotor actions) are quite common and can be found among up to 10% of all school children. These tics often disappear and exhibit significant overlap with ADHD and Asperger syndrome (Gillberg, 1995). A persistent pattern of multiple motor and vocal tics – that is, Gilles de la Tourette's syndrome – is much less common and affects 0.2-1% of all school-aged children, with a clear dominance for boys (Kadesjö & Gillberg, 2000). Significant co-morbidity, including full syndromes or substantial symptoms of ADHD and autism spectrum disorders were found among approximately two-thirds of individuals with Tourette's syndrome. Kadesjö and Gillberg (2000) concluded that co-occurring symptoms and syndromes are likely to cause more suffering than the tics *per se*. The clinical symptoms that bring individuals suffering from Tourette's syndrome to the attention of professionals are often different kinds of obsessions, disinhibition, and impulsivity. Other co-morbid problems (studies cited in Gillberg, 1995) are conduct-disordered behavior and sexually exhibitionistic behaviors (among less than 10%). However, several recent studies indicate that conduct-disordered behavior including aggression and sexual problem behaviors in clinic-referred children may be due rather to co-morbid ADHD than to Tourette's syndrome in itself (Budman, Bruun, Park, Lesser, & Olson, 2000; Carter et al., 2000; Stephens & Sandor, 1999; Spencer et al., 1998). Viewed from another perspective, among ADHD children, a co-morbid diagnosis of Tourette's syndrome implies a much more problematic clinical picture associated also with other forms of co-morbidity (Pierre, Nolan, Gadow, Sverd, & Sprafkin, 1999).

8. Autism Spectrum Disorders

Autistic disorder or childhood autism is a neurodevelopmental disorder, defined by the DSM-IV (American Psychiatric Association, 1994) with criteria reflecting abnormal development of communication and social interaction and a markedly restricted repertoire of activity and interests. By definition, autistic disorder should be clearly discernible before age 3 years and is part of a spectrum of disorders, also referred to as *autism spectrum disorders.* The prevalence among children in the general population is approximately .5% for all autism spectrum disorders, and .1% for childhood autism (see Gillberg, 1998 for an overview). The psychosocial outcome for childhood autistic disorder varies, but is often poor. Individuals often remain dependent on help and support from the environment also as adults. The strongest predictive factor for poor long-term outcome in autism is low IQ and low language skills when starting school. In practice, it is most likely Asperger syndrome (see below) and so-called high-functioning autism that may affect the development of criminal behaviors and would require differential diagnostic skills and decisions.

9. Asperger Syndrome

Asperger syndrome is often conceptualized as a form of *high-functioning autism* characterized by normal intelligence and no history of language delay as compared to individuals suffering from autistic disorder. Still, it is not clear whether Asperger syndrome truly is part of the continuum of autism spectrum disorders or represents a specific nosologic entity. Swedish population-based studies suggest that Asperger syndrome occurs in approximately 0.4% of 8-16 year-olds (Gillberg, 1998). In contrast to individuals suffering from autistic disorder, often recognized during childhood because of major problems with social interactions, it is likely that individuals suffering from Asperger syndrome could remain undiagnosed into adulthood. Certain syndrome-specific difficulties reduce the likelihood that affected individuals will consult mental health professionals. Thus, follow-up studies of clinical cases will probably give us a worse picture of the long-term outcome than what is generally the case. It is fair to assume that the basic problems will be expressed differently during different developmental phases. Therefore, we have reason to believe that individuals with Asperger syndrome, once consulting professionals, sometimes have been diagnosed with disorders such as schizophrenia and atypical depression. Likewise, schizotypal, schizoid, paranoid, antisocial, borderline and obsessive-compulsive personality traits or disorders are often encountered among people fulfilling criteria for Asperger syndrome (Gillberg, 1998). Two differential diagnostic signs are noteworthy. Due to cognitive as well as affective empathic problems (Gillberg, 1998) or deficient "theory of mind", individuals with Asperger syndrome very seldom are able to fool or manipulate others. Further, they usually display stability of relationships and behaviors over time. Notably, the former characteristic stands in contrast to more circumscribed affective empathy problems seen in individuals that may often more are more appropriately diagnosed with ASPD or psychopathic personality disorder. Such individuals may very well have a good sense of the "inner world" of other people, but are not particularly emotionally concerned about cheating or manipulating, if such behavior serves their own purposes.

A significant and important overlap between different neuropsychiatric syndromes, not the least for differential diagnostic decisions, has also been underlined by Gillberg. Thus, autistic traits can be found among about 50% of all children suffering from severe ADHD with developmental coordination disorder (DCD) and it is sometimes almost impossible to clearly differentiate between severe ADHD + DCD and Asperger syndrome (Gillberg, 1995, for an overview see Gillberg & Billstedt, 2000). In addition, data suggest that attention deficits and other co-morbidity are common in Asperger syndrome. Thus, neuropsychological studies (Ehlers et al., 1997; Nydén, Gillberg, Hjelmquist, & Heiman, 1999) have demonstrated that high-functioning autism and Asperger syndrome are quite often characterized by similar types and degrees of inattention and executive functioning deficits that were previously been identified mainly in children with ADHD only. Such comorbid syndromes or symptoms may affect the risk for development of antisocial behaviors and require specific treatment.

No study so far has systematically studied the long-term outcome for individuals diagnosed with Asperger syndrome. However, the prognosis appears to be much better than for autistic disorder (Gillberg, 1998). The prospective studies that exist suggest that the disturbances of communication and socialization will remain for many years and that the diagnosis is very stabile over time. Very good outcomes concerning the ability for appropriate self-care, educational achievements and professional careers can be seen in many individuals. A few case studies have suggested that some individuals that have committed crimes related to extreme special interests in gunpowder, poison, or fire (Gillberg, 1995; Kohn, Fahum, Ratzoni, & Apter, 1998), also fulfill criteria for Asperger syndrome. In a study of adult male inmates in a British maximum-security prison, 1.5% of the subjects were diagnosed with Asperger syndrome (Scragg & Shaw, 1994). These authors suggested that Asperger syndrome and high-functioning autism should be considered more often by forensic psychologists and -psychiatrists. Finally, a Swedish study suggested that Asperger syndrome and high-functioning autism disorders could sometimes be found in young adult forensic psychiatric patients if the analysis was made with the particular focus to find psychiatric and developmental disorders with onset in childhood (Siponmaa, Kristiansson, Nydén, Jonson, & Gillberg, 2001). However, it does not seem as if criminality *per se* is especially common among individuals with Asperger syndrome (Ghaziuddin, Tsai, & Ghaziuddin, 1991, see also Kohn et al., 1998).

References

AACAP (1997). Practice parameters for the assessment and treatment of children and adolescents with conduct disorder. *Journal of the American Academy of Child and Adolescent Psychiatry, 36, Supplement,* 122S-139S.

af Klinteberg, B., Andersson, T., Magnusson, D., & Stattin, H. (1993). Hyperactive behavior in childhood as related to subsequent alcohol problems and violent offending: a longitudinal study of male subjects. *Personality and Individual Differences, 15,* 381-383.

Alm P.-O. (1996). *Juvenile and adult criminality: Relationships to platelet MAO-activity, triiodothyronine, ADHD, conduct disorder and psychopathy.* Ph.D. thesis. Stockholm, Sweden: Karolinska Institutet. ISBN 91-628-1999-2.

American Psychiatric Association. (1994). *Diagnostic and statistical manual of mental disorders* (4th ed.). Washington DC, USA: Author.

Andershed, H., Kerr, M., & Stattin, H. (2001). *The importance of callous, unemotional traits as opposed to hyperactivity-impulsivity-attention problems in identifying a particularly violent subgroup of conduct-problem youths.* Submitted manuscript.

Babinski, L. M., Hartsough, C. S., & Lambert, N. M. (1999). Childhood conduct problems, hyperactivity-impulsivity, and inattention as predictors of adult criminal activity. *Journal of Child Psychology and Psychiatry, 40,* 347-355.

Barkley, R. A., Fischer, M., Edelbrock, C. S., & Smallish, L. (1990). The adolescent outcome of hyperactive children diagnosed by research criteria: I. An 8-year prospective follow-up study. *Journal of the American Academy of Child and Adolescent Psychiatry, 29,* 546-557.

Biederman, J, Wilens, T., Mick, E., Spencer, T., & Faraone, S. V. (1999). Pharmacotherapy of attention-deficit/hyperactivity disorder reduces risk for substance use disorder. *Pediatrics, 104(2),* e20. URL: *http://www.pediatrics.org/cgi/content/full/104/2/e20.*

Biederman, J., Faraone, S. V., Mick, E., Williamson, S., Wilens, T. E., Spencer, T. J., Weber, W., Jetton, J., Kraus, I., Pert, J., & Zallen, B. (2000). Clinical correlates of ADHD in females: findings from a large group of girls ascertained from pediatric and psychiatric referral sources. *American Journal of Psychiatry, 157,* 1077-1083.

Biederman, J., Faraone, S. V., Milberger, S., Curtis, S., Chen, L., Marrs, A., Ouelette, C., Moore, P., & Spencer, T. (1996). Predictors of persistence and remission of ADHD into adolescence: Results from a four-year prospective follow-up study. *Journal of the American Academy of Child and Adolescent Psychiatry, 35,* 343-351.

Biederman, J., Mick, E., & Faraone, S. V. (1998). Normalized functioning in youths with persistent attention-deficit/hyperactivity disorder. *Journal of Pediatrics, 133,* 544-551.

Biederman, J., Mick, E., & Faraone, S. V. (2000). Age-dependent decline of symptoms of attention deficit hyperactivity disorder: impact of remission definition and symptom type. *American Journal of Psychiatry, 157,* 816-818.

Biederman, J., Milberger, S., Faraone, S. V., Kiely, K., Guite, J., Mick, E., Ablon, S., Warburton, R., & Reed, E. (1995a). Family-environment risk factors for attention-deficit hyperactivity disorder. *Archives of General Psychiatry, 52,* 464-470.

Biederman, J., Milberger, S., Faraone, S. V., Kiely, K., Guite, J., Mick, E., Ablon, S., Warburton, R., Reed E., & Davis S. G. (1995b). Impact of adversity on functioning and co-morbidity in children with attention-deficit hyperactivity disorder. *Journal of the American Academy of Child and Adolescent Psychiatry, 34,* 1495-1503.

Brennan, P. A., & Raine, A. (1997). Biosocial bases of antisocial behavior: psychophysiological, neurological, and cognitive factors. *Clinical Psychology Review, 17,* 589-604.

Budman, C. L., Bruun, R. D., Park, K. S., Lesser, M., & Olson, M. (2000). Explosive outbursts in children with Tourette's disorder. *Journal of the American Academy of Child and Adolescent Psychiatry, 30,* 383-387.

Carlson, C. L., Tamm, L., & Gaub, M. (1997). Gender differences in children with ADHD, ODD, and co-occurring ADHD/ODD identified in a school population. *Journal of the American Academy of Child and Adolescent Psychiatry, 36,* 1706-1714.

Carter, A. S., O'Donnell, D. A., Schultz, R. T., Scahill, L., Leckman, J. F., & Pauls, D. L. (2000). Social and emotional adjustment in children affected with Gilles de la Tourette's syndrome: Associations with ADHD and family functioning. *Journal of Child Psychology and Psychiatry, 41,* 215-223.

Caspi, A., Moffitt, T. E., Newman, D. L., & Silva, P. A. (1996). Behavioral observations at age 3 years predict adult psychiatric disorders: Longitudinal evidence from a birth cohort. *Archives of General Psychiatry, 53,* 1033-1039.

Chilcoat H. D., & Breslau, N. (1999). Pathways from ADHD to early drug use. *Journal of the American Academy of Child and Adolescent Psychiatry, 38,* 1347-1354.

Dalteg, A., & Levander, S. (1998). Twelve thousand crimes by 75 boys: A 20-year follow up study of childhood hyperactivity. *Journal of Forensic Psychiatry, 9,* 39-57.

Disney, E. R., Elkins, I. J., McGue, M., & Iacono, W. G. (1999). Effects of ADHD, conduct disorder, and gender on substance use and abuse in adolescence. *American Journal of Psychiatry, 156,* 1515-1521.

Ehlers, S., Nyden, A., Gillberg, C., Sandberg, A. D., Dahlgren, S. O., Hjelmquist, E., & Oden A. (1997). Asperger syndrome, autism and attention disorders: a comparative study of the cognitive profiles of 120 children. *Journal of Child Psychology and Psychiatry, 38,* 207-217.

Elander, J., Simonoff, E., Pickles, A., Holmshaw, J., & Rutter, M. A (2000). A longitudinal study of adolescent and adult conviction rates among children referred to psychiatric services for behavioural or emotional problems. *Criminal Behaviour & Mental Health, 10,* 40-59.

Faraone, S. V., Biederman, J., Mennin, D., Russell, R., & Tsuang, M. T. (1998a). Familial subtypes of attention deficit hyperactivity disorder: A 4-year follow-up study of children from antisocial-ADHD families. *Journal of Child Psychology and Psychiatry, 39,* 1045-1053.

Faraone, S. V., Biederman, J., Weber, W., & Russell, R. L. (1998b). Psychiatric, neuropsychological, and psychosocial features of DSM-IV subtypes of attention-deficit/hyperactivity disorder: results from a clinically referred sample. *Journal of the American Academy of Child and Adolescent Psychiatry, 37,* 185-93.

Farrington, D. P., Loeber, R., & Van Kammen, W. B. (1990). Long-term criminal outcomes of hyperactivity-impulsivity- psychosocial-attention deficit and conduct problems in childhood. In L Robins & M. Rutter (Eds.), *Straight and devious pathways from childhood to adulthood* (pp. 169-188). New York: Cambridge University Press.

Fergusson, D. M., Lynskey, M. T., & Horwood, L. J. (1997). Attentional difficulties in middle childhood and psychosocial outcomes in young adulthood. *Journal of Child Psychology and Psychiatry, 38,* 633-644.

Frick, P. J., Barry, C. T., & Bodin, S. D. (2000a). Applying the concept of psychopathy to children: Implications for assessment of antisocial youth. In C. B. Gacono (Ed.), *The clinical and forensic assessment of psychopathy: A practitioner's guide* (pp. 3-24). Mahwah, NJ: Erlbaum.

Frick, P. J., Bodin, S. D., & Barry, C. T. (2000b). Psychopathic traits and conduct problems in community and clinic-referred samples of children: Further development of the Psychopathy Screening Device. *Psychological Assessment, 12,* 382-393.

Ghaziuddin, M., Tsai, L., & Ghaziuddin, N. (1991). Violence in Asperger syndrome: a critique. *Journal of Autism and Developmental Disorders, 21,* 349-354.

Gillberg, C., & Billstedt, E. (2000). Autism and Asperger syndrome: coexistence with other clinical disorders. *Acta Psychiatrica Scandinavica, 102,* 321-330.

Gillberg, C. (1995). *Clinical child neuropsychiatry.* Cambridge, UK: Cambridge University Press.

Gillberg, C. (1998). Asperger syndrome and high-functioning autism. *British Journal of Psychiatry, 172,* 200-

209.

Gittelman, R., Manuzza, S., Schenker, R., & Bonagura, N. (1985). Hyperactive boys almost grown up. I. Psychiatric status. *Archives of General Psychiatry, 42,* 937-947.

Gray, A., Pithers, W. D., Busconi, A., & Houchens, P. (1999). Developmental and etiological characteristics of children with sexual behavior problems: treatment implications. *Child Abuse & Neglect, 23,* 601-621.

Greene, R. W., Biederman, J., Faraone, S. V., Sienna, M., & Garcia-Jetton, J. (1997). Adolescent outcome of boys with attention-deficit/hyperactivity disorder and social disability: results from a 4-year longitudinal follow-up study. *Journal of Consulting and Clinical Psychology, 65,* 758-767.

Hellgren, L., Gillberg, I. C., Bågenholm, A., & Gillberg, C. (1994). Children with deficits in attention, motor control and perception (DAMP) almost grown up: psychiatric and personality disorders at age 16 years. *Journal of Child Psychology and Psychiatry, 35,*1255-1271.

Hill, J. C., & Schoener, E. P. (1996). Age-dependent decline of Attention Deficit Hyperactivity Disorder. *American Journal of Psychiatry, 153,* 1143-1146.

Jensen, P. S., Hinshaw, S. P., Kraemer, H. C., Lenora, N., Newcorn, J. H., Abikoff, H. B., March, J. S., Arnold, L. E., Cantwell, D. P., Conners, K. C., Elliott, G. R., & Greenhill, L. L. (2001). ADHD comorbidity findings from the MTA study: comparing comorbid subgroups. *Journal of the American Academy of Child and Adolescent Psychiatry, 40,* 147-158.

Jensen, P., Martin, C., & Cantwell, D. (1997). Co-morbidity in ADHD: implications for research, practice, and DSM-IV. *Journal of the American Academy of Child and Adolescent Psychiatry, 36,* 1065-1079.

Kadesjö, B., & Gillberg, C. (2000). Tourette's disorder: epidemiology and co-morbidity in primary school children. *Journal of the American Academy of Child and Adolescent Psychiatry, 39,* 548-555.

Kafka, M. P., & Prentky, R. A. (1998). Attention-Deficit/Hyperactivity Disorder in males with paraphilias and paraphilia-related disorders: A co-morbidity study. *Journal of Clinical Psychiatry, 59,* 388-396.

Klein, R. G., & Manuzza, S. (1991). Long-term outcome of hyperactive children: A review. *Journal of the American Academy of Child and Adolescent Psychiatry, 30,* 383-387.

Kohn, Y., Fahum, T., Ratzoni, G., & Apter, A. (1998). Aggression and sexual offense in Asperger's syndrome. *The Israel Journal of Psychiatry and Related Sciences, 35,* 293-299.

Kratzer, L., & Hodgins, S. (1997). Adult outcomes of child conduct problems: A cohort study. *Journal of Abnormal Child Psychology, 25,* 65-81.

Lahey, B. B., Waldman, I. D., & McBurnett, K. (1999). Annotation: The development of antisocial behavior: An integrative causal model. *Journal of Child Psychology and Psychiatry, 40,* 669-682.

Lilienfield, S. O., & Waldman, I. D. (1990). The relation between childhood attention-deficit hyperactivity disorder and adult antisocial behavior reexamined: The problem of heterogeneity. *Clinical Psychology Review, 10,* 699-725.

Loeber, R., & Farrington, D. P. (1998). *Serious & violent juvenile offenders: Risk factors and successful interventions.* Thousand Oaks, CA, USA: Sage Publications.

Loeber, R., & Farrington, D. P. (2000). Young children who commit crime: Epidemiology, developmental origins, risk factors, early interventions, and policy implications. *Development and Psychopathology, 12,* 737-762.

Loeber, R., & Hay, D. (1997). Key issues in the development of aggression and violence from childhood to early adulthood. *Annual Review of Psychology 48,* 371-410.

Loeber, R., Burke, J. D., Lahey, B. B., Winters, A., & Zera, M. (2000). Oppositional defiant and conduct disorder: a review of the past 10 years, part I. *Journal of the American Academy of Child and Adolescent Psychiatry, 39,* 1468-1484.

Loeber, R., Green, S. M., Keenan, K., & Lahey, B. B. (1995). Which boys will fare worse? Early predictors of the onset of conduct disorder in a six-year longitudinal study. *Journal of the American Academy of Child and Adolescent Psychiatry 34,* 499-509.

Loeber, R., Stouthamer-Loeber, M., & White, H. R. (1999). Developmental aspects of delinquency and internalizing problems and their association with persistent juvenile substance use between ages 7 and 18. *Journal of Clinical Child Psychology, 28,* 322-332.

Lynam, D. R. (1997). Pursuing the psychopath: Capturing the fledgling psychopath in a nomological net. *Journal of Abnormal Psychology, 106,* 425-438.

Lynam, D. R. (1998). Early identification of the fledgling psychopath: Locating the psychopathic child in the current nomenclature. *Journal of Abnormal Psychology, 107,* 566-575.

MacDonald, V. M., & Achenbach, T. M. (1996). Attention problems versus conduct problems as six-year predictors of problem scores in a national sample. *Journal of the American Academy of Child and Adolescent Psychiatry, 35,* 1237-46.

Manuzza, S., Klein, R. G., Bessler, A., Malloy, P., & LaPadula, M. (1993). Adult outcome of hyperactive boys. *Archives of General Psychiatry, 50,* 565-576.

Manuzza, S., Klein, R. G., Bessler, A., Malloy, P., & LaPadula, M. (1998). Adult psychiatric status of hyperactive boys grown up. *American Journal of Psychiatry, 155,* 493-498.

Manuzza, S., Klein, R. G., Bonagura, N., Malloy, P., Giampino ,T. L., & Adalli, K. A. (1991). Hyperactive boys almost grown up. V. Replication of psychiatric status. *Archives of General Psychiatry, 48*, 77-83.

Milberger, S., Faraone, S. V., Biederman, J., Chu, M. P., &Wilens, T. (1998). Familial risk analysis of the association between attention-deficit/hyperactivity disorder and psychoactive substance use disorders. *Archives of Pediatrics and Adolescent Medicine, 152*, 945-995.

Moffitt, T. E. (1990). Juvenile delinquency and attention deficit disorder: boys' developmental trajectories from age 3 to age 15. *Child Development, 61*, 893-910.

Moffitt, T. E. (1993). Adolescence-limited and life-course-persistent antisocial behavior: A developmental taxonomy. *Psychological Review, 100*, 674-701.

Nagin, D., & Tremblay, R. E. (1999). Trajectories of boys' physical aggression, opposition, and hyperactivity on the path to physically violent and nonviolent juvenile delinquency. *Child Development, 70*, 1181-1196.

Nydén, A., Gillberg, C., Hjelmquist, E., & Heiman, M. (1999). Executive function/attention deficits in boys with Asperger syndrome, attention disorder and reading/writing disorder. *Autism, 3*, 213-228.

Patterson, G. R., DeGarmo, D. S., & Knutson, N. (2000). Hyperactive and antisocial behaviors: Comorbid or two points in the same process? *Development and Psychopathology 12*, 91-106.

Patterson, G. R., Forgatch, M. S., Yoerger, K. L., & Stoolmiller, M. (1998). Variables that initiate and maintain an early-onset trajectory for juvenile offending. *Development and Psychopathology, 10*, 531-547.

Pierre, C. B., Nolan, E. E., Gadow, K. D., Sverd, J., Sprafkin, J. (1999). Comparison of internalizing and externalizing symptoms in children with attention-deficit hyperactivity disorder with and without comorbid tic disorder. *Journal of Developmental and Behavioral Pediatrics, 20*, 170-176.

Raine, A., Brennan, P., & Mednick, S. A. (1994). Birth complications combined with early maternal rejection at age 1 year predispose to violent crime at age 18 years. *Archives of General Psychiatry, 51*, 984-988.

Raine, A., Brennan, P., & Mednick, S. A. (1997). Interaction between birth complications and early maternal rejection in predisposing individuals to adult violence: Specificity to serious, early-onset violence. *American Journal of Psychiatry, 154*, 1265-1271.

Raine, A., Brennan, P., Mednick, B., & Mednick, S. A. (1996). High rates of violence, crime, academic problems, and behavioural problems in males with both early neuromotor deficits and unstable family environments. *Archives of General Psychiatry, 53*, 544-549.

Rasmussen, P., & Gillberg, C. (2000). Natural outcome of ADHD with developmental coordination disorder at age 22 years: A controlled, longitudinal, community-based study. *Journal of the American Academy of Child and Adolescent Psychiatry, 39*, 1424-1431.

Rey, J. M., S., Morris-Yates, A., Singh, M., Andrews, G., & Stewart, G. W. (1995). Continuities between psychiatric disorders in adolescents and personality disorders in young adults. *American Journal of Psychiatry, 152*, 895-900.

Robins, L. N., & Price, R. K. (1991). Adult disorders predicted by childhood conduct problems: results from the NIMH Epidemiologic Catchment Area project. *Psychiatry, 54*, 116-32.

Rutter, M. (1995). Relationships between mental disorders in childhood and adulthood. *Acta Psychiatrica Scandinavica, 91*, 73-85.

Satterfield, J., & Schell, A. (1997). A prospective study of hyperactive boys with conduct problems and normal boys: adolescent and adult criminality. *Journal of the American Academy of Child and Adolescent Psychiatry, 36*, 1726-1735.

Satterfield, J., Swanson, J., Schell, A., & Lee, F. (1994). Prediction of antisocial behavior in attention-deficit hyperactivity disorder boys from aggression/defiance scores. *Journal of the American Academy of Child and Adolescent Psychiatry 33*, 185-90.

Scragg, P., & Shaw, A. (1994). Prevalence of Asperger's syndrome in a secure hospital. *British Journal of Psychiatry, 165*, 679-682.

Silverthorn, P., & Frick, P. J. (1999). Developmental pathways to antisocial behavior: The delayed-onset pathway in girls. *Development and Psychopathology, 11*, 101-126.

Siponmaa, L., Kristiansson, M., Nydén, A., Jonson, C., & Gillberg, C. (2001). *Young forensic psychiatric patients: the role of child neuropsychiatric disorders*. Submitted manuscript.

Spencer, T., Biederman, J., Harding, M., O'Donnell, D., Wilens, T., Faraone, S., Coffey, B., & Geller, D. (1998). Disentangling the overlap between Tourette's disorder and ADHD. *Journal of Child Psychology and Psychiatry, 39*(7), 1037-44.

Stattin, H., & Magnusson, D. (1991). Stability and change in criminal behaviour up to age 30. *British Journal of Criminology, 31*, 327-346.

Stephens, R. J., & Sandor, P. (1999). Aggressive behaviour in children with Tourette syndrome and co-morbid attention-deficit hyperactivity disorder and obsessive-compulsive disorder. *Canadian Journal of Psychiatry, 44*, 1036-1042.

Taylor, E., Chadwick, O., Heptinstall, E., & Danckaerts, M. (1996). Hyperactivity and conduct disorder as risk factors for adolescent development. *Journal of the American Academy of Child and Adolescent Psychiatry, 35,* 1213-1226.

Toone, B. K., & van der Linden, G. J. H. (1997). Attention deficit hyperactivity disorder or hyperkinetic disorder in adults. *British Journal of Psychiatry, 170,* 489-491.

Weiss, G., Hechtman, L., Milroy, T., & Perlman, T. (1985). Psychiatric status of hyperactives as adults: a controlled prospective 15-year follow-up of 63 hyperactive children. *Journal of the American Academy of Child and Adolescent Psychiatry, 24,* 211-220.

Wilens, T. E., Biederman, J., & Mick, E. (1998). Does ADHD affect the course of substance abuse? Findings from a sample of adults with and without ADHD. *American Journal of Addiction, 7,* 156-163.

Willcutt, E. G., Pennington, B. F., Chhabildas, N. A., Friedman, M. C., & Alexander, J. (1999). Psychiatric co-morbidity associated with DSM-IV ADHD in a nonreferred sample of twins. *Journal of the American Academy of Child and Adolescent Psychiatry, 38,* 1355-1362.

World Health Organization. (1992). International statistical classification of diseases and related health problems (10th revision). Geneva, Switzerland: Author.

Multi-Problem Violent Youth
R.R. Corrado et al. (Eds.)
IOS Press, 2002

105

Family Impact on Youth Violence

Vladislav Ruchkin

From the first breath to the last, our lives begin, proceed and are dominated by personal interaction. And perhaps the most important of these interactions is the first; family provides the context for loving attention and care and the foundation for development. Family interactions shape our experiences throughout life and can determine an individual's functioning and mental health prognosis. Hence, family is generally recognized to be the "primary arena" for socialization, where parents are the central characters (Maccoby, 1984). A long line of research moreover, links the development of competence and adaptive adjustment in children with a loving, understanding, autonomy-supporting style of parental rearing, and, conversely, suggests that dysfunctional rearing practices might lead to low competence and maladaptive personality patterns (Perris, 1994). Parents frequently have been implicated as the principal causal agents in their children's behavioral, emotional, personality and cognitive development (Holden & Edwards, 1989), influencing a wide array of developmental processes (Baumrind, 1980; Sears, Maccoby, & Levin, 1957).

Although the first attempts to understand the influences of child rearing on psychosocial development can be dated back to 17th century educational philosophy (see Grant & Tarcov, 1996; Rousseau, 1762), it was only in the 20th century that the first structured measures were developed to investigate the relations between aggressive behavior and parental rearing. In the early 20th century, all major fields of psychology, to a certain degree, implicated the role of parental attitudes in determining child outcome. The main sources of inspiration in this field were stimulated by psychoanalytic theory, especially its attention to the role of past experiences in shaping current behavior; early behaviorism and its attempts to understand the learning process; social psychology, suggesting that attitudes affect parental behavior and thus, influence child's environment; mental hygiene movement with its emphasis on the ways in which interpersonal relationships shape children's behavior; and finally clinical findings of dysfunctional parenting in children with behavioral and emotional problems (Holden & Edwards, 1989; Perris, 1994; Rutter, 1999 for more detailed reviews). The conclusion that emerges from these different sources is that early experiences with parents can be precursors in the development of maladaptive personality characteristics and psychopathology (Perris, 1994). These first theoretical approaches placed the cornerstone for later research and in the 50s and 60s the field flourished. At this time, numerous psychiatric case studies reported cruel punishment and severe discipline in the background of serious offenders (Duncan, Frazier, Litin, Johnson, & Barron, 1958; Easson & Steinhilber, 1961; Kempe, Silverman, Steele, Droegemuller, & Silver, 1962; Silver, Dublin, & Lourie, 1969). Since aggressive parents tend to produce aggressive children (Bandura & Walters, 1959; Eron, Walder, & Lefkowitz, 1971), it was suggested that violence breeds violence (Curtis, 1963; Silver et al., 1969) – an idea that has received a close consideration in more recent research (Kaufman & Zigler, 1987; Widom, 1989).

About the same time, the first monographs describing the role of parental rearing in the development of antisocial behavior were published, such as the classic writings by Glueck and Glueck (1950) and Robins (1966). Although the role of family background such as socio-economic status, parental criminality, and parental mental health had been seen as related to criminal behavior for some time (e.g., Zuckerman, Barrett, & Bragiel, 1960), these were the first studies that were able to demonstrate in a systematic way the role of parental affection, supervision, and discipline on later aggression and antisocial behavior.

Thus, the first major studies also delineated the major components of parental rearing, providing the basis for more structured research approaches. Most of the studies on parental rearing reveal a set of similar rearing practices, which are consistent across different cultures. The classic research of Baumrind (1967, 1971), for example, resulted in the identification of three major types of child rearing styles: authoritative (warm and non-punitive, but disciplining and consistent), authoritarian (cold, rigid and often physically punitive), and permissive (inconsistent and permissive). Maccoby and Martin (1983) further divided the latter type into indulgent (allowing the child to do what he or she wants) and neglectful types. Rohner (1986) identified two major factors – parental acceptance and parental rejection, with the latter subdivided into hostility/aggression, undifferentiated rejection and indifference/neglect. A large international study, conducted in 14 nations, revealed three similar factors of parental rearing: rejection, emotional warmth and overprotection (Arrindell et al., 1994), whereas Parker, Tupling, and Brown (1979) also identified two main factors: care-indifference and autonomy-overprotectiveness.

Several decades of intensive research that followed have further increased our understanding of the factors that underlie behavior problems in youth in general, and aggressive behavior in particular. Recent reviews have summarized the accumulated knowledge and have described in detailed and structured way the relationship between child rearing and later violent and antisocial behavior (Haapasalo & Pokela, 1999; Malinosky-Rummel & Hansen, 1993; Widom, 1989). Independent of the approach adopted by different authors, they reveal a fair level of agreement that dysfunctional rearing practices, especially those defined as hostile, punitive, shaming, rejecting or overcontrolling, significantly relate to the development of different patterns of aggression (e.g., Becker, 1964; Jacobsson, Lindstrom, von Knorring, Perris, & Perris, 1980; Perris, 1994; Sears, Whiting, Nowlis, & Sears, 1953). The evidence for this relationship comes from three major sources (see Malinosky-Rummel & Hansen, 1993; Haapasalo & Pokela, 1999 for a detailed review). First, adolescents who reveal aggressive and violent behaviors generally tend to demonstrate higher rates of maltreatment than in general population (Garbarino & Plantz, 1986; Malinosky-Rummel & Hansen, 1993). Second, retrospective file reviews of incarcerated violent adolescent boys reveal higher rates of harsh parenting than in less violent controls (Haapasalo & Pokela, 1999; Lewis, Shanok, Pincus, & Glaser, 1979). Finally, among the children who receive psychiatric treatment, those who have been physically abused, exhibit more aggressive behavior than their non-abused peers (Blount & Chandler, 1979; Cavaiola & Schiff, 1988; Malinosky-Rummel & Hansen, 1993).

In addition to abusive parental rearing practices per se, non-violent family dysfunction also has been linked to problems in child development. For instance, the occurrence of antisocial behavior has been related to psychiatric disorders among parents (e.g., Ge et al., 1996), as well as to parental substance use (e.g., Kandel, Rosenbaum, & Chen, 1994) and parental criminality (Farrington, Gundry, & West, 1975; Farrington, 1978). The latter was particularly emphasized in the development of antisocial behavior, as a family variable in which both heritability and shared environment play important roles (see Wambolt & Wambolt, 2000 for a review of family role in childhood disorders). Finally, the role of domestic violence has been emphasized (Suh & Abel, 1990; Widom, 1989), with some authors suggesting that witnessing family violence could be almost as influential for future aggressive behavior in children, as being a victim of abusive rearing practices (Hughes, Parkinson, & Vargo, 1989).

In their recent review of the major longitudinal studies in the field, Haapasalo and Pokela (1999) grouped parental influences in five main factors that have been repeatedly identified as crucial for the development of aggressive and antisocial behavior. These factors include deviant parental characteristics (criminality, substance abuse, mental problems), family disruption (separations, divorce, instability, marital conflict), punitive practices and attitudes (corporal punishment, authoritarian attitudes, strict discipline), lack of love (rejection), and lax parenting (poor monitoring, lack of supervision) (Haapasalo & Pokela, 1999).

Despite the definite role played by these major factors, it is often difficult to separate the individual effect of each particular factor, since they commonly co-occur and mutually influence each other's development. Low socio-economic status tend to be related to substance abuse, increased rates of criminality, and family disruption, which in turn are often related to poor parental control, punitive rearing practices and neglect. In addition, children with different types of dysfunctional parenting tend to have similar behavior outcomes. For example, physically abused children demonstrate similar behavior problems to those children, who witness domestic violence (Jaffe, Wolfe, Wilson, & Zak, 1986), or to those children, who experience general problems in their relations with parents (Wolfe & Mosk, 1983). Thus, it was suggested that the developmental effects of child maltreatment tend to be non-specific and are largely independent of the type of maltreatment. Whether the child develops negative consequences depends on the severity, frequency and lifetime duration of maltreatment, developmental level and age of the child, and the presence of intrinsic strengths or vulnerabilities of the child (Haapasalo & Pokela, 1999). Furthermore, the interactive influence of different family variables may be much more important for the psychological development of the child than the effect of any single variable and thus, their combination might have a cumulative effect on the functioning of the child (Hotaling, Straus, & Lincoln, 1990; Kashani, Daniel, Dandoy, & Holcomb, 1992; O'Keefe, 1994).

For the development of effective preventive strategies, it is crucial to understand the mechanisms that underlie the development of antisocial behavior and thus, I will briefly review parental influences from a developmental perspective and will demonstrate the specific mechanisms, identified in the recent research, which may, or may not lead to aggressive and violent behavior in youth.

From the recent research it has become clear that the way parents construct their relationships with their infants is highly influenced by parental characteristics, and potential risk and buffering conditions are detectable even before the child is born (Zeanah et al., 1997). Heinecke et al. (1986) for example, demonstrated that child's modulation of aggression at age 2 years could be predicted by the pre-birth measures of maternal adjustment, maternal competence and maternal warmth.

At the time of birth the development of the child's brain is still incomplete. However, the child enters the world equipped with a set of inborn neural patterns and associated information-processing programs that develop through a genetically determined course (Perris, 1994) and interact with the environment such that the infant's sensory experiences actually shape the development of his brain (Shatz, 1990). As the child's development is extremely sensitive toward environmental influences, both biological and social, these inborn capacities allow the child to deal adaptively with all the influences to which he is exposed. How adults engage and respond to an infant determines the child's ability to learn, to relate to people, to manage his emotions. Thus, development proceeds through a continuous interaction between nature and nurture. For example, as demonstrated by Raine, Brennan, and Mednick (1997), birth complications become manifest in violent behavior later in life only when combined with maternal rejection, such as attempts to terminate pregnancy.

A great deal of development and functional adjustment to the environment occurs in the first years of life and one of the most important tasks of the child's development is a successful adaptation to the environment. Singer (1986) has drawn attention to the fact that the self-organization of the brain depends on experience and must be considered as an active dialogue between the brain and environment. The structures of the brain that support the cognitive functions require sensory stimulation for maturation and environmental input is necessary for normal growth and maturation of the brain. During this period different environmental factors can affect the brain development through modulation of neuronal activity by sensory signals (Shatz, 1990). Relevant in this present context is Bowlby's postulate of genetic readiness to develop certain types of cognitive maps or "working model" of self and of the social world (Bowlby, 1969; 1973), which are in turn modified by the information that is received. As research shows, those brain functions that can potentially develop, but are not "requested" at certain critical periods of development, will regress (Hubel & Wiesel, 1970; Lombroso & Pruett, in press). Presumably, parents in these circumstances serve not only as a "secure base" (Ainsworth et al., 1978), but also exemplify successful adaptation to the unfamiliar environment. They will serve as a model, since this model, supposedly, has been successful in adaptation and survival, and the child naturally copies this model to become as successfully adapted as his parents. Both theory and research in cognitive development, largely influenced by Vygotsky (1978), have similarly suggested that the development of cognitive skills primarily occurs during parent-child interactions, as the child is guided by an adult, who structures and models the ways to solve the problem. Indeed, from the very beginning infants try to adjust and communicate with adults, look and behave like their parents, as well as copy their sounds, intonations, and gestures. These early patterns of behavior, developed in childhood, are often very persistent and difficult to modify.

In this context, the impact of the absence of both, or even of one (especially of the same sex) parent is hard to overestimate. In situations, where the child is not provided with the model for developing empathic, warm and caring attitudes towards others, the child may not only experience a lag in emotional and cognitive development, as demonstrated by the numerous studies of adoptees from an institutional environment (e.g., Verhulst & Versluis-den Bileman., 1994), but may not even develop these feelings in the future. Being reared in a public care institution in the first year of life has been shown as another key aspect that interacts with birth complications towards predisposing an individual to violence (Raine et al., 1997). Similarly, the inability to maintain close relationships with other people observed in psychopaths has been attributed, among other factors, to a lack of experiences with normal relationships and a paucity of parental identification (Widom, 1989). Thus, although earlier claims that prolonged separation from parents *usually* result in affectionless psychopathy were not supported by later research (Rutter, 1999), early loving relationships with parents are considered as crucial for the adequate development of the child.

It has been demonstrated that early abuse and neglect are related to disorganized attachment (e.g., Carlson et al., 1989), which in turn was connected to increased aggression (Solomon et al., 1995) and hostility (Lyons-Ruth et al., 1993). The above mentioned changes in attachment and in child's behavior are often attributed to biological changes, which in the long run, may lead to antisocial behavior. As brain functions develop "on demand", absence of parents or lack of parental warmth and neglect towards the needs of the child may lead to serious developmental sequelae. Studies on nonhuman primates who have experienced abnormal rearing practices have consistently demonstrated emotional and social impairment (Harlow, Doddsworth, & Halrlow, 1965; Hinde & McGinnis, 1977; Hinde, Leighton-Shapiro, & McGinnis, 1978; Suomi, 1991). As suggested by Porter (1996), repeated abusive experiences in a childhood may lead to psychopathy later in life due to the child's detachment from the painful emotions caused by such experiences.

Children might also become desensitized to painful or anxiety provoking experiences and thus, become "less emotionally and physiologically responsive to the need of others, callous and non-empathic, without remorse or guilt" (Widom, 1997). Similarly, inconsistent care, unreliability of the main attachment figure or rejection may result in an insecure-avoidant style of attachment, leading the child to interpret neutral or even friendly behavior as hostile and evoking inappropriately aggressive responses (Perris, 1994; Widom, 1997). The developmental period in which maltreatment occurred is particularly important (Barnett, Manly, & Cicchetti, 1993), as neurobiological sequelae of abnormal environmental experiences usually have more profound effects at earlier stages (Kandel, 1985).

The long-lasting effects of above-mentioned changes in responsivity as a result of early interactions between child and parents are often ascribed to the acute stress followed by the neurophysiological changes, such as pathology in neurotransmission, or hormonal changes, which in turn modify a child's behavior. For example, Galvin et al. (1995) reported a relations between early maltreatment and extremely low levels of dopamine-β-hydroxylase, an enzyme playing an important role in catecholamine metabolism that has been consistently related to severe conduct problems with inability to establish close interpersonal contacts (e.g. Gabel, Stadler, Bjorn, Shindledecker, & Bowden, 1993; Galvin et al., 1995; Rogeness, Hernandez, Macedo, & Mitchell, 1982). High levels of stress hormones such as cortisol and epinephrine are at least in part responsible for laying down abnormal connections in the brain of the child, which in turn set up aberrant networks that promote the learning of violent responses to benign stimuli. Changes in reactivity also served as a basis for the posttraumatic model of violence, which suggests that child's traumatic reactions to maltreatment may result in addictive physiological changes, which later will lead to the search of activities that would cause similar physiologic effects, such as, for example, committing a violent crime (Hodge, 1992).

Once developed, the structured features of personality have their own means of self-regulation that determine "what is perceived and what is ignored, how a new situation is constructed, and what plan of action is likely to be constructed to deal with it" (Bowlby, 1973, p. 417). As human beings tend to follow patterns, it might be that early negative experiences, either through parental abuse, or other types of maltreatment, lead to the development of persistent dysfunctional "working models" (Bowlby, 1973), or dysfunctional coping styles (Widom, 1997), which may persist through the life, leading to ongoing problems in interpersonal relations.

Since children learn behavior, at least in part, by imitating the behavior of important others, it has been suggested that the child can model parental abusive and punitive behaviors, resulting in further aggressive behavior patterns (Bandura, 1973; Eron, Huesman, & Zelli, 1991; Huesmann & Eron, 1992). In this context, aggressiveness is perceived by the child as a normal and justified expressive behavior, and is understood as a permissible tool in getting others to do what one wishes (McCord, 1988). Several studies (Strassberg, Dodge, Pettit, & Bates, 1994; Straus, Sugarman, & Giles-Sims, 1997), for example, have demonstrated that when parents use spanking to correct child's behavior, the child learns to use aggression as a tool in obtaining a desirable effect in his relationships with other children. More recently Huesmann, Moise, and Podolski (1997) suggested that children learn "scripts" that determine how they perceive and interpret the world. It has been suggested that maltreated children develop deficits in the processing of socially provocative information (Dodge, Bates, & Pettit, 1990; Dodge, Pettit, Bates, & Valente, 1995), called hostile attributional biases, which represent a form of cognitive distortion or misperception of social cues, resulting in the tendency to overattribute hostile intent to a provocation in social situations regardless of actual intent (e.g., benign or ambiguous) (Dodge, 1986).

Aggressive behavior, developing as a result of coercive family interactions, can be further supported through the negative reinforcement mechanisms to the behavior of the child (Patterson, 1982; Patterson, Reid, & Dishion, 1992). For example, when parents withdraw limit setting in response to child's temper tantrum, the child's tendency to throw tantrums is considered negatively reinforced by the removal of the unpleasant parental discipline.

Developmental science, however, has moved from a dominant view of parenting to a more interactive view of parent child-processes that recognizes the necessity of reciprocity and cooperation. Bandura (1978), for example, suggested the concept of "reciprocal determinism", which takes into account the influence that the child might have on parenting, in terms of development occurring in a context of active interaction between the parents and the child. The child is not a *tabula rasa* and his temperament and behavior to a large degree define the behavior of his parents, who have to adjust or to modify their approaches (Lytton, 1990; Rowe, 1994). It was further suggested that a child with a "difficult" temperament can cause negative reactions from parents, leading in turn to excessively harsh discipline, rejection or punishment (e.g. Bates, 1989, Friedrich & Boriskin, 1976; Hubert & Wachs, 1985). From somewhat different perspective, Shaw et al. (1994) demonstrated that mother-directed attention seeking behavior in infant boys and maternal unresponsiveness at 12 months predicted children's non-compliance at 18 months and aggression at age 24 months. In addition, aggressive, antisocial, or non-compliant behavior by children and adolescents has been shown to elicit coercive parental behavior. Repeated experiences of high rates of coercive exchanges between parents and children may mutually reinforce patterns of negative interactions (Forehand, 1990; Patterson, 1986). Finally, a child's temperament might determine how the developmental consequences of harsh parenting will unfold (Widom, 1989) and either prevent or potentiate harmful influences from the environment. Thus, the role of a child's personality should always be kept in mind, when considering family interaction.

According to the data now available, it is reasonable to conclude that dysfunctional parental rearing can significantly contribute to maladaptive psychosocial adjustment. However, linear causal models to explain the linkage between adverse experiences of parental rearing and child's behavior are untenable (Perris, 1994). First, not all the children, who experience maltreatment or negative family interactions become criminals (Widom, 1991); thus there has probably been some space for maneuvering, when, in spite of all negative influences, harmful outcomes do not occur. Thus, the issue of resiliency should be taken into account. Also, early intervention can improve behavioral adjustment and reduce the effects of maltreatment (Fisher, Gunnar, Chamberlain, & Reid, 2000), implying at least some malleability and opportunity for rehabilitative efforts.

Also, it has been suggested that early experiences, including maladaptive family influences, have mostly indirect chain effects to the negative behavior outcomes in adolescence (Rutter, 1999), which are mediated by individual vulnerability and dysfunctional self-schemata (Perris, 1994). Once the primary influence on childhood, defining the individual's views, values, and interactions, the role of parents changes in adolescence. During this period there is a continuous shift of authorities and important others and the significance of parental rearing per se in the life of an adolescent decreases. The role of family functioning and parental monitoring, however, remains extremely important, though with changing accents in the family dynamics, where the dominant role of parents slowly changes towards a role of support and trust. In this context a dominant control may play a negative role. A longitudinal analysis of factors predicting violence among arrested adolescents revealed that increased parental control and decreased support were associated with increased delinquency in early adulthood (Scholte, 1999).

The role of parental control, however, is not necessarily negative. Although the effectiveness of direct control as a proactive strategy for parents was originally questioned, based on the assumption that due to increasing independence of youth, they spend most of their time in places where parents cannot directly control them (Nye, 1958), many later studies have shown that parental control and monitoring could play an important role in limiting antisocial behavior (Loeber & Stouthammer-Loeber, 1986; Snyder & Patterson, 1987; see also Stattin & Kerr, 2000 for a more detailed discussion). Parental monitoring implies parental knowledge about their children's activities outside the home, which comes partly from parent's efforts to find out what their children are doing and partly from the child's spontaneous and willing disclosure of information, with the latter being the strongest negative predictor for normbreaking behavior (Stattin & Kerr, 2000). This raises the issue of trustful relationships in the family, which, although depend a wide range of factors, nevertheless play an important role.

Indeed, trusting family relationships has been increasingly reported as a mediator between family dysfunction and the child's delinquency. In a recent study by Kerr, Stattin, and Trost (1999), family dysfunction from the child's perspective was based on whether they believed that parents trusted him or her, whereas parental perceptions of family dysfunction were based on their own trust in the child. In dysfunctional family relations, this beneficial factor is replaced by suspicion and blaming, thus further potentiating the intrafamilial crisis of relationships. When several aspects of parenting are considered, the findings often suggest that parent-child communication is more beneficial than surveillance and control (Stattin & Kerr, 2000). As poorly monitored youth tend to have deviant friends, they may also become delinquent because of peer pressure (Fridrich & Flannery, 1995). Lack of parental support and/or inadequate monitoring could further promote distancing from the family and affiliation with delinquent peers, and thus potentiate various types of antisocial activities (Reid & Patterson, 1989; Snyder & Patterson, 1987). Furthermore, parents and family may define aspects of peer relationships (Durbin, Darling, Steinberg, & Brown, 1993), where adolescents with more authoritative parents are more likely to have peers that accept both adult and youth values and norms (such as academic achievement/school success and athletic/social popularity), whereas youth with uninvolved parents tend to have peers that did not support adult norms or values, and boys with indulgent parents were in peer groups that stressed fun and partying (Durbin et al., 1993). In this context general family functioning might be a more substantial source of problems, since the behaviors of youth often develop as a result of the close interactions between him, his parents and his peers (Ruchkin, Eisemann, Koposov, & Hagglof, 2000).

Parental rearing, although an important etiological factor for behavioral problems, is one part of a larger, complex system of a daily pattern of family interactions that either serves as a source of confidence and positive reinforcement of an individual's activity or, alternatively, increases his or her vulnerability (Perris, 1994). Parental characteristics associated with juvenile delinquency, such as aggression, hostility, cruel and neglecting attitudes toward the child; negativism and permissiveness of aggression are all behaviors that tend to discourage trust and warmth in family interactions.

As might be concluded from this brief review, even within the family there is a myriad of factors and mechanisms that might lead to the development of antisocial behavior. Rarely if ever do any of these factors act in isolation. Rather they occur in a close interaction between the biological characteristics of the individual and his social environment, leading to the development of individual vulnerability, with the development of psychopathology related to the interaction between negative experiences and the level of dysfunctionality of the self-schemata, to which dysfunctional family interactions greatly contribute (Perris, 1994)

References

Ainsworth, M. D. S., Blehar, M. C., Waters, E., & Wall, S. (1978). *Patterns of attachment: A psychological study of the strange situation.* Hillsdale, NJ: Erlbaum.

Arrindell, W. A., Perris, C., Eisemann, M., van der Ende, J., Gaszner, P., Iwawaki, S., Maj, M., & Zhang, J-e. (1994). Parental rearing behavior from a cross-cultural perspective: A summary of data obtained in 14 nations. In C. Perris, W.A. Arrindell, & M. Eisemann (Eds.), *Parenting and psychopathology* (pp.145-172). Chichester: Wiley.

Baumrind, D. (1967). Child care practices anteceding three patterns of the preschool behavior. *Genetic Psychology Monographs, 75,* 43-88.

Baumrind, D. (1971). Current patterns of parental authority. *Developmental Psychology Monographs, 4* (1, Pt.2), 1-103.

Baumrind, D. (1980). New directions in socialization research. *American Psychologist, 35,* 639-652.

Bandura, A., & Walters, R. H. (1959). *Adolescent aggression.* New York: Ronald Press.

Bandura, A. (1973). *Aggression: A social learning analysis.* Englewood Cliffs, NJ: Prentice Hall.

Bandura, A. (1978). The system of reciprocal determinism. *American Psychologist, 33,* 344-358.

Barnett, D., Manly, J. T., & Cicchetti,D. (1993). Defining child maltreatment: The interface between policy and research. In D. Cichetti & S.L. Toth (Eds.), *Child abuse, development and social policy. Advances in applied and developmental psychology* (pp.7-73). Norwood, NJ: Aplex.

Bates, J. E. (1989). Concepts and measures of temperament. In G. A. Kohnstamm, J. E. Bates, & M. K. Rothbart (Eds.), *Temperament in childhood* (pp.3-26). New York: Wiley.

Becker, W.C. (1964). Consequences of different kinds of parental discipline. In Y. L. Hoffman & L. W. Hoffman (Eds.), *Review of child development research* (Vol.1, pp. 509-535). New York: Russell Sage.

Blount, H.R., & Chandler, T.Z. (1979). Relationships between childhood abuse and assaultive behavior in adolescent male psychiatric patients. *Psychological Reports, 44,* 1126.

Bowlby, J. (1969). *Attachment and loss,* Vol.1: Attachment. New York: Basic.

Bowlby, J. (1973). *Attachment and loss,* Vol.2: Separation, anxiety and anger. New York: Basic.

Bowlby, J. (1980). *Attachment and loss,* Vol.3: Loss: sadness and depression. New York: Basic.

Carlson, V., Chicchetti, D., Barnett, D, and Braunwald, K. (1989). Finding order in disorganization: lessons from maltreated infant's attachment to their caregivers. In D. Cicchetti & V. Carlson (Eds.), *Child maltreatment: theory and research on causes and consequences of child abuse and neglect* (pp.494-528). Cambridge: Cambridge University Press.

Cavaiola, A. A., & Schiff, M. (1988). Behavioral sequelae of physical and/or sexual abuse in adolescents. *Child Abuse and Neglect, 12,*181-188.

Curtis, G. C. (1963). Violence breeds violence – perhaps? *American Journal of Psychiatry, 120,* 386-387.

Dodge, K. A. (1986). A social information processing model of social competence in children. In M. Permutter (Ed.), *Minnesota symposium in child psychology* (Vol.18, pp. 77-125). Hillsdale, NJ: Erlbaum.

Dodge, K. A., Bates, J. E., & Pettit, G. S. (1990). Mechanisms in the cycle of violence. *Science, 250,* 1678-1683.

Dodge, K. A., Pettit, G. S., Bates, J. E., & Valente, E. (1995). Social information-processing patterns partially mediate the effect of early physical abuse on later conduct problems. *Journal of Abnormal Psychology, 104,* 632-643.

Duncan, G. M., Frazier, S. H., Litin, E. M., Johnson, A. M., & Barron, A. J. (1958). Etiological factors in first-degree murder. *Journal of American Medical Association, 168,* 1755-1758.

Durbin, D. L., Darling, N., Steinberg, L., & Brown, B. B. (1993). Parenting style and peer group membership among European-American adolescents. *Journal of Research on Adolescence, 3,* 87-100.

Easson, W. M., & Steinhilber, R. M. (1961). Murderous aggression by children and adolescents. *Archives of General Psychiatry, 4,* 27-35.

Eron, L. D., Walder, L. O., & Lefkowitz, M. M. (1971). *Learning aggression in children.* Boston: Little Brown.

Eron, L. D., Huesman, R. L., & Zelli, A. (1991). The role of parental variables in the learning of aggression. In D. J. Pepler & K. M. Rubin (Eds.), *The development and treatment of childhood aggression* (pp.169-188). Hillsdale, NJ: Erlbaum.

Farrington, D. P., Gundry, G., & West, D. J. (1975). The familial transmission of criminality. *Medical Science and Law, 15,* 177-186.

Farrington, D. P. (1978). The family backgrounds of aggressive youths. Book Supplement to the *Journal of Child Psychology and Psychiatry, 1,* 73-93.

Fisher, P. A., Gunnar, M. R., Chamberlain, P., & Reid, J. B. (2000). Preventive intervention for maltreated preschool children: Impact on children's behavior, neuroendocrine activity, and foster parent functioning. *Journal of American Academy of Child and Adolescent Psychiatry, 39,* 1356-1364.

Fridrich, A. H., & Flannery, D. J. (1995). The effects of ethnicity and acculturation on early adolescent delinquency. *Journal of Child and Family Studies, 4,* 69-87.

Friedrich, W., & Boriskin, J. A. (1976). The role of the child in abuse: A review of the literature. *American Journal of Orthopsychiatry, 46,* 580-590.

Forehand, R. (1990). Families with a conduct problem child. In G. H. Brody & I. E. Sigel (Eds.), *Methods of family research: Biographies of research projects* (Vol.2, pp.1-30). Hillsdale, NJ: Erlbaum.

Gabel, S., Stadler, J., Bjorn, J., Shindledecker, R., & Bowden,C. (1993). Dopamine-beta-hydroxylase in behaviorally disturbed youth. *Biological Psychiatry, 34,* 434-442.

Galvin, M., Ten Eyck, R., Sheckhar, A., Stilwell, B., Fineberg, N., Laite, G., & Karwisch, G. (1995). Serum dopamine beta hydroxylase and maltreatment in psychiatrically hospitalized boys. *Child Abuse and Neglect, 19,* 821-832.

Garbarino, J., & Plantz, M. C. (1986). Child abuse and juvenile delinquency: What are the links? In J. Garbarino, C. J. Schellenbach, & J. M. Sebes (Eds.), *Troubled youth, troubled families* (pp.27-39). New York: Aldine de Gruyter.

Ge, X., Conger, R. D., Cadoret, R. J., & Neiderhiser, J. M. (1996). The developmental interface between nature and nurture: A mutual influence model of child antisocial behavior and parent behaviors. *Developmental Psychology, 32,* 574-589.

Glueck, S., & Glueck, E. T. (1950). *Unraveling juvenile delinquency.* New York: Commonwealth Fund.

Haapasalo, J., & Pokela, E. (1999). Child-rearing and child abuse antecedents of criminality. *Aggression and Violent Behavior, 4,* 107-127.

Harlow, H. F., Doddsworth, R. O., & Halrlow, M. K. (1965). Total social isolation in monkeys. *Proceedings of the National Academy of Sciences of the United States of America, 54,* 90-97.

Heinicke, C. M., Diskin, S. D., Ramsey-Klee, D. M., and Oates, D. S. (1986). Pre- and postbirth antecedents of 2-year-old attention, capacity for relationships and verbal expressiveness._*Developmental Psychology, 22,* 777-787.

Hinde, R. A., & McGinnis, L. (1977). Some factors influencing the effect of temporary mother-infant separation: Some experiments with rhesus monkeys. *Psychological Medicine, 7,* 197-212.

Hinde, R. A., Leighton-Shapiro, M. E., & McGinnis,L. (1978). Effects of various types of separation experience on rhesus monkeys 5 months later. *Journal of Child Psychology and Psychiatry, 19,* 199-211.

Hodge, J. E. (1992). Addiction to violence: A new model of psychopathy. *Criminal Behaviour and Mental Health, 2,* 212-223.

Holden, G. W., & Edwards, L. A. (1989). Parental attitudes toward child rearing: Instruments, issues, and implications. *Psychological Bulletin, 106,* 29-58.

Hotaling, G. T., Straus, M. A., & Lincoln, A. J. (1990). Intrafamily violence and crime and violence outside the family. In M. A. Straus & R. J. Gelles (Eds.), *Physical violence in American families* (pp.431-470). New Brunswick, NJ: Transaction Publishers.

Hubel, D. H., & Wiesel, T. N. (1970). The period of susceptibility to the physiological effects of unilateral eye closure in kittens. *Journal of Physiology, 206,* 419-436.

Hubert, N. C., & Wachs, T. D. (1985). Parental perception of the behavioral components of infants easiness/difficultness. *Child Development, 56,* 1525-1537.

Huesmann, L. R & Eron, L. D. (1992). Childhood aggression and adult criminality. In J. McCord (Ed.), *Facts, frameworks and forecasts: Advances in criminological theory* (Vol.3, pp.137-156). New Brunswick: Transaction Publishers.

Huesmann, L. R., Moise, J. F., & Podolski, C. L. (1997). The effects of media violence on the development of antisocial behavior In D. M. Stoff, J. Breiling, & J. D. Maser (Eds.), *Handbook of antisocial behavior* (pp. 181-193). NY: Wiley.

Hughes, H., Parkinson, D., & Vargo, M. (1989). Witnessing spousal violence and experiencing physical abuse: a "double whammy"? *Journal of Family Violence, 4,* 197-209.

Jacobsson, L., Lindstrom, H., von Knorring, L., Perris, C., & Perris, H. (1980). Perceived parental behaviour and psychogenic needs. *Archiv fur psychiatrie und nervenkrankheiten, 228,* 21-30.

Jaffe, P., Wolfe, D., Wilson, S., & Zak, L. (1986). Similarities in behavioral and social maladjustment among child victims and witnesses to family violence. *American Journal of Orthopsychiatry, 56,* 142-146.

Kandel, E. R. (1985). Early experience, critical periods and developmental fine-tuning of brain architecture. In E. R. Kandel & J. H. Schwarz (Eds.), *Principles of neural science* (pp.757-770). New York: Elsevier Science.

Kandel, D. B., Rosenbaum, E., & Chen, K. (1994). Impact of maternal drug use and life experiences on preadolescent children born to teenage mothers. *Journal of Marriage and the Family, 56,* 325-340.

Kashani, J., Daniel, A., Dandoy, A., & Holcomb, W. (1992). Family violence: Impact on children. *Journal of American Academy of Child and Adolescent Psychiatry, 31,* 181-189.

Kaufman, J., & Zigler, E. (1987). Do abused children become abusive parents? *American Journal of*

Orthopsychiatry, 57, 186-192.

Kempe, C. H., Silverman, F. N., Steele, B. F., Droegemuller, W., & Silver, H. K. (1962). The battered child syndrome. *Journal of American Medical Association, 181,* 17-24.

Kerr, M., Stattin, H., Trost, K. (1999). To know you is to trust you: Parent's trust rooted in child disclosure of information. *Journal of Adolescence, 22,* 737-752.

Lewis, D. O., Shanok, S. S., Pincus, J. H., & Glaser, G. H. (1979). Violent juvenile delinquents: Psychiatric, neurological, psychological, and abuse factors. *Journal of American Academy of Child and Adolescent Psychiatry, 18,* 307-319.

Lyons-Ruth, K., Alpern, L., Repacholi, B. (1993). Disorganized infant attachment classification and maternal psychosocial problems as predictors of hostile-aggressive behavior in preschool classroom. *Child Development, 64,* 572-585.

Grant, R.W., & Tarcov, N. (Eds.). (1996). *Some thoughts concerning education and of the conduct of the understanding by John Locke (1693).* New York: Hackett.

Loeber, R., & Stouthamer-Loeber, M. (1986). Family factors as correlates and predictors of juvenile conduct problems and delinquency. In M. Torny & N. Morris (Eds.), *Crime and justice: An annual review of research* (Vol.7, pp.29-149), Chicago: Chicago University Press.

Lombroso, P., & Pruett, K. (in press). Critical periods in CNS development. In: E. Zigler & S. Styfco (Eds.), *The Headstart debate (friendly and otherwise).* Yale University Press.

Lytton, H. (1990). Child and parent effects in boys' conduct disorder: A reinterpretation. *Developmental Psychology, 26,* 683-697.

Maccoby, E. E. (1984). Socialization and developmental change. *Child Development, 55,* 317-328.

Maccoby, E. E., & Martin, J. (1983). Socialization in the context of the family: Parent-child interaction. In E. M. Hetherington (Ed.), *Handbook of child psychology: Socialization, personality, and social development* (Vol. 4, pp. 1-101). NY: Wiley.

Malinosky-Rummel, R., & Hansen, D. J. (1993). Long-term consequences of childhood physical abuse. *Psychological Bulletin, 114,* 68-79.

McCord, J. (1988). Parental behavior in the cycle of aggression. *Psychiatry, 51,* 14-53.

Nye, F. I. (1958). *Family relationships and delinquent behavior.* New York: Wiley.

O'Keefe, M. (1994). Linking marital violence, mother-child/father-child aggression, and child behavior problems. *Journal of Family Violence, 9,* 63-78.

Parker, G., Tupling, H., & Brown,L. B. (1979). A parental bonding instrument. *British Journal of Medical Psychology, 52,* 1-10.

Patterson, G. R. (1982). *Coercive family process.* Eugene, OR: Castalia.

Patterson, G. R. (1986). Performance models for antisocial boys. *American Psychologist, 41,* 432-444.

Patterson, G. R., Reid, J. B., & Dishion, T. J. (1992). *Antisocial boys.* Eugene, OR: Castalia.

Perris, C. (1994). Linking the experience of dysfunctional parental rearing with manifest psychopathology: a theoretical framework. In C. Perris, W. A. Arrindell, & M. Eisemann (Eds.), *Parenting and psychopathology* (pp. 3-32). Chichester: Wiley.

Porter, S. (1996). Without conscience or without active conscience? The etiology of psychopathy revisited. *Aggression and Violent Behavior, 1,* 179-189.

Raine, A., Brennan, P., & Mednick, S. A. (1997). Interaction between birth complications and early maternal rejection in predisposing individuals to adult violence: Specificity to serious, early-onset violence. *American Journal of Psychiatry, 154,* 1265-1271.

Reid, J. B., & Patterson, G. (1989). The development of antisocial behavior patterns in childhood and adolescence: Personality and aggression. *European Journal of Personality, 3,* 107-119.

Robins, L. (1966). *Deviant children grown up.* Baltimore, MD: Williams and Wilkins.

Rohner, R. P. (1986). *The warmth dimension: Foundations of parental-acceptance rejection theory.* Newbury Park, CA: Sage.

Rogeness, G. A., Hernandez, J. M., Macedo, C. A., & Mitchell, E. L. (1982). Biochemical differences in children with conduct disorder socialized and undersocialized. *American Journal of Psychiatry, 139,* 307-311.

Rousseau, J. J. (1762). *Emile or education.* Bloom A (Transl.). New York: Basic Books, 1979.

Rowe, D. C. (1994). *The limits of family influence.* New York: Guilford Press.

Ruchkin, V. V., Eisemann, M., Koposov, R. A., & Hagglof, B. (2000). Family functioning, parental rearing and behavioral problems in delinquents. *Clinical Psychology and Psychotherapy, 7,* 310-319.

Rutter, M. L. (1999). Psychosocial adversity and child psychopathology. *British Journal of Psychiatry, 174,* 480-493.

Scholte, E. M. (1999). Factors predicting continued violence into young adulthood. *Journal of Adolescence, 22,* 3-20.

Sears, R. R, Whiting, J. W. M., Nowlis, V., & Sears, P. S. (1953). Some child-rearing antecedents of aggression and dependency in young children. *Genetic Psychology Monographs, 47,* 135-234.

Sears, R. R., Maccoby, E. E., & Levin, H. (1957). *Patterns of child rearing.* Evanston: Harper and Row.

Shaw, D. S., Keenan, K., Vondra, J. I. (1994). Developmental precursors of externalizing behavior; ages 1 to 3. *Developmental Psychology, 30*, 355-364.

Singer, W. (1986). The brain as self-organizing system. *European Archives of Psychiatry and Neurological Sciences, 236*, 4-9.

Silver, L. B., Dublin, C. C., & Lourie, R. S. (1969). Does violence breed violence? Contributions from a study of the child abuse syndrome. *American Journal of Psychiatry, 126*, 404-407.

Snyder, J., & Patterson, G. (1987). Family interaction and delinquent behavior. In H. C. Quay (Ed.), *Handbook on juvenile delinquency* (pp.216–243). New York: Wiley.

Solomon, J., George, C., & DeJong, A. (1995). Children classified as controlling at age six: evidence of disorganized representational strategies and aggression at home and at school. *Development and Psychopathology, 7*, 447-464.

Stattin, H., & Kerr, M. (2000). Parental monitoring: A reinterpretation. *Child Development, 71*, 1072-1085.

Strassberg, Z., Dodge, K. A., Pettit, G. S., & Bates, J. E. (1994). Spanking in the home and children's subsequent aggression toward kindergarten peers. *Development and Psychopathology, 6*, 445-461.

Straus, M. A., Sugarman, D. B., & Giles-Sims, J. (1997). Spanking by parents and subsequent antisocial behavior of children. *Archives of Pediatrics and Adolescent Medicine, 151*, 761-767.

Suomi, S. J. (1991). Early stress and adult emotional reactivity in rhesus monkeys. *Ciba Foundation Symposium, 156*, 171-188

Verhulst, F. C., & Versluis-den Bileman, H. J. M. (1994). Developmental course of problem behaviors in adolescent adoptees. *Journal of American Academy of Child and Adolescent Psychiatry, 34*, 151-159.

Vygotsky, L. S. (1978). *The mind in society: the development of higher psychological processes.* Cambridge, MA: Harvard University Press.

Widom, C. S. (1989) Does violence beget violence? A critical examination of the literature. *Psychological Bulletin, 106*, 3-28.

Widom, C. S. (1997). Child Abuse, neglect and witnessing violence. In D. M. Stoff, J. Breiling, & J. D. Maser (Eds.), *Handbook of antisocial behavior* (pp.159-170). NY: Wiley.

Wolfe, D. A., & Mosk, M. D. (1983). Behavioral comparison of children from abusive and distressed families. *Journal of Consulting and Clinical Psychology, 51*, 702-708.

Zeanah, C. H., Boris, N. W., & Scheeringa, M. S. (1997). Psychopathology in infancy. *Journal of Child Psychology & Psychiatry, 38*, 81-99.

Zuckerman, M., Barrett, B. H., & Bragiel, R. M. (1960). The parental attitudes of parents of child guidance cases: I. Comparisons with normals, investigations of socio-economic and family constellation factors, and relations to the parents' reaction to the clinic. *Child Development, 31*, 401-417.

Multi-Problem Violent Youth
R.R. Corrado et al. (Eds.)
IOS Press, 2002

116

Aggressive and Violent Girls: Prevalence, Profiles and Contributing Factors

Marlene Moretti and Candice Odgers

One of the most consistent findings throughout youth violence research and literature is that males are more heavily involved in serious forms of violence than females. According to official charge statistics, males are far more likely to be involved in both serious (homicide, assault causing bodily harm, aggravated assault) and minor (Level 1 assaults, intimidation) forms of violence during adolescence (Dell & Boe, 1998; Duffy, 1996; Totten, 2000; U.S. Department of Health and Human Services, 2001). This relationship has held true across time (Corrado, Cohen & Odgers, 2000; Rowe, Vazsonyi, & Flannery, 1995) and across cultures (Budnick & Shields-Fletcher, 1998; Department of Justice Canada, 1998; Tanner, 1996). In addition, self-report data has consistently shown higher rates of violence among adolescent males when traditional measures of physical aggression and violence are employed (U.S. Department of Health and Human Services, 2001). Support for unequal rates of antisocial behavior between young males and females is also evident within psychiatric literature where an approximate 4:1 male to female prevalence rate of pre-adolescence conduct disorder diagnosis has been reported (Butts et al., 1995; Cohen et al., 1993; Shaffer et al., 1995).

Despite this widely acknowledged sex difference in serious forms of violence, the involvement of females in aggressive and violent behavior has recently captured the attention of a number of individuals working in mental health, youth justice, and educational settings (Artz, 1998a; Budnick & Shields-Fletcher, 1998; Reitsma-Street, 1999). The growing recognition that there are a significant number of young women who engage in behaviors that are highly aggressive, both overtly and relationally (Moretti, Holland, & McKay, 2001), has brought forth a myriad of important research and policy questions. However, despite the immediate demand for answers to these questions and the creation of gender specific programming (Budnick & Shields-Fletcher, 1998) we still know relatively little about prevalence rates, psychosocial profiles, risk factors, and developmental trajectories of violent girls.

The inclusion of young women as a footnote, subset, or minor variation of behavior among males (Artz, 1998a; Bergsmann, 1989; Calhoun, Jurgens, & Chen, 1993; Chesney-Lind & Sheldon, 1998; Figueria-McDonough, 1992) limits our capacity to develop comprehensive theories of female violence (Horowitz & Pottieger, 1991; Kruttschnitt, 1994; Levine & Rosich, 1996).A substantial degree of confusion surrounding how we should best understand and respond to violence among girls continues to exist. In this chapter we summarize the existing research on prevalence, profiles, and developmental trajectories. Limitations of research and challenges to the field are discussed.

1. Rates of Aggression and Violence in Girls: Characteristics and Trends

Many argue that violence among girls is rising. This is not a new observation; in fact, Freda Alder voiced this concern over 25 years ago in her seminal publication *Sisters in*

Crime (1975). At that time, Alder stated that "the phenomenon of female criminality is but one wave in the rising tide of female assertiveness- a wave which has not yet crested and may even be seeking its level uncomfortably close to the high-water mark set by male violence" (p.14). Although Alder was speaking more generally about criminality, the underlying concern was that the behavior of young women was becoming more serious in nature and warranted immediate attention. Since that time, violence among young women has continued to rise, and although it has not reached the "high-water mark" set by males, recent media portrayals of girl gangs, swarmings, and brawls throughout the Western world (Burman, Tisdall, & Brown, 1998; Chisholm, 1997; Hennington, Hughes, Cavell, & Thompson, 1998) have contributed to the impression that female violence is increasing exponentially and may be transforming into a more vicious phenomenon (Schissel, 1997).

The question, then, is whether the impression that female violence is on the rise can be supported by what we currently know. This issue can be addressed by first examining what the long-term trends in female violence are, and then examining how the rates of female and male violence compare over time. In North America, female violence has risen substantially over the last decade. Statistics Canada has reported a 127% increase in charges for violent crimes among females over this period (Savioe, 2000). Similarly, the Violent Crime Index arrest rate[1] in the United States more than doubled for females between 1987 and 1994; although it has decreased consistently since that time (between 1995 and 1999), the rate still remains 74% above the 1980 rate, while the rate for males has dropped to 7% below its 1980 rate (OJJDP Statistical Briefing Book, 2000). The female arrest rate for simple assault in the United States, however, has not followed a similar pattern of rapid escalation and moderate decrease. Instead, it has risen over 250% since 1981 and is continuing to rise sharply (Snyder & Sickmund, 1999).

Comparing male and female charges for violence over time reveals that violent crime has increased at a greater rate among girls. In Canada, violent crime has been increasing at twice the rate for female as compared to male youth over the last decade (Statistics Canada, 1999). Similarly, between 1987 and 1994, the Violent Crime Index arrest rate in the United States more than doubled for females while increasing 64% for males (OJJDP Statistical Briefing Book, 2000). Measures of simple assault in the US are even more telling where arrest rates have risen 260% for females versus 148% for males between 1981 and 1987.

Admittedly, official statistics only capture a portion of the profile of violence and aggression among girls. While they are an essential source of standardized data in the analysis of prevalence and long term trends, self report data aids in the approximation of the actual prevalence of aggression and violence. Overall, self-report measures of aggression are also supporting the notion that female youth may be "closing the gap". According to the Surgeon General's recent report on youth violence, the ratios of self-report male to female violent have decreased from 7.5:1 to 3.5:1 between 1983 and 1999 (U.S. Department of Health and Human Services, 2001).

Although there are consistent indications that female violence is on the rise, and that the ratio of male to female violence has decreased over time, a couple of cautionary notes should be considered. First, the percentage increase of female violence is somewhat misleading due to low initial base rates. For instance, although there was a 125% increase in Violent Index Offence arrest rates between 1985 and 1994 in the United States (Snyder & Sickmund, 1999), the arrest rate for males remained 5.8 times the rate for females. Second, males still tend to be more heavily involved in the most serious types of violent crimes, such as robbery and major assault, whereas females are far more likely to be charged with common assaults. For example, Statistics Canada arrest data indicates that in 1999, two-thirds of female youths were charged with common assault compared to just under half (46%) of male youths (Savioe, 2000).

Another important methodological issue concerns the forms of aggression and violence that are measured. For example, when overt aggression is measured, which includes acts of physical aggression, significantly more boys than girls report engaging in violence (Bjorkqvist, Osterman, & Kaukiainen, 1992; Cotton et al., 1994; Ryan, Matthews, & Banner, 1993; Saner & Ellickson, 1996; U.S. Department of Health and Human Services, 2001), although a handful of studies have reported comparable rates (Finkelstein, Von Eye, & Preece, 1994; Snyder & Sickmund, 1999). However, when the traditional definition of violence is expanded, to include indirect or relational forms of aggression, the disparity between males and females decreases (Crick, 1995; Crick & Grotepeter, 1995; Everett & Price, 1995).

Overall, then, official and self-report data indicate that girls' aggression has consistently risen across the past two decades. It is important to keep in mind, however, that female violence is not skyrocketing and girls continue to be underrepresented as perpetrators of serious forms of overt aggression.

2. Do Girls Express Aggression Differently than Boys?

A recent body of literature suggests that girls may be as aggressive as boys if gender specific forms of aggression are considered. There is little question that in early childhood boys are more physically or overtly aggressive then are girls (Fabes & Eisenberg, 1992; Maccoby & Jacklin, 1980; Parke & Slaby, 1983). Crick and Grotpeter (1995) argue, however, that girls are just as aggressive as are boys if gender differences in the expression of aggressive behavior are recognized. Gender differences in aggression arise because of fundamental differences between males and females in social goals: males' social goals emphasize instrumentality and physical dominance while females' goals are more focused on interpersonal issues. The bilateral model of aggression captures gender differences in aggressive behavior, according to the specific focus or goal to which aggressive acts are directed. Two forms of aggressive behavior are differentiated. Overt aggression includes physical acts and verbal threats toward others, such as hitting or threatening to hit others. In contrast, relational aggression which is intended to harm others through damage to personal and social relationships, such as spreading rumors and excluding others from social groups. In studies of pre-school children (Crick, Casas, & Ku, 1999; Crick, Casas, & Mosher, 1997); middle-age children (Crick, 1996; Crick & Bigbee, 1998; Cunningham, et al., 1998; Rys & Bear, 1997) and young adults (Werner & Crick, 1999), relational aggression has been associated with greater loneliness and less social satisfaction, independently of level of overt aggression. While both relational and overt aggression are viewed as equally hostile, relationally aggressive acts have been shown to be particularly distressing for girls (Crick & Bigbee, 1998; Paquette & Underwood, 1999).

How consistently are gender differences in the expression of aggression found? The results of pre-school studies with children as young as three to five years of age indicate that teachers and peers readily distinguish relational from overt aggression. Even at this young age, girls display a significantly higher level of relationally aggressive behavior than do boys (Crick et al., 1997), and girls are more likely to experience relational victimization than are boys (Crick & Bigbee, 1998; Crick et al., 1999). By middle childhood, the distinction between the gender specific forms of aggressive behavior appears relatively well-established; although the percentage of aggressive girls and boys is comparable (27% of boys vs. 21.7% of girls; Crick and Grotpeter, 1995), girls tend to display this aggressive behavior through covert, relational acts and boys through overt, physical acts.

Yet not all research supports the view that girls and boys express aggression differently. Some studies have found, for example, that girls and boys engage in relational aggression to the same extent (Crick & Grotpeter, 1995; Rys & Bear, 1997). Indeed in some studies (Craig, 1998; Henington, Huges, Cavell, & Thompson, 1998; Roecker, Caprini, Dickerson, Parks, & Barton, 1999; Wolke, Woods, Bloomfield, & Karstadt, 2000) boys are found to engage in even higher levels of relational aggression than are girls. There are several factors that play a role in these diverse findings including diverse assessment procedures (self-report, teacher report, behavior observation) and the age of children across various studies. Nonetheless, it is simply not the case that relational aggression is exclusively a female form of aggressive behavior at any developmental level. Girls and boys both engage in relational aggression across development. Girls, however, generally engage in higher levels of relational than overt aggression and boys generally engage in higher levels of overt than relational aggression.

3. How Important is Relational Aggression?

An important question to ask, then, is whether relational aggression is of any particular significance in understanding severe aggression and violence in girls. There are two types of information that are relevant in this regard. First, clinically elevated or severe levels of relational aggression may be a 'marker' of other forms of aggressive behavior that are present at the current time. The evidence pertaining to the role of relational aggression as a 'marker' of other forms of aggressive behavior is limited but some trends can be extrapolated from existing data. A close look at published studies shows that the correlation between relational and overt aggression is typically very high. For example, in a study of 245 third to sixth grade children, Crick (1996) found a correlation of .77 between relational and overt aggression. Although studies show that relational aggression has unique consequences on social-emotional functioning in girls and boys, independent of overt aggression (Crick & Bigbee, 1998; Paquette & Underwood, 1999), the high correlation indicates that these two forms of aggression often co-occur. Similar results were found in our study of conduct disordered adolescents (Moretti, Holland, & McKay, 2001). Girls engaged in significantly higher rates of relational aggression than did boys; however, they did not engage in lower levels of overt aggression and assaultive behavior. Furthermore, the correlation between these two forms of aggression was high for both girls and boys, $r=.62$ and $r=.54$ respectively. More importantly, a robust correlation emerged between relational aggression and engagement in assaultive behavior for girls, $r=.47$, $p=.001$, but not for boys, $r= -.12$, ns. These results suggest that very high levels of relational aggression in girls may be a marker for serious overt aggressive behavior. These girls are often highly controlling and manipulating of their social networks (i.e., relationally aggressive), and at the same time, physically aggressive toward others. This is consistent with observations of other researchers (Artz, 1998b; Campbell, 1984; Chesney-Lind & Sheldon, 1998) who have used diverse methods to understand the lives of aggressive and violent girls. For example, Artz (1998b) describes the social relationships of violent girls as focused on issues of power and dominance designed to secure their position within a tenuous social milieu.

Relational aggression may also be important as a predictor of future violent behavior even if such behavior is not present at the current time. Unfortunately, longitudinal evidence of such a relationship is limited. In one study, however, Crick (1996) found that level of relational aggression is related to peer rejection in girls and that peer rejection increases over time (6 month follow-up) for relationally aggressive girls. Clearly, further research is required to assess the predictive significance of relational aggression to later violent behavior.

In sum, although research shows that relational aggression is generally more pronounced in girls than is overt aggression, relational aggression is not restricted to girls. It is clear that relational aggression is linked with increased levels of psychological problems and social relations difficulty at least concurrently and in the short term. However, research findings are insufficient to conclude that relational aggression, independent of physical aggression, is predictive of the development of severe aggressive behavior or violence in girls or boys. Nonetheless, preliminary findings show that very high relational aggression typically co-occurs with overt aggression and assaultive behavior in high-risk girls but not high-risk boys. Thus, relational aggression may define the social context in which serious acts of overt aggression occur for girls.

4. Risk Factors, Mental Health Profiles and Developmental Trajectories

Based on our previous discussion it is clear that various forms of aggression are associated with a myriad of social-emotional difficulties in both girls and boys. However, there are very few studies that have made the distinction between minor and serious forms of aggression among girls. Instead, the majority of research has treated girls that engage in antisocial behavior or delinquency as a homogeneous group. For example, a meta-analysis of 60 studies conducted by Simourd and Andrews (1994) concluded that the risk factors that are important for male delinquency are also important for female delinquency. The majority of outcome measures employed in these studies, however, failed to distinguish between minor forms of antisocial behavior (ie. skipping school, drinking, lying, shoplifting) and more serious measures of aggression (physical fights, use of weapons, robbery, etc.).

The second limitation throughout this body of research relates to the tendency to rely heavily on normative or low-risk populations. For instance, in arriving at the conclusion that there are no significant differences in the correlates of delinquency for males and females, Rowe, Vazonyi, and Flannery (1995), relied on a sample ($n= 836$) of predominately middle class, Caucasian youth, from intact families (89%), who reported relatively minor involvement in delinquency. Similarly, the most recent review of the research on female adolescent aggression (Leschied, Cummings, Van Brunschot, Cunningham, & Saunders, 2000) was based on studies where the majority—over 70%—of samples were drawn from normative or high school populations.

Arguably, there are two problems with relying on these types of summaries for the purposes of profiling girls who engage in serious forms of violence. First, we know from previous research that highly aggressive youth are not likely to be found in school populations due to high rates of expulsion and dropping out (Corrado, Odgers & Cohen, 2000; Figueria-McDonough, 1986). Second, although the preceding meta–analyses and literature reviews concluded that the factors associated with aggression for males and females were remarkably similar, it is not clear whether this relationship holds true at more extreme ends of the continuum of violence.

Although it is important that researchers understand the significance of even moderately elevated levels of aggression for psychological adjustment in girls, there are limitations in generalizing from research based largely on normative populations and relatively non-serious definitions of violence to highly aggressive and violent girls. In other words, there is a need to examine separately the factors that contribute to very severe aggressive and violent behavior. Not surprisingly, the information on these very high-risk girls is extremely scant. There are, however, two sources of relevant information on severely aggressive and violent girls, namely, juvenile delinquency and conduct disorder (CD) research. Although not all female offenders and conduct disordered youth are violent, most are either overtly or covertly aggressive and thus findings from these studies are of relevance here.

With respect to risk factors, there is a reasonably large body of juvenile delinquency research profiling female offenders. Overall, these studies (Bergsmann, 1989; Chesney-Lind & Sheldon, 1998; Corrado, Odgers & Cohen, 2000; Crawford, 1988; Lewis, Yeager, Cobham-Portorreal, & Klein, 1991; Rosenbaum, 1989; Warren & Rosenbaum, 1986) report similar findings pointing to high levels of physical and sexual victimization, family dysfunction, substance use, and psychological distress. A review of these studies indicated that 45% to 75% of incarcerated girls have been sexually abused, as compared to approximately 2% to 11% of incarcerated males (Bergsmann, 1989; Corrado, Odgers, & Cohen, 2000; U.S.Bureau of Justice Statistics, 1997; Viale–Val & Sylvester, 1993). Reported levels of physical abuse are also extremely high among girls in jail. For example, Corrado et al.(2000), in a Canadian study of 460 incarcerated youth, reported that 70% of females (n=110) versus 38% of males (n= 360) reported exposure to physical abuse. Similarly, other studies (Bergsman, 1989; Calhoun et al., 1993; Viale-Val & Sylvester, 1993) show rates of physical abuse among girls as ranging between 40% and 62%.

Familial dysfunction (Calhoun, et al., 1993; Chesney-Lind & Sheldon, 1998; Corrado, Odgers & Cohen, 2000), psychopathology (Bergsmann, 1989; Rosenbaum, 1989), and family violence (Heimer & deCoster, 1999) are also extremely common among girls in jail. For instance, Rosenbaum (1989) reported that 97% of girls committed to the California Youth Authority came from non-intact families, and that 76% had family members with previous records of arrest. Likewise, Corrado, Odgers & Cohen (2000) found significantly higher levels of familial dysfunction among girls in custody; 70% had a family member with a criminal record, 76% had a family member of their family with significant substance abuse problem, and 78% reported that a family member had been physically abused. Levels of family conflict among the females were also significantly higher, with 88% of girls leaving home, and 57% reporting that they had been kicked out of their homes. Moreover, in accordance with previous research (see Bergsmann, 1989; Shaw & Dubois, 1995; Smith & Thomas, 2000), the levels of family dysfunction and level of conflict experienced within the home was significantly higher among the female, as compared to the male offenders.

CD research has produced mixed findings with respect to the effects of exposure to maltreatment. In a study of early-onset CD, Webster-Stratton (1996) found no difference between girls and boys in a host of family variables including parental drug and alcohol abuse and depression, disconfirming the hypothesis that it takes a worse environment to produce conduct problems in girls than boys. In other studies, however, conduct disordered girls are found to be more likely to be placed outside the home in foster care or other such facilities, to be removed from the home earlier than boys, and to be exposed to sexual abuse (Moretti, Holland, & McKay, 2001; Moretti, Wiebe, Brown, & Kovacs, 2000).

With respect to mental health profiles, studies of youth in detention centers have confirmed the view that girls are more likely than boys to have a broad array of problems. In particular, high rates of suicide ideation and attempts (Bergsmann, 1989; Lewis et al. 1991) have been reported in these samples. In a self report study conducted by the American Correctional Association Task Force on the Female Offender, over half of the girls reported attempting suicide (Crawford, 1988), while a seven-year follow up study of female offenders, conducted by Lewis et al. (1991), found that close to 90% of these girls had attempted suicide. Numerous studies have also highlighted the presence of depression and low levels of self-esteem among female young offenders (Chesney-Lind & Sheldon, 1998; Crawford, 1988). In addition, higher rates of substance use disorders (SUDs; see Ellickson, Saner, & McGuigan, 1997; Jasper, Smith, & Bailey, 1998; Kingery, Mirzaee, Pruitt, Hurley, & Heuberger 1991) and hard drug use have consistently been found among incarcerated girls (Corrado et al., 2000; Crawford, 1988; Horowitz & Pottieger, 1991).

Studies examining the mental health profiles of conduct disordered girls is limited, but findings typically confirm a pattern of pervasive psychopathology which exceeds that found for conduct disordered boys. In one of the first papers to address this issue, Loeber and Keenan (1994) selectively reviewed studies on comorbidity with CD and specifically examined the effects of age and gender. Where possible, general population studies were selected but studies using high-risk and clinical samples were noted as well. Odds ratios showed that girls with CD were more likely to suffer from comorbid conditions of attention-deficit hyperactivity disorder (ADHD), anxiety disorder, depression and substance use disorder (SUD) than were their male counterparts. Similar results were found in our recent study of gender differences in rates of comorbidity among 70 adolescents diagnosed with conduct disorder (Moretti, Lessard, Weibe, & Reebye, 2001). Girls and boys in this sample were found to show highly similar patterns of conduct disordered behavior; for example, girls were as likely as boys to be involved in violent or aggressive behaviors such as mugging, cruelty to others, and use of weapons. Despite the comparability in CD symptoms, comorbid psychiatric disorders were much more prevalent among girls than boys. For boys, 16.1% met criteria for CD alone and approximately 80% were diagnosed with between one to three additional disorders. In contrast, all girls in our sample were diagnosed with a comorbid condition; in fact, 37% of conduct disordered girls met criteria for between one and three additional disorders and a further 63% were diagnosed with four or more additional disorders. Most commonly, girls met criteria for at least one internalizing, one externalizing, and a substance use disorder. Similar findings were found regardless of whether analyses focused on results from diagnostic interviews or from independent caregiver reports.

A few studies have specifically examined comorbidity between conduct disorder and post-traumatic stress disorder (PTSD) as a test of the hypothesis that exposure to trauma is associated with both delinquent behavior and PTSD. Cauffman, Feldman, Waterman, and Steiner (1998) found that approximately 60% of incarcerated female juvenile offenders met partial (12%) or full (49%) criteria for PTSD. These rates were significantly higher than those noted for male juvenile delinquents. Furthermore, compared to males, females were more likely to report being victims of violent acts (15% vs. 51% for males and females respectively) rather than witnesses to such acts (48% vs. 17% for males and females respectively). Similar findings were reported by Reebye, Moretti, Wiebe, and Lessard (2001). Girls diagnosed with conduct disorder met criteria for PTSD more frequently than did boys. Girls more frequently reported exposure to sexual assault while boys were more likely to report exposure to physical assaults, being involved in accidents and witnessing the death of a loved one.

In summary, the typical delinquent and conduct disordered girl has generally experienced more severe maltreatment and trauma than boys with comparable behavior problems. There is some evidence that the type of maltreatment and trauma experienced by delinquent and conduct disordered girls is different than that experienced by boys; girls are more likely to be victims of sexual abuse than are boys. Finally, there is consistent evidence that girls have a far greater scope of mental health problems, beyond their aggressive behavior, than do boys.

Although these findings are provocative in suggesting that aggressive and violent behavior in girls is linked to distinct risk profiles, there are several notable limitations. First, existing research is almost exclusively descriptive in nature. Most has focused on assessing the relative *level* of risk factors in girls and boys rather than the *relationship* between the risk factors and aggressive behavior. All or some of these risk factors may be more or less strongly related to aggressive behavior in girls than boys. Second, existing research is typically retrospective. The findings provide a good picture of the types of events that have transpired in the lives of these girls, and the scope of the mental health problems with which they contend. However, they do not provide a test of the causal relationships between risk factors and the development of later aggressive behavior. These are just some of the challenges for future research in this area.

5. Developmental Trajectories

Identification of risk for violent behavior in girls requires knowledge of the developmental pathway to persistent and serious aggressive behavior in girls. A classic and now common distinction in the literature on youth aggression is the one between early-onset, life course persistent delinquents versus adolescent-onset delinquents (Moffitt, 1993). This distinction often mistakenly leads to the conclusion that all or most boys with early childhood aggressive behavior or CD show greater persistence in aggressive behavior than adolescent-onset boys. Interestingly, roughly 20-25% of boys with highly aggressive behavior in early childhood do *not* persist in aggressive behavior. Persistence is more common for boys with severely aggressive behavior in early childhood; approximately 95% of these boys continue to show aggressive behavior lending some degree of credence to the model (Loeber & Stouthamer-Loeber, 1998; Tremblay, 2000). Whether or not there is one pathway (Patterson, Forgatch, Yoerger, & Stoolmiller, 1998) or multiple pathways (Loeber, Keenan, & Zhang, 1997) from early aggressive behavior to later delinquency and violence is an issue of debate. It is generally accepted, however, that early and serious aggressive behavior is a marker of risk for later aggressive problems. It is also assumed that adolescent-onset delinquency and aggressive behavior carries a less severe prognosis in terms of continuity of aggression to adulthood and pervasiveness of social-emotional impacts.

Only recently have we questioned whether these developmental trajectories truly apply to girls and preliminary findings are mixed. There are two fundamental questions that are relevant in the assessment of whether developmental trajectories modeled on aggressive boys are relevant for girls. The first question is whether the classic early-onset, life-course persistent versus adolescent-onset distinction is applicable for girls. In general, the developmental slope of overtly aggressive behavior is different for girls than boys. While boys show more physical aggression early in childhood (Maccoby & Jacklin, 1980) followed by a consistent decrease in aggression over development (Haapsalo & Tremblay, 1994), girls show lower physical aggression in childhood and increasing levels of aggression over time (Boothe, Bradley, Flick, Keough, & Kirk, 1993; Cameron, deBruijne, Kennedy, & Morin, 1994; Dobb & Sprott, 1998) These gender specific developmental trends are reflected in shifts in the prevalence of CD for preadolescent versus adolescent girls: generally, prevalence rates of CD in girls increase in adolescence and, in some studies, reach levels that are comparable to the prevalence in boys (Zoccolillo, 1993). This has led some researchers to conclude that the classic distinction between early versus adolescent onset is not applicable to girls since most girls show an adolescent-onset pattern (Loeber & Loeber-Stouthamer, 1998; Silverthorn & Frick, 1999).

Although adolescent-onset aggression or CD is frequently time-limited in boys, many adolescent-onset girls show persistence in aggression and criminality into adulthood. These girls are also likely to have a multitude of mental health problems in adulthood, including substance dependence, involvement in abusive relationships, antisocial personality disorder and social welfare dependence (Bardone, Moffitt, Caspi, Dickson, & Silva, 1996; Robins, 1986; Silverthorn & Frick, 1999). Such findings have led Silverthorn and Frick (1999) to propose that girls show a 'delayed-onset' pattern that is distinctive from both the early-onset and adolescent-onset patterns characteristic of boys. Specifically, the delayed-onset pattern in girls emerges in adolescence but persists into adulthood.

Although the majority of girls develop aggressive behavior or CD in adolescence, there are girls who develop these problems in childhood. In our research of adolescent girls and boys with highly similar profiles of CD symptoms (Moretti, Lessard, et al., 2001), the same proportion of girls and boys were characterized as early onset, but retrospective evaluation of their histories showed they were quite distinctive groups. Disorders such as ADHD, oppositional defiant disorder (ODD), overanxious disorder (OAD) and separation anxiety disorder (SAD) typically predated the onset of CD for *both* girls and boys, but girls had a greater number of these disorders in their developmental history than did boys. Girls were also significantly more likely than boys to have pre-existing SUDs prior to the onset of CD than were boys and were more likely to have a pre-existing depression, although this difference did not reach significance. These results suggest that the 'delayed-onset' pattern proposed by Silverthorn and Frick (1999) may only be delayed with respect to severe aggressive behavior problems; it is likely that these girls suffered from a wide range of mental health problems prior to the emergence of aggressive behavior.

It is also important to determine whether the developmental trajectory to aggression in girls and boys is influenced by the same risk factors and to the same extent. In Webster-Stratton's (1996) study of very early onset (4-7 years) CD, the persistence of girls', but not boys', aggression over a two-year follow-up was significantly influenced by family functioning variables (e.g., parenting characteristics, attitude of parents). These findings are provocative in suggesting that the persistence of girls' aggression may be more strongly linked to the quality of family functioning. Further research is required to determine whether the factors that influence developmental trajectories for girls and boys are comparable.

In sum, researchers are only beginning to grapple with the question of the comparability of developmental trajectories that account for the development of aggression and violence in girls versus boys. Nonetheless, preliminary research suggests that onset patterns and developmental consequences may be quite different for girls than boys. In addition, the developmental precursors to aggression may be distinctive for girls. Although models based on research with boys do not appear to fit the data on development of aggression in girls, it is a mistake to disregard the progress that has been made in research with boys. This research is of substantial value in elucidating methodologies that will speed the development of appropriate models for girls.

6. Challenges and Future Directions

Despite the fact that official police charge statistics and self-report data indicate significant increases in violence among girls, there are a number of individuals who question whether we are witnessing an *actual* rise in female violence (Schramm, 1998). Alternative explanations include the presence of a "moral panic" driven by hysteria and the sensationalism of isolated incidents of "girl violence" (Chisholm, 1997; Schissel, 1997), an increased willingness to charge young women due to changing gender role expectations and zero tolerance polices within schools (Gabor, 1999; Totten, 2000), the possibility of discriminatory processing of young women for minor forms of assault (Horowitz & Pottieger, 1991; Reitsma-Streeet, 1993, 1999), and an increasing reliance by family members on juvenile justice and mental health systems to resolve familial conflict and control their "unruly" daughters (Totten, 2000). More generally, there is the concern that the recent attention surrounding violence and girls is excessive due to the relatively low numbers of actual new violent offence charges that have been laid against female youth (Schramm, 1998). It is important that we recognize sociopolitical pressures that may unjustly criminalize or pathologize girls, yet we cannot dismiss consistent trends of increasing aggression and violence in girls that are evident in juvenile justice statistics and self-report data. At the same time, a tempered perspective needs to prevail: the magnitude of increase in girls' aggression is significant but not alarming. Even if we dismiss recent trends, it still remains the case that research on aggression has consistently neglected these problems in girls. There have always been – and there continues to be – a significant number of highly aggressive girls. It is imperative that we understand why these girls become aggressive and how we can best introduce preventative or remedial programs.

What are the major limitations and most pressing challenges to researchers in the field? In our review, we have pointed to the need for clearer definitions and focused measurement of aggressive and violent behavior. Regardless of what population is sampled, researchers need to rely on standardized assessment procedures that differentiate rebellious and antisocial behavior from truly aggressive and violent behavior. One solution may be to develop a diagnostic protocol or rating system that selectively draws items tapping aggressive behavior from the DSM criteria for CD and other measures of aggression and violence. Attempts should also be made to ensure that such protocols contain developmentally appropriate items.

Other limitations of current research include the failure of investigators to examine both the *level* and the *relation* of risk factors to aggressive behavior. Gender differences may be present in the level of exposure to risk factors associated with the development of aggression and violence; however, these risk factors may be similarly related to aggressive behavior. Research differentiating these two aspects of how risk influences aggression is necessary to clarify developmental models of aggression in girls and boys. Similarly, although retrospective reports have been helpful in understanding the life experiences of highly aggressive girls, prospective studies are sorely needed.

Two fundamental challenges in the field include the need to understand issues related to sample selection and the necessity of systematic research guided by developmentally informed models. It is clear that researchers are struggling to understand which samples are best suited to investigating the development of aggression and how findings from normative and high-risk samples can be integrated. This is a common problem in the study of psychopathology. It may be a more serious problem in the study of aggressive girls, however, because of the relatively lower base rate of these problems in girls and consequent discontinuities in the relationships between risk factors and mildly versus seriously aggressive behavior. Thus, risk factors that are related to mildly aggressive behavior may be quite different than those related to highly aggressive and violent behavior. In addition, research with high-risk or clinical samples may skew estimates of risk exposure and vulnerability. Research shows clearly that rates of comorbidity, for example, are typically overestimated in high-risk or clinical samples (Caron & Rutter, 1991; Goodman, Lahey, Fielding, Dulcan, & Regier, 1997). Ideally, research from both normative and high-risk or clinical samples can be productively integrated but this will require careful thought about how best to understand divergent findings from these two sources of information.

The need to articulate clear developmental models is also a pressing issue. Although the relevance of existing developmental models of aggression in boys should be recognized, researchers also need to consider research on gender differences in developmental psychopathology more broadly. We know, for example, that prevalence rates for the vast majority of *childhood* psychological disorders are higher for boys than girls. In adolescence, however, prevalence rates shift so that many disorders become more common among girls than boys. A good example is the fact that prior to age 12, few reliable gender differences are found in prevalence rates of depression. By age 16, girls are twice as likely to be diagnosed with this disorder (Hankin et al., 1998). Thus, a broad understanding of gender differences in developmental psychopathology raises the question of whether the emergence of aggression and CD in girls during adolescence is unique and specific or reflective of a general developmental process. Although provocative, Silverthorn and Frick's (1999) delayed-onset model of girls' antisocial behavior fails to address such issues. Further work is required to develop a more comprehensive and developmentally informed model of girls' aggression.

In closing, we recognize that there is a pressing need to develop methods of risk identification and intervention, yet the limitations of our current understanding of girls aggression cannot be underscored. Further research is required to ensure accuracy in risk identification and efficacy of programming.

Notes

[1] Violent Crime Index includes the offences of murder, manslaughter, forcible rape, robbery, and aggravated assault.

References

Alder, F. (1975). *Sisters in crime.* New York: McGraw-Hill.

Artz, S. (1998a). *Sex, power, and the violent school girl.* Toronto: Trifolium Books.

Artz, S. (1998b). Where have all the school girls gone? Violent girls in the school yard. *Child & Youth Care Forum, 27,* 77-109.

Bardone, A.M., Moffitt, T., Caspi, A., Dickson, N., & Silva, P.A. (1996). Adult mental health and social outcomes of adolescent girls with depression and conduct disorder. *Development and Psychopathology, 8,* 811-829.

Bergsmann, I.R. (1989). The forgotten few: Juvenile female offenders. *Federal Probation, 12,* 73-78.

Bjorkqvist, K., Osterman, K., & Kaukiainen, A. (1992). The development of direct and indirect aggressive strategies in males and females. In K. Bjorkqvist & P. Niemela (Eds.), *Of mice and women: Aspects of female aggression* (pp. 51-64). San Diego: Academic Press.

Boothe, J., Bradley, L., Flick, M., Keough, K., & Kirk,S. (1993). The violence at your door. *The Executive Educator, February 1993,* 16-21.

Budnick, J., & Shields-Fletcher, E. (1998, September). *What about girls?* (Office of Juvenile Delinquency and Prevention Fact Sheet, 84). Washington, DC: U.S. Department of Justice.

Burman, M., Tisdall, K., & Brown, J. (1998, July). *Researching girls and violence: Searching for a suitable methodology.* Paper presented at the 14th World Congress of Sociology.

Butts, J., Snyder, H., Finnegan, T, Aughenbaugh, A., Tierney, N., Sullivan, D., & Poole, R. (1995). *Juvenile Court Statistics: 1992.* Washington, DC: Office of Juvenile Justice and Delinquency Prevention.

Calhoun, G., Jurgens, J., & Chen, F. (1993). The neophyte female delinquent: A review of the literature. *Adolescence, 28,* 461-471.

Cameron, E., deBruijne, L., Kennedy, K., & Morin, J. (1994). *British Columbia Teachers' Federation Task Force on Violence in Schools Final Report.* Vancouver, B.C: British Columbia Teachers' Federation.

Campbell, A. (1984). *The girls in the gang.* New York: Basil Blackwell.

Caron, C., & Rutter, M. (1991). Comorbidity in child psychopathology: Concepts, issues and research strategies. *Journal of Child Psychology and Psychiatry and Allied Disciplines, 32,* 1063-1080.

Cauffman, E., Feldman, S.S., Waterman, J., & Steiner, H. (1998). Posttraumatic stress disorder among female juvenile offenders. *Journal of the American Academy of Child and Adolescent Psychiatry, 37,* 1209-1216.

Chesney-Lind, M. & Sheldon, R. (1998). *Girls, delinquency, and juvenile justice* (2nd ed.). Pacific Grove, CA: Brooks/Cole.

Chisholm, P. (1997, December 8). Bad girls. *Macleans,* 13-15.

Cohen, P., Cohen, J., Kasen, S., Velez., C., Hartmark, C., Johnson., J., Rojas, M., Brook, J., & Stenning, E. (1993). An epidemiological study of disorders in late childhood and adolescence: Age and gender specific prevalence. *Journal of Child Psychology and Psychiatry, 34,* 851-867.

Corrado, R., Odgers, C., & Cohen, I. (2000). The use of incarceration for female youth: Protection for whom? *Canadian Journal of Criminology, 42,* 189-206.

Corrado, R., Cohen, I., & Odgers, C. (2000). Teen violence in Canada. In A.M. Hoffman & R.W. Summers (Eds.), *Teen violence: A global view.* Westport, CT: Greenwood Press.

Cotton, N. U., Resnick J., Browne, D. C, Martin, S. L., McCarraher, D. R., & Woods, J. (1994). Aggression and fighting behavior among African-American adolescents: Individual and family factors. *American Journal of Public Health, 84,* 618-622.

Craig, W.M. (1998). The relationship among bullying, victimization, depression, anxiety, and aggression in elementary school children. *Personality and Individual Differences, 24*, 123-130.

Crawford, J. (1988). *Tabulation of nationwide survey of female inmates*. Phoenix, AZ: Research Advisory Services.

Crick, N. R. (1995). Relational aggression: The role of intent attributions, feelings of distress, and provocation type. *Development and Psychopathology, 7*, 313-322.

Crick, N. R. (1996). The role of overt aggression, relational aggression, and prosocial behavior in the prediction of children's future social adjustment. *Child Development, 67*, 2317-2327.

Crick, N. R., & Bigbee, M. A. (1998). Relational and overt forms of peer victimization: A multiinformant approach. *Journal of Consulting and Clinical Psychology, 66*, 337-347.

Crick, N. R., Casas, J. F., & Ku, H. (1999). Relational and physical forms of peer victimization in preschool. *Developmental Psychology, 35*, 376-386.

Crick, N. R., Casas, J. R., & Mosher, M. (1997). Relational and overt aggression in preschool. *Developmental Psychology, 33*, 579-588.

Crick, N. R., & Grotpeter, J. K. (1995). Relational aggression, gender, and social-psychological adjustment. *Child Development, 66*, 710-722.

Cunningham, C. E., Cunningham, L. J., Martorelli, V., Tran, A., Young, J., & Zacharias, R. (1998). The effects of primary division, student-mediated conflict resolution programs on playground aggression. *Journal of Child Psychology and Psychiatry, 39*, 653-662.

Dell, A., & Boe, R. (1998). *Female young offenders in Canada* (rev. ed.). Ottawa: Correctional Service of Canada Research Branch.

Department of Justice Canada (1998). *A strategy for the renewal of youth justice*. Ottawa: Department of Justice Canada.

Dobb, A.N., & Sprott, J.B. (1998). Is the "quality" of youth violence becoming more serious? *Canadian Journal of Criminology, 40*, 185-194

Duffy, A. (1996). Bad girls in hard times: Canadian female juvenile offenders. In G. O'Bireck (Ed.), *Not a kid anymore* (pp. 203-220) Scarborough, ON: Nelson Canada.

Ellickson, P., Saner, H., & McGuigan, K. A. (1997). Profiles of violent youth: Substance use and other concurrent problems. *American Journal of Public Health, 87*, 985-991.

Everett, S. A., & Price, J. H. (1995). Students' perceptions of violence in the public schools: The Metlife survey. *Journal of Adolescent Health, 17*, 345-352.

Figueria-McDonough, J. (1986). School context, gender, and delinquency. *Journal of Youth and Adolescence, 15*, 79-97.

Figueria-McDonough, J. (1992). Community structure and female delinquency rates. *Youth and Society, 24*, 3-30.

Finkelstein, J. W., Von Eye, A., & Preece, M. A. (1994). The relationship between aggressive behavior and puberty in normal adolescents: A longitudinal study. *Journal of Adolescent Health, 15*, 319-326.

Fabes, R.A., & Eisenberg, N. (1992). Young children's coping with interpersonal anger. *Child Development, 63*, 116-128.

Gabor, T. (1999). Trends in youth crime: Some evidence pointing to increases in the severity and volume of violence on the part of young people. *Canadian Journal of Criminology, 41*, 385-92.

Goodman, S.H., Lahey, B.B., Fielding, B., Dulcan, W.N., & Regier, D. (1997). Representativeness of clinical samples of youths with mental disorders: A preliminary population-based study. *Journal of Abnormal Psychology, 106*, 3-14.

Haapsalo, J., & Tremblay, R.E. (1994). Physically aggressive boys from age 6 to 12: Family background, parenting behavior, and prediction of delinquency. *Journal of Consulting and Clinical Psychology, 62*, 1044-1052.

Hankin, B.L., Abramson, L.Y., Moffitt, T.E., Silva, P.A., McGee, R., & Angell, K.E. (1998). Development of depression from preadolescence to young adulthood: Emerging gender differences in a 10-year longitudinal study. *Journal of Abnormal Psychology, 107*, 128-140.

Heimer, K., & de Coster, S. (1999). The gendering of violent delinquency. *Criminology, 37*, 277-317.

Hennington, C., Hughes, J. N., Cavell,T.A., & Thompson,B. (1998). The role of relational aggression in identifying aggressive boys and girls. *Journal of School Psychology, 36*, 57-477.

Horowitz, R., & Pottieger, A. E. (1991). Gender bias in juvenile justice handling of seriously crime-involved youths. *Journal of Research in Crime and Delinquency, 28*, 75-100.

Jasper, A., Smith, C., & Bailey, S. (1998) One hundred girls in care referred to an adolescent forensic mental health service. *Journal of Adolescence, 21*, 555-568.

Kingery, P. M., Mirzaee, E., Pruitt, B. E., Hurley, R. S., & Heuberger, G. (1991). Rural communities near large metropolitan areas: Safe havens from adolescent violence and drug use? *Health Value, 15*, 39-48.

Kruttschnitt, C. (1994). Gender and interpersonal violence. In A.J. Reiss, Jr. & J.A. Roth (Eds.), *Understanding and preventing violence* (Vol. 3). Washington, DC: National Academy Press.

Leschied, A., Cummings, A., Van Brunschot, M., Cunningham, A., & Saunders, A. (2000). *Female adolescent aggression: A review of the literature and the correlates of aggression* (User Report No. 2000-04). Ottawa: Solicitor General Canada.

Levine, F., & Rosich, K. (1996). *Social causes of violence: Crafting a scientific agenda.* Washington, DC: American Sociological Association.

Lewis, D.O., Yeager, C.A., Cobham-Portorreal, C.S., & Klein, N. (1991). A follow-up of female delinquents: Maternal contributions to the perpetuation of deviance. *Journal of the American Academy of Child and Adolescent Psychiatry, 30,* 197-201.

Loeber, R., & Keenan, K. (1994). Interaction between conduct disorder and its comorbid conditions: Effects of age and gender. *Clinical Psychology Review 14,* 497-523.

Loeber, R., Keenan, K., & Zhang, Q. (1997) Boys' experimentation and persistence in developmental pathways toward serious delinquency. *Journal of Child and Family Studies, 6,* 321-357.

Loeber, R., & Stouthamer-Loeber, M. (1998). Development of juvenile aggression and violence: Some common misconceptions and controversies. *American Psychologist, 53,* 242-259.

Maccoby, E. E., & Jacklin, C. N. (1980). Sex differences in aggression: A rejoinder and reprise. *Child Development, 51,* 964-990.

Moffitt, T.E. (1993). Adolescence-limited and life course-persistent antisocial behavior developmental taxonomy. *Psychological Review, 100,* 674-701.

Moretti, M.M., Holland, R., & McKay, S. (2001). Self-other representations and relational and overt aggression in adolescent girls and boys. *Behavioral Sciences and the Law, 19,* 109-126.

Moretti, M.M., Lessard, J.C., Wiebe, V.J., & Reebye, P. (2001). *Comorbidity in adolescents with conduct disorder: The gender paradox.* Unpublished manuscript. Simon Fraser University. Burnaby, B.C., Canada.

Moretti, M.M.,Wiebe, V.J., Brown, C., & Kovacs, S. (2000, October). *Maltreatment and attachment in aggressive adolescents.* Paper presented at the 47th annual meeting of the American Academy of Child and Adolescent Psychiatry, New York.

OJJDP   Statistical   Briefing   Book.   (2000,   December).   Online.   Available: *http://ojjdp.ncjrs.org/ojstatbb/qa253.html.*

Paquette, J.A., & Underwood, M.K. (1999) Gender differences in young adolescents' experiences of peer victimization: Social and physical aggression. *Merrill-Palmer-Quarterly, 45,* 242-266.

Parke, R. D., & Slaby, R.G. (1983). The development of aggression. In P. H. Mussen (Ed.), *Handbook of child psychology* (Vol. 4, pp.547-641). New York: Wiley.

Patterson, G.R., Forgatch, M.S., Yoewrger, K.L., & Stoolmiller, M. (1998). Variables that initiate and maintain an early-onset trajectory for juvenile offending. *Development and Psychopathology, 10,* 531-547.

Reebyee, P., Moretti, M.M., Wiebe, V.J., & Lessard, J.C. (2000). Symptoms of posttraumatic stress disorder in adolescents with conduct disorder: Sex differences and onset patterns. *Canadian Journal of Psychiatry, 45,* 746-751.

Reitsma-Street, M. (1993). Canadian youth court charges and dispositions for females before and after implementation of the Young Offenders Act. *Canadian Journal of Criminology, 35,* 437-458.

Reitsma-Street, M. (1999). Justice for Canadian girls: A 1990s update. *Canadian Journal of Criminology, 41,* 335-364.

Roecker, C.E., Caprini, J., Dickerson, J., Parks, E., & Barton, A. (1999, March). *Children's responses to overt and relational aggression.* Paper presented at the Biennial Meeting of the Society for research in Child Development, Albuquerque.

Rosenbaum, J. L. (1989). Family dysfunction and female delinquency. *Crime & Delinquency, 35,* 31-44.

Rowe, D., Vazsonyi, A., & Flannery, D. (1995). Sex differences: Do means and within-sex variation have similar causes? *Journal of Research in Crime and Delinquency, 32,* 84- 100.

Ryan, C., Matthews, F., & Banner, J.(1993). *Student perceptions of violence.* Toronto: Central Toronto Youth Services.

Rys, G. S., & Bear, G. G. (1997). Relational aggression and peer relations: Gender and developmental issues. *Merrill-Palmer Quarterly, 43,* 87-106.

Saner, H., & Ellickson, P. (1996). Concurrent risk factors for adolescent violence. *Journal of Adolescent Health, 19,* 94-103.

Savioe, J. (1999). *Youth violent crime* (Statistics Canada, Catalogue no. 85-002-XPE, Vol. 19, no. 3) Ottawa: Canadian Centre for Justice Statistics.

Schissel, B. (1997). *Blaming children: Youth crime, moral panics, and the politics of hate.* Halifax: Fernwood.

Schramm, H. (1998). *Young women who use violence: Myths and facts.* Calgary, AB: Elizabeth Fry Society of Calgary.

Shaffer, D., Fisher, P., Dulcan, M., Davies, M., Piacentini, J., Schwab–Stone, M. E., Lahey, B. B., Bourdon, K., Jensen, P. S., Bird, H. R., Rubio-Stipec, M., & Rae, D. S. (1995). The NIMH Diagnostic Interview Schedule for Children Version 2.3: Description, acceptability, prevalence rates, and performance in the MECA study. *Journal of the American Academy of Child and Adolescent Psychiatry, 35*, 865–877.

Shaw, M., & Dubois, S. (1995) *Understanding violence by women: A review of the literature.* Ottawa: Correctional Service of Canada.

Silverthorn, P., & Frick, P. (1999). Developmental pathways to antisocial behavior: The delayed-onset pathway in girls. *Development and Psychopathology, 11*, 101-126.

Simourd, L., & Andrews, D.A. (1994). Correlates of delinquency: A look at the gender differences. *Women in Prison, 6*, 28-44.

Smith, H., & Thomas, S.P. (2000). Violent and nonviolent girls: Contrasting perceptions of anger experiences, school, and relationships. *Issues in Mental Health and Nursing, 21*, 547-575.

Snyder, H., & Sickmund, M. (1999). *Juvenile offenders and victims: 1999 National Report.* Washington, DC: Office of Juvenile Justice and Delinquency Prevention.

Statistics Canada. (1999). *Canadian dimensions: Youth and adult crime rates.* Online. Available: http://www.statcan/english/Pgdb/State/Justice/legal.

Tanner, J. (1996). *Teenage troubles: Youth and deviance in Canada.* Toronto: Nelson Canada.

Totten, M. (2000). *The special needs of females in Canada's youth justice system: An account of some young women's experiences and views.* Ottawa: Department of Justice Canada, Services Bureau of Ottawa-Carleton.

Tremblay, R.E. (2000). The development of aggressive behaviour during childhood: What have we learned in the past century. *International Journal of Behavioral Development, 24*, 129-141.

U.S. Bureau of Justice Statistics. (1997). *Privacy and juvenile justice records: A mid-decade status report.* Annapolis Junction, MD: Bureau of Justice.

U.S. Department of Health and Human Services. (2001). *Youth violence: A report of the Surgeon General.* Rockville, MD: U.S. Department of Health and Human Services, Centers for Disease Control and Prevention, National Center for Injury Prevention and Control; Substance Abuse and Mental Health Services Administration, Center for Mental Health Services; and National Institutes of Health, National Institute of Mental Health.

Viale-Val, G., & Sylvester, C. (1993). Female delinquency. In M. Sugar (Ed.), *Female adolescent development* (pp.169-191) New York: Brunner-Mazel.

Warren, M.Q., & Rosenbaum, J.L. (1986). Criminal careers of female offenders. *Criminal Justice and Behavior, 13*, 393-418.

Webster-Stratton, C. (1996). Early-onset conduct problems: Does gender make a difference? *Journal of Consulting and Clinical Psychology, 64*, 540-551.

Werner, N. E., & Crick, N. R. (1999). Relational aggression and social-psychological adjustment in a college sample. *Journal of Abnormal Psychology, 108*, 615-623.

Wolke, D., Woods, S., Bloomfield, L., & Karstadt, L. (2000). The association between direct and relational bullying and behavior problems among primary school children. *Journal of Child Psychology and Psychiatry, 41*, 989-1002.

Zoccolillo, M. (1993). Gender and the development of conduct disorder. *Development and Psychopathology, 5*, 65-78.

Multi-Problem Violent Youth
R.R. Corrado et al. (Eds.)
IOS Press, 2002

Neurological Aspects of Violence, Particularly in Youth

Anneliese A. Pontius

This chapter will emphasize certain still neglected neurological aspects of aggression and violence, with the goal of contributing to an integration of psychoneurological aspects of the study of youth violence. Another purpose of this chapter is to encourage the use of clinical neuropsychiatric and neuropsychological research methods. Initial theorizing and research about psychoneurological factors relevant to youth violence was conducted decades ago, but has been more recently supported by research relying on technological advances, including invasive brain imaging techniques (such as Positron Emission Tomography (PET) and by functional Magnetic Resonance Imaging (fMRI). For example, it had been hypothesized that frontal lobe system dysfunctioning was associated with attention deficit/hyperactivity disorder (ADHD; Pontius, 1973 a, b) two decades before Zametkin et al's. (1990) congruent PET findings. Similarly, basic distinctions between types of murderers, one with reduced frontal and increased subcortical brain functioning (Raine et al., 1995), as well as reduced frontal lobe gray matter in antisocial personality (Raine et al., 2000) were both congruent with the earlier mostly clinical determinations regarding homicides (Pontius, 1981, 2000), and frontal lobe dysfunctioning in delinquents and adult criminals (Pontius & Ruttiger, 1976; Pontius & Yudowitz, 1980).

Technological advances in neuropsychiatry have facilitated controversial extensions of certain forms of violence. The neuroscientist Fried (1997), for instance, proposed a model for ethnic mass murderers whose symptomatology and implicated brain areas were a virtual mirror image to the symptomatology of Limbic Psychotic Trigger Reaction (LPTR). This model and data provided support for each one of the two hypothesized opposite homicidal syndromes (Syndrome E and LPTR; see Pontius, 2000).

Neuropsychiatric contributions to understanding youth violence are potentially very important for evaluating risk for violence. It is proposed, in this chapter, that this perspective should be included in any diagnostic instrument used to identify risk for violence and interventions that might reduce future violence. The initial step is to review the research and theories about these factors.

1. General Neuro-Anatomical Aspects

Postnatally the human brain undergoes phases of anatomical maturation, corresponding with specific behavioral phases. With regard to aggression, two brain systems are of particular interest: the evolutionarily old limbic system (LS) mediates all functions related to survival of the species (sex) and of the individual (aggression). The frontal lobe system (FLS) functions as a counterpart to the LS in order to facilitate social interaction between individuals. This socialization process includes a "civilizing" effect with modification of raw drive expression, including the consideration of the consequences and implications of action. The anatomical maturation of the FLS extends, however, at least beyond puberty, and it still continues over decades into the 90s (Pontius, 1972; Yakovlev &

Lecours, 1967).

A balance between FLS and LS, whereby the FLS is normally dominant over the LS, is required for socialized behavior. The fronto-limbic balance is subject to various kinds of interference, likely more frequent during youth, since they are characterized by relative frontal lobe immaturity. An extreme example of the destructive impact even of a fleeting fronto-limbic imbalance, in which the LS is fleetingly dominant (probably due to a partial seizures), will be discussed in more detail. A violent outcome may be explained by the aforementioned LPTR (Pontius, 1981-2000). An important theme is the developmental sequences that are identified with different neurological impacts on aggression.

1.1 Three Sets of Neurological Impact on Aggression

Late frontal lobe system (FLS) maturation is an initial problem associated with youth violence. The normally late FLS maturation is of special interest since the developmental phases can be used for a neuro-anatomically based classification by the age groupings of youth. Thus, an appropriate evaluation of such groups' action behavior requires awareness of their neuro-developmentally based differences (Pontius, 1974). Further, each one of the FLS maturational phases corresponds specifically to certain kinds of delinquent behaviors. For example, impulsivity is normally characteristic for children up to ages 2 to 3; lack of planning—up to ages 3 to 4; and the inability to switch or reset the principle or plan of an ongoing activity persists beyond puberty and not uncommonly even into young adulthood (Pontius, 1972; Pontius & Ruttiger, 1976; Pontius & Yudowitz, 1980).

Normally the neuro-anatomical maturation of the prefrontal cortex, FLS in short, extends beyond puberty (Pontius, 1972), then FLS maturation attains a certain plateau which can be delayed even into young adulthood (Pontius & Yudowitz, 1980). Finally, FLS maturation extends slowly into old age (Yakovlev & Lecours, 1967). In effect, given that the FLS influences patterns of socialized behaviors, various kinds of behavior problems can be expected during the phases of FLS immaturity. Specific FLS tests have been employed to identify phase-specific behavior problems, including the Trail Making Test-B (TM-B) and the new clinical Narratives Test (NT) (Pontius, 1976, Pontius & Ruttiger, 1976; Pontius & Yudowitz, 1980). For all these instruments, a consistent phase-specific congruency was shown between three sets of data: 1) just cited results of the NT and/or the TMT-B; 2) anatomical FLS maturational phases (Burr, 1960; Conel, 1959; Yakovlev & Lecours, 1967); and 3) behavioral test results from young children (Luria & Homskaya, 1964). Thus, the NT and other test results were based on an analysis of the form (not of the content) of young children's spontaneously told stories (Pitcher & Prelinger, 1963). Such a form analysis (which was subsequently used as a basis for a new clinical Narratives Test (Pontius, 1974), revealed that the typical action behavior of children between 2 and 3 years of age was disjointed and impulsive. Impulsivity of behavior still characterized the behavior of the majority of the children between about 3 and 4 years of age. Between ages 4 and 6, a reflection of the neuro-anatomically known first important spurt in FLS maturation appeared in the form of planned behavior. Yet, only 13% of the children between ages 5 and 6 were able to switch the principle or program of action during an ongoing activity. By contrast, stopping an action for a while before starting a new program of action presented no problem. This fact had been used in remedial attempts to improve the behavior of these young children (Palkes, 1968).

1.2 Disabilities Corresponding to FLS Maturational Phases

As briefly mentioned above, already two decades before intrusive PET-scan brain imaging supported FLS dysfunctioning in Attention Deficit/Hyperactivity Disorder

(ADHD) (Zametkin et al., 1990), clinical behavioral-observational studies had proposed FLS immaturity in ADDH, then called "Minimal Brain Dysfunction" (MBD) (Pontius, 1973 a, b). Such novel findings were, however, widely ignored during a time when the existence of the central nervous system was "resented." Next, a blind form analysis of a group of juvenile delinquents revealed FLS dysfunction on the Narratives Test (Pontius, 1974), in significant contrast to the NT results in control youth without delinquent behavior (Pontius & Ruttiger, 1976). Similar results were obtained in a group of young adult criminals, based on the NT and supplemented by corresponding TMT-B test results as an added mechanism of control (Pontius & Yudowitz, 1980). In these two sets of studies the adolescent participants had narrated invented action stories, while the young adult participants had narrated their actual criminal acts.

The congruence between actual action behavior and the narration of action has been supported by anatomical evidence (Pontius, 1974; Pontius & Ruttiger, 1976; Pontius & Yudowitz, 1980). The specific FLS dysfunction in the criminal groups was the specific inability to switch or reset their program of action during an ongoing activity. FBI crime files, showing that the highest rates of felonies and of fatal car accidents occur up to age 24, are also congruent with late FLS maturation (Pontius & Ruttiger, 1976).

1.3 Clinical and Neuro-psychological Non-intrusive Test Instruments for Future Research

The Narratives Test (Pontius, 1974; Pontius & Ruttiger, 1976; Pontius & Yudowitz, 1980; Pontius, 1995), possibly in combination with the Trail Making Test-B (Armitage, 1946), are suggested as instruments for future research on delinquent behaviors associated with FLS immaturity. In such cases, delinquent behavior may escalate into felonies, even into homicide. For example, when a young male committing a burglary is surprised by the owner, he may just eliminate the interfering person as part of his action principle of stealing, and in continuation of this principle, the young burglar is neuro-anatomically incapable of reprogramming his plan of stealing.

1.4 Fostering or Hindering FLS Functioning

As is the case with any late evolving ability, FLS functioning can to some extent be improved by exercises and practice, or delayed by the lack of them. With regard to fostering FLS functioning, specific exercises that mimic the above used test instruments could be employed. For example, the Narratives Test could be used to let children actively recall and critically discuss the form of their own actions behavior, the way it progresses or changes, or the form of action as reflected in stories heard or seen on film or on TV. In contrast, it would be counter-productive to "spoil" youngsters by sheltering them from actively assuming age-appropriate responsibilities, e.g., for their own safety and decisions, or to "seduce" them to passivity, e.g., by extensive TV watching (aside from the possible emotional impact of viewing explicitly violent content).

With regard to teaching rule-governed behavior mediated by the FLS, a caveat is in order: practice has to be a hands-on type of action (including the use of external speech as well as "inner" thought-based speech). Although cognitive instruction of rules of behavior is necessary, there is a "dissociation between knowing and doing" (Teuber, 1964). Thus, moral teaching addressing cognition exclusively, is not sufficient to produce actual moral acts.

1.5   Fleetingly Reversed Fronto-limbic Balance in "Limbic Psychotic Trigger Reaction" (LPTR)

A second neurological impact on aggression is the Limbic Psychotic Trigger Reaction. Although rare, LPTR may be more common in non-homicidal, albeit bizarre, motiveless, unplanned kinds of aggression. These individuals are typically not court-referred for extensive neuro-psychiatric examination in maximum-security facilities, so such bizarre acts may remain an undetected source of mild forms of LPTR, especially in youth whose FLS has not yet sufficiently matured and whose hormonal level may stimulate the LS, possibly facilitating a fronto-limbic imbalance, as hypothesized to exist in LPTR.

LPTR has been delineated in over two decades of neuro-psychiatric forensic research. It has been proposed as a diagnosis in 21 otherwise unexplainable cases of bizarre aggressive acts by white, mostly young middle class males. All 21 males had not acted under the influence of alcohol or of other drugs (according to police tests and as reported by frequently present witnesses of such unplanned acts). They were physically healthy social loners with otherwise unremarkable socio-economic and medical background. The 21 cases included one bank robber (in his 50s) (Pontius, 2001), three juvenile firesetters (ages 8 - 15) (Pontius, 1999) and 17 homicidal men (mostly between age 18 and their early 30s) (Pontius, 1981, 2000). The 13 symptoms and signs of LPTR were essentially present in all 21 cases. LPTR is characterized by sudden, completely out of character acts in previously non-violent men, most of whom had no prior mental health history. All their acts were committed without plan and motive and with flat affect. The acts occurred quite suddenly upon a chance encounter with a highly individualized trigger stimulus (a symbol or perception of any sensory modality). The trigger stimulus had vividly revived the memories of past merely mild to moderate stresses, which these loners had not shared with others. Thus, they had addressed the sources of their stress, but likely had ruminated on them in their memory. All their bizarre acts occurred with brief autonomic arousal (e.g., ice-cold or strange epigastric sensations, vertigo, incontinence, or nausea). Also present was a brief de novo psychosis (unformed or formed hallucinations of various modalities and/or delusions, often of grandeur). The LPTR symptomatology occurred in three phases—aura, ictus and post-ictal (with particular indications of frontal lobe dysfunctioning). Thus, the symptomatology was strongly suggestive of a partial seizure, whereby all LPTR patients retained their "core consciousness" (Damasio, 1999). Therefore, they could remember the essential details of their puzzling acts about which they felt deep remorse, assumed full responsibility for them, and typically asked to be punished.

The underlying neuro-physiological mechanism of the proposed partial seizures (without convulsions) is limbic seizure kindling (Goddard, 1967; Goddard & McIntyre, 1986). Kindling occurs by intermittently applied subthreshold stimuli ultimately eliciting convulsions in lower mammals, but typically nonconvulsive "behavioral seizures" in primates (Wada, 1986). Some inadvertent human cases of kindling have been reported (Heath et al., 1955; Sramka et al., 1984)

Differential diagnoses of LPTR, based on 16 inclusion and 13 exclusion criteria, has been reviewed elsewhere (Pontius, 1981-2000). About half of the LPTR cases had a known history of paranatal traumata, infantile febrile convulsions or closed head injury; about half had an abnormal "objective" test at some time during their lives, e.g., scalp EEG, CT-scan or MRI.

1.6 Potential Detection of Mild Forms of LPTR Cases Requiring Future Research

So far, only felonious cases of LPTR have received attention through the court system that initiated their extensive work-up. Non-felonious and milder degrees of LPTR may, however, occur virtually in everyday life. Thus, it is common knowledge, that at times persons may suddenly act so much out of character that there can ensue long lasting damage to themselves and/or to others. The literature may provide additional examples of such acts, as sensitively described by Proust in the story of "M. Swan" (Pontius, 1993).

Senseless acts are particularly recounted repeatedly in anecdotes about puzzling acts of youths without any influence of alcohol or other drugs. It appears reasonable to expect an as yet undetected reservoir of LPTR in youths, given the late FLS maturation together with their hormonal changes. Thus, future research could benefit from being alert to possible LPTR symptoms in previously non-violent youths (with histories of head injury) and with otherwise unexplainable sudden motiveless, unplanned violence and other similarly bizarre antisocial act, especially if associated with autonomic arousal and by a fleeting psychosis. In terms of interventions, given that social loners are the most vulnerable to a proposed FLS-LS imbalance, certain remedial measures could be considered, such as providing opportunities such as "Big Brothers" or "Big Sisters" to these children.

1.7 Research Instrument for the Detection of LPTR

The 16 inclusion and exclusion criteria listed in Table 1described in Pontius (1997) can form the core of future research about LPTR. Inquiries must be indirect, nonsuggestive, embedded in and supplemented by detailed history taking and by sensitive interviews. Such patients are so perplexed by their involuntary acts that they search for some "explanation" based on common sense, e.g. that one "must have felt rage" to kill somebody. Therefore they may readily respond in the affirmative to suggestive questions about rage. It is essential to inquire in great detail, particularly about the 24 hours preceding a bizarre act.

1.8 Link Between Fear and Subcortical Information Processing

The third area relates to the link between fear, crude subcortical information processing and "unblinkingly" prompt violence. It is hypothesized that crude visuo-spatial test performances under the fear-inducing impact of inter-tribal warfare linked with their "unblinking" violence, may have counterparts in poor literacy skills of violent Westerners exposed to fear-inducing gang warfare or abusive families (Pontius, in press). The hypothesis is based on the following: The brain can process information in two ways, using neocortical or subcortical structures and pathways. Fast processing occurs via subcortical routes (mainly involving limbic structures) with quick but only globally correct, crude stimulus evaluation, but making possible a prompt, unthinking motor reaction. In contrast, neocortical stimulus processing is accurate and detailed, albeit painstakingly "slow" and deliberate, requiring 250 msec more time than required by the shortcut via subcortical pathways.

Fast subcortical information processing is resorted to under the impact of fear for life. Crude evaluation of potentially life-threatening stimuli allows fast, life-saving "unblinking" motor reaction. By contrast, a subtle full neocortical evaluation of stimuli can constitute a life-threatening luxury, since it loses precious milliseconds. Subcortical pathways resorted to under fear involve particularly the amygdala, mediating aggression as a survival mechanism under life-threatening circumstances. The expectation of fear is ever present in warfare-like environments, be it an abusive family or a gang-ridden environments. Such reasoning is deduced from repeatedly consistent results from visuo-spatial tests (with emphasis on intra-pattern, not on inter-object, map-like spatial relations). The test results were obtained in remote areas on three continents (Amazonian Auca-Indians, SW Ethiopian Karo, and West-New Guinean Asmat and Dani. As "controls" served peaceful comparison groups living nearby). A modification (Pontius, 1993, Figure 2) of the Kohs (1923) block design test was used, and traditional face drawing tests were rendered more subtle with emphasis on intra-pattern spatial-relations, called "Draw-A-Person-With-the-Face-in-Front" (e.g., Pontius, 1982, 1997).

Significant differences were consistently found between the test performance of the three peaceful comparison groups, as against those three tribes that were involved in persistent warfare and suffered pervasive fear for life. In these warring tribes, there was a significant prevalence of the telltale signs of subcortical processing. Their test performances showed crudeness of intra-pattern spatial-relational detail, while the basic shapes within the target designs were grossly preserved. Such specifically crude test performances are the hallmark of subcortical processing, which largely bypasses detailed neocortical stimulus evaluation as well as deliberate "slow" motor reaction to the stimuli.

Refined intra-pattern spatial relational evaluation is not necessary for survival under the impact of fear for life, where fast reaction is required. The processing of subtle spatial relations are, however, necessary for literacy skills. Thus, poor literacy skills, especially in the form of spatial or visual dyslexia and dysgraphia, could be indicative of subcortical processing particularly if the youth is living in a fear-inducing environment. The deduction that visuo-spatial test results reflect neocortical parietal bypassing is strongly supported by reports on patients with parietal brain damage (Friedman et al., l995; Mattingley et al., 1997). Such patients showed similarly crude, only globally correct visuo-spatial stimulus processing, as did the warring, fear-ridden hunter-gatherers in contrast to their peaceful controls living nearby (Pontius, 1989; 1993; 1997; 2001).

1.9 Instruments to Test Hypothesized Link Between Specifically Distorted Test Performances and Fear

Suggested future research of a subgroup of youths habitually resorting to "unblinking" violence and living in fear-inducing environments could begin by evaluating their academic performance, with particular emphasis on indications of visual or spatial dyslexia or dysgraphia (Pontius, 1997), or of specific problems in other courses requiring spatial abilities, such as geometry.

To recall, it is hypothesized that a positive correlation (analogous to that consistently found in the specific test performances by fear-ridden hunter-gatherers on three continents involved in persistent inter-tribal warfare and known to react with unthinking, "unblinking" violence can also be expected in Westerners, e.g., in unthinkingly violent youths living in fear-inducing environments and showing specifically poor academic performance. As an additional step of testing the hypothesis, the two visuo-spatial tests (the modified Kohs Block Design Test (Draw-A-Person-With-Face-In-Front Test. If the hypothesis (Pontius, 1998) is supported, a new neurologically-based understanding of unthinking and "unblinking" violence, likely of a subcortical kind, could be gained. The latter differs from the deliberate violence with substantial neocortical participation, showing goal directedness and planning. Further, such a study of Western violent youths could encourage a new kind of remediation, which could occur in the form of encouraging slow but accurate information processing, particularly in reading and geometry. This would amount to an attempt at reprogramming students' habit of resorting to fast but crude, unthinking subcortical processing short-cuts by superimposing slow, deliberate neocortical participation.

2. Summary and Conclusion

The neurological aspects of violence can no longer remain neglected in research designs and instruments without risking skewed explanations. Thus, as a contribution toward the goal of a more comprehensive model of violence, examples of novel conceptualizations and research approaches to the study of violent youth are suggested, with emphasis on neglected neurological factors. As an illustration, three types of aggresssive and/or violent acts, especially in youth, can be distinguished, each one calling for a first-line approach based on potential neurological factors to be ruled out:

1) With regard to impulsivity and/or lack of planning associated with aggressive behavior, rule out in accordance with age level: a) normal FLS immaturity in children up to age 6; b) FLS maturational lag in youths up to young adulthood.

2) With regard to bizarre, motiveless acts without planning in previously non-violent persons (especially in loners), rule out a fleeting partial limbic seizure (with preserved core consciousness and memory for the acts), such as in the proposed Limbic Psychotic Trigger Reaction.

3) With regard to "unblinkingly" prompt violence in youths with poor reading skills and living in fear-inducing environments, rule out habitual subcortical processing, which bypasses full neocortical participation of visuo-spatial stimulus evaluation.

References

Armitage, S.G. (1946). An analysis of certain psychological tests for the evaluation of brain injury. *Psychological Monographs, 60*, (whole no. 277).

Burr, H.S. (1960). *The neural basis of human behavior*. Springfield, Ill.: Thomas.

Conel, J.L. (1959). *The postnatal development of the human cerebral cortex*. Cambridge, MA: Harvard University Press.

Damasio, A. (1999). *The feeling of what happens. Body and emotion in the making of consciousness*. New York: Harcourt Brace.

Davis, M. (1992). The role of the amygdala in conditioned fear. In J.P. Aggleton (Ed.), *The amygdala* (pp. 255-306). New York: Wiley.

Fried, I. (1997). Syndrome E. *Lancet, 350*, 1845-1848.

Goddard, G.V. (1967). Development of epileptic seizures through brain stimulation of low intensity. *Nature, 214*, 1020-1021.

Goddard, G.V., & McIntyre, D.C. (1986). Some properties of a lasting, epileptogenic trace kindled by repeated electrical stimulation of the amygdala in mammals. In K.B. Doane & K.E. Livingston (Eds.), *The limbic system: Functional organization and clinical disorders* (pp. 95-105). New York: Raven.

Heath, R.G., Monroe, R.R., & Mickle, W. (1955). Stimulation of the amygdaloid nucleus in a schizophrenic patient. *American Journal of Psychiatry, 11*, 862-863.

Kohs, S.C. (1923). *Manual for Kohs block design test*. Chicago: Stoelting.

Luria, A.R., & Homskaya, E.D. (1964). Disturbance in the regulative role of speech with frontal lobe lesions. In J.M. Warren & K. Akert (Eds.), *The frontal granular cortex and behavior* (pp. 353-371). New York: McGraw Hill.

Mattingley, J.B., Davis, G., & Driver, J. (1997). Preattentive filling-in of visual surfaces in parietalextinction. *Science, 272*, 671-674.

Palkes, H., Stears, M., & Kohan, B. (1968). Porteus maze performances of hyperactive boys after training in self-directed verbal commands. *Child Development, 39*, 817-826.

Pontius, A.A. (1972). Neurological aspects in some types of delinquency, especially among juveniles: Toward a neurological model of ethical action. *Adolescence, 7*, 289-308.

Pontius, A.A. (1973a). A conceptual neuro-psychiatric model of Minimal Brain Dysfunction. *Annual of New York Academy of Sciences, 205*, 61-63.

Pontius, A.A. (1973b). Dysfunction patterns analogous to frontal lobe system and caudate nucleus syndromes in some groups of minimal brain dysfunction. *Journal of the American Women's Association, 28*, 285-290.

Pontius A.A. (1974). Basis for a neurological test of frontal lobe system functioning up to adolescence. *Adolescence, 9,* 221-232.

Pontius, A.A. (1981). Stimuli triggering violence in psychoses. Journal of Forensic Sciences, 26, 123-128.

Pontius, A.A. (1984). Specific stimulus-evoked violent action in psychotic trigger reaction, a seizure-like imbalance between frontal lobe and limbic systems? *Perceptual & Motor Skills, 59,* 299-333. (Monograph Supplement, I-V59).

Pontius, A.A. (1989). Color and spatial error in block design in stone-age Auca Indians: Ecological underuse of occipital-parietal system in men; of frontal lobes in women. Brain & Cognition, 10, 54-671.

Pontius. A.A. (1993a). Overwhelming remembrance of things past: Proust portrays limbic kindling by external stimulus. *Psychological Reports, 73,* 615-621.

Pontius. A.A. (1993b). Spatial representation, modified by ecology: From hunter-gatherers to city-dwellers in Indonesia. *Journal of Cross-Cultural Psychology, 24,* 399-413.

Pontius, A.A. (1995). Neurodevelopmental aspects of adolescence. Bulletin of New York Academy of Medicine, 72, 157-161.

Pontius, A.A. (1996). Forensic significance of the limbic psychotic trigger reaction. *Bulletin of American Academy of Psychiatry and the Law, 24,* 125-134.

Pontius, A.A. (1997a). Homicide, linked to moderate repetitive stress, kindling limbic seizures in 14 cases of "Limbic Psychotic Trigger Reaction." *Aggression and Violent Behavior, 2,* 125-141.

Pontius, A.A. (1997b). Dyslexia-dysgraphia. In R. Dulbecco (Ed.), *Encyclopedia of human biology* (2nd ed.). San Diego: Academic Press.

Pontius, A.A. (1997c). Impact of literacy training on spatial representation in SW Ethiopia. *International Journal Intercultural Relations, 21,* 299-304.

Pontius, A.A. (1999). Motiveless firesetting implicating partial seizure kindling by reviving memories of fire in "Limbic Psychotic Trigger Reaction." *Perceptual & Motor Skills, 85,* 970-982.

Pontius, A.A. (2000). Comparison between two opposite syndromes with homicide (Syndrome E vs. Limbic Psychotic Trigger Reaction). *Aggression & Violent Behavior,5,* 423-428

Pontius, A.A. (2001). Two bankrobbers with 'antisocial' and 'schizoid-avoidant' personality disorders, comorbid with partial seizures: Temporal lobe epilepsy and Limbic Psychotic Trigger Reaction. *Journal of Developmental & Physical Disabilities, 13,* 191-197.

Pontius, A.A. (in press). Impact of fear-inducing violence on neuropsychological visuo-spatial tests in warring hunter-gatherers: Analogies to Western enviroments. *Aggression & Violent Behavior.*

Pontius, A.A., & Ruttiger, K.F. (1976). Frontal lobematurational lag in juvenile delinquents shown in Narratives Test. *Adolescence, 11,* 509-518.

Pontius, A.A., & Yudowitz, B.S. (1980). Frontal lobe system dysfunction in criminal action, as shown in Narratives Test. *Journal Nervous & Mental Diseases, 168,* 111-117.

Sramka, M., Sedlak, F., Nadvornik, P. (1984). Observation of kindling phenomenon in treatment of pain by stimulation of the thalamus. In W.H. Sweet, S. Abrador, & J. Martin-Rodriguez (Eds.), *Neurosurgical treatment in psychiatry* (pp. 651-654). New York: Elsevier.

Teuber, K.L. (1964). The riddle of frontal lobe function in man. In J.M. Warren & K. Akert (Eds.), *The frontal granular cortex and behavior.* New York: McGraw Hill.

Wada, J.A. (1978). The clinical relevance of kindling: Species, brain sites and seizure susceptibility. In K.E. Livingston & O. Horykiewicz (Eds.), *Limbic mechanisms, the continuing evolution of the limbic system concept* (pp. 369-388). New York: Plenum.

Zametkin, A.J., Nordahl, T.E., Gross, M. et al. (1990).Cerebral glucose metabolism in adults with hyperactivity of childhood onset. *New England Journal of Medicine, 323,* 1361-1366.

Yakovlev, P.I., & Lecours, A.R. (1967). The myelogenetic cycles of regional maturation of the brain. In A. Minkowski (Ed.), *Regional development of the brain in early life.* Oxford: Blackwell.

Multi-Problem Violent Youth
R.R. Corrado et al. (Eds.)
IOS Press, 2002

Underlying Vulnerability Influencing Outcome Factors/Behaviors in Psychosocial Disturbances

Britt af Klinteberg[*]

Developmental research results support the notion that underlying biological mechanisms, like sympathetic-adrenal excretion and transmitter turnover functioning, are influencing behaviors like hyperactive behavior, aggressiveness and impulsivity, that have shown to be early indicators of different forms of later disinhibitory psychosocial disturbances, such as criminal activity, psychopathy, abuse and violence (Gorenstein & Newman, 1980; Oreland, 1993). Disinhibition is expressed in an impulsive lifestyle characterized by talkativeness, acting before thinking, fast decision-making and inability to learn from mistakes. From a neuropsychological perspective, the concept of disinhibition, comprising both motor and cognitive functions, seems to be related to a reduced frontal lobe functioning (Stuss, Eskes, & Foster, 1994). Early behavior indicators (antecedents) seem, in turn, to be differently expressed during different age periods over the life path. This is of interest in the light of results reported on Attention Deficit Hyperactivity Disorder (ADHD) according to DSM IV (American Psychiatric Association APA, 1994), to be significantly related to genetic markers, i.e., degree of loading for markers and the distribution of polymorphisms of genes involved in aminergic mechanisms (Ebstein et al., 1996; Comings et al., 1996, Comings, 1997). The individual differential expression of early behavior indicators, possibly initiated by related underlying biological mechanisms, is even more interesting and comprehensible, since we know that the genetic expressions are importantly influenced by individual hormonal excretion patterns.

An increasing knowledge of the functioning and development of the individual in interaction with the environment is continuously enriching this field (Loeber, Wei, Stouthamer-Loeber, Huizinga, & Thornberry, 1999; Stattin & Magnusson, 1996; Tremblay, 2000). There are findings supporting the suggestion that biological factors also interact with the modulating effect of the environment and cooperate in the development of a personality pattern making *some* individuals more impulsive and aggressive than others (Pine et al., 1996; Young & Chiland, 1994). Would it then be possible, already at an early age, to discover risk indicators, that is hyperactive behavior, aggressiveness and impulsivity, for possible psychosocial disturbances in the child? This chapter will discuss different research results which together suggest that there might be some possibilities. It will also describe the content of some of these risk indicators and report on how often they occur within one and the same individual. These research results might form, and in the long run possibly constitute, a basis for a conceivable extra support and help during the development of children at risk.

1. Research Approaches

The goal in psychological research is more and more aimed at studying the interaction of factors within the individual; and between the individual and the situation (af Klinteberg, 1988; Magnusson, 1999). The research approach in the study of risk indicators is an interactionistic personality-psychological perspective, where, in studies of the individuals' behavior, *the interplay* between the individual abilities and the environment or situation is emphasized. Individual differences in how to understand and to interpret situations and phenomena are closely associated with differences in managing situations and in responding to what is happening in the specific context. With such an approach, and on the basis of earlier research results, it was assumed that there are differences in level of 'vulnerability' for developing disinhibitory psychosocial disturbances, and that this vulnerability is expressed in the interaction with the environment. Thus, a special situation, like a stressful situation or a negative life event seems to trigger an individual with such a 'vulnerability' to behave in a destructive manner toward others.

In a psychobiological attempt, when studying personality syndromes and testing a vulnerability model, personality inventories and neuropsychological tests are used to explore how personality characteristics and problem solving (cognitive style) characterize behavior over time. This approach implies that we study risk indicators in relation to outcome variables, theoretically assumed to be influenced by environmental stress. A more specific purpose is to understand when normal development turns into problem behaviors or psychiatric disturbances. Hyperactive behavior (for a recent overview of subgroups, see Brown, 2000), aggressiveness and impulsivity are seen as important risk indicators in the development of disinhibitory psychosocial disturbances. The impulsivity concept, that include a lack of self-control and deficient ability to foresee the consequences of own behavior, is repeatedly found to characterize young adults with externalizing behavior, e.g., violence. One way of studying this developmental change into a potential disinhibitory psychosocial disturbance is to observe the risk behaviors hyperactive behavior, aggressiveness and impulsivity at an early age among individuals as evidenced in their daily behavior over time. Consequently, for a more thorough understanding of individual differences, we examined connections between different factors and then focused the interest on the individual, studying patterns of factors (here a combination of risk indicators) in the same individuals.

2. The Biological Basis of Personality Indicates Whether an Individual is at Risk for Developing Disinhibitory Psychosocial Disturbances

Different experiences – same situation. What are the neurobiological factors, which make us experience the same situation in such different ways? There are theories suggesting how developmental events in the brain could relate to the development of behaviors (Carlson & Earls, 2000; van der Molen & Molenaar, 1994; Suomi, in press). The level of neuropsychological functioning, e.g., in terms of perception and cognitive ability, are considered as underlying the unique individual experiences. Whether or not differences in temperament is associated with individual perceptual ability characteristics of the child as concerns personality development is still an unresolved issue. Acquisition of motor and cognitive abilities, and emotional control during childhood appear, during critical periods, to be related to neuronal changes within the central nervous system (CNS) as a result of cell death and reorganization of those cells that survive. The neural changes of most interest when dealing with the interplay between the environment and the CNS are those taking place in the frontal parts of the cerebral cortex especially during the period just before and after birth.

Another important issue pertains to the various effects of sexual hormones on cerebral maturation as possibly reflected in differences in personality. Hence, there are findings showing that an immature or disturbed function of the frontal lobes, perhaps due to hormonal influence, is associated with the personality trait *impulsivity*. The concept of impulsivity include behavior characteristics like 'acting on the spur of the moment', 'deciding quickly and functioning in the present', 'carefreeness without thinking about the future' (Shapiro, 1965). Impulsivity is considered one of the core features in disinhibitory behavior and will be described more in detail later on in this chapter.

3. What Characterizes the Disinhibited-Aggressive-Externalizing Group?

Some of the most important evidences in the research of disinhibitory psychosocial disturbances, especially studies on aggressive and violent children, are the following: these children have not learned to handle strong aggressive feelings and have difficulties in controlling their impulses. They are unable to recognize most of their feelings and experiences. Their impulsive-aggressive acting-out behavior becomes an habitual manner in responding to the environment (Young & Chiland, 1994). These behavior characteristics might be initiated and worsened by long and unhealthy stress in the environment, as, for instance, addiction in the family or unpredictability in everyday life. If there is an extension of problem behaviors over the life span, including criminal acts, it was reported that those who commit violent crimes show personality characteristics like an impulsive lifestyle, lack of behavior control, promiscuous sexual life, lack of responsibility for own behavior, and often alcohol and drug abuse (Farrington, Loeber, & Van Kammen, 1990). These characteristics remind us of the essential features in the concept of psychopathy (Hare, 1991) – the impulsive lifestyle, lack of consideration for others, low empathy, shallow contacts, early antisocial behavior and impulsive, random driven crimes.

Early indicators of disinhibitory psychosocial disturbances. The most important predictors for the development of different forms of disinhibitory psychosocial disturbances, found in extensive studies, are reported to be hyperactive behavior in childhood, early impulsivity and antisocial behavior. These predictors partly overlap the risk indicators (i.e., hyperactive behavior, aggressiveness and impulsivity) found in studies of antisocial personality disorder (APD, according to DSM IV, characterized by irresponsible and antisocial behavior, inability to feel guilt, and with behavioral disturbances present since early adolescence) (American Psychiatric Association APA, 1994), psychopathic personality (a personality that is corresponding to the main traits in psychopathy) and psychopathy (a personality disorder that according to current psychiatric diagnostics in principle can be equal to APD but is more well-defined concerning personality characteristics: high impulsivity, need for new and strong experiences, deficient inhibitory functioning, insufficient sense of responsibility, low empathy and a callous or superficial emotional life).

Hyperactive behavior risk subjects. The finding that a subgroup of hyperactive children has shown to be a distinct risk group in developing APD and psychopathy, is important when it comes to forming intervention and support to make possible a more favorable development of those children (Lynam, 1996; af Klinteberg, 1997; Satterfield, 1978). The hyperactive syndrome could be described as a subtle handicap that during the development is causing difficulties in the interaction with the environment, thus constituting the basis for different kinds of problems over time. When everyday-life demands are increasing, these deficient capabilities become more obvious. Childhood hyperactive behavior diagnoses have been found to predict not only adult psychiatric disorders like antisocial and drug abuse disorders, but also relative risk for educational and vocational disadvantage (Lahey & Loeber, 1997; Mannuzza, Klein, Bessler, Malloy, & LaPadula, 1993). Psychopathy and APD have in several studies shown to be related to violent behavior and violent criminality (Hare, 1999; Williamson, Hare, & Wong, 1987; Grann, Långström, Tengström, & Kullgren, 1999). The high impulsivity in individuals with psychopathic personality, often result in violent acts. In the light of various studies, impulsivity seems to be closely connected to the aggressive expression – or more exactly to the destructive aspect – of violent behavior toward others. This finding indicates that individuals with high impulsivity and vulnerability for disinhibitory tendencies often act destructively in stressful situations. Even if there is strong evidence of a common biological basis for this destructiveness – it can still be expressed in different ways: it can be externalizing in the form of interpersonal violence, or more internalizing as against the self in terms of, for instance, suicide.

4. Impulsivity – A Risk Factor in the Development of Disinhibitory Psychosocial Disturbances

Impulsivity is a characteristic personality trait in individuals with different kinds of 'disinhibitory syndromes', like alcohol problems, suicide, hyperactivity and psychopathy. The psychobiological correlates of impulsivity are well documented, not only in groups of patients and criminals but also in groups of normals (Barratt & Patton, 1983; for an overview, see Schalling, 1993). The findings of connections between impulsivity and psycho-biological indicators even in normal adolescents and adults, support the hypothesis that a biological vulnerability in an individual might be existing, *but* that the individual do not necessarily develop any kind of psychosocial disturbance if there are protecting resources in the individual or in his/her environment.

Psychobiological correlates of impulsivity. Most importantly, impulsivity has shown to be connected to low turnover of the serotonin transmitter system. However, since it is not possible to measure the existing quantity of serotonin directly in the CNS, assessments of indirect measurements are used: (1) level of the serotonin metabolite 5HIAA in the cerebrospinal fluid, and (2) activity from the enzyme monoamine oxidase (MAO) in blood platelets. It is assumed that platelet MAO activity reflects the central serotonin turnover. Indications of a weak or unstable serotonergic system have shown to be related to both aggressive behavior (Brown et al., 1982; Schalling, Edman, Åsberg & Oreland, 1988) and disinhibitory syndromes (Bongioanni, 1991; Soubrié, 1986). The genetic expression of low MAO activity is, however, dependent on the hormonal excretion of testosterone and oestrogen over time. Disinhibitory symptoms are, as described earlier, characterized predominantly by deficient control of behavior, a tendency to act in an impulsive way, and an inability to foresee the consequences of own acting.

A strong negative connection between impulsive-aggressive behaviors and serotonergic activity has been found, as well as signs of a specific connection with the impulsivity aspects of violent behavior (Linnoila et al., 1983; Virkkunen et al., 1994). When groups of individuals with these kind of disturbances have been studied (as, for instance, hyperactive children and psychopaths) not only the above-mentioned frontal cortical deviations in development, and disturbances of neurochemical functions in CNS have been found, but also a certain imbalance in the sympathetic-adrenergic and neuroendocrine systems (LaPierre, Braun, & Hodgins, 1995; Lewis, 1991; Satterfield, 1987; Schalling, 1978). Those results have partly shown to be replicable even in groups of normals (af Klinteberg, 2000; Magnusson, af Klinteberg, & Stattin, 1994).

5. Early Hyperactive-Aggressive Behavior Problems are related to Subsequent Disinhibitory Psychosocial Disturbances

Projects. Here below are some general research results in this field presented, and some results obtained from studies in two Swedish longitudinal projects: the 'Individual Development and Adaptation' (IDA) (Magnusson, 1988), a research program in which a representative group of normal children has been studied from a social-psychological-biological aspect since the age of 10 years and up to early adult age (26-27 yrs); and the 'Young Lawbreakers as Adults' (YLA) (af Klinteberg, Humble, & Schalling, 1992), a follow-up project of boys at risk for developing antisocial behavior from the age of 11-15 years and up to adult age (38-40 yrs); as well as some experimental studies of a group of normal adolescents (17-19 yrs) and pilot recruits (17-24 yrs).

Results from the studies presented here in more detail are focusing on hyperactivity-related behaviors (motor restlessness and concentration difficulties). Most hyperactive boys (boys with high ratings on both motor restlessness and concentration difficulties), in these studies, displayed high ratings also on aggressiveness, such as being actively obstructive, aggressive against classmates and teachers, disturbing and quarrelling. However, it was shown that hyperactive behavior more than aggressiveness *per se* was the important factor in the development of disinhibitory syndromes, such as alcohol problems, antisocial behavior and violence. Earlier studies of the specific content of the connections between, for instance, hyperactivity and aggressiveness versus risk factors as high impulsivity and low sympathetic-adrenergic activity/reactivity had shown that the variable hyperactivity was strongly connected to those risk factors, which was not the case with the variable aggressiveness (af Klinteberg & Magnusson, 1989; af Klinteberg, Magnusson, & Schalling, 1989). In this connection it should be mentioned that low activity/reactivity in the sympathetic-adrenal system, in terms of adrenaline excretion, was found in boys with high hyperactive behavior at 13 years of age, and interpreted as indicating reduced capability to communicate effectively with the environment (Magnusson et al., 1994).

Furthermore, *hyperactive behavior in childhood* and later *antisocial behavior* showed a connection, as did hyperactive behavior and alcohol problems. Problem behaviors are often acting together in the same individual, as evidenced in several follow-up studies. It has been clearly shown that hyperactivity, antisocial behavior and alcohol problems have some sort of common psychobiological basis. Results have also shown that *impulsivity is connected to both alcohol problems and antisocial behavior*. Biological links are again especially obvious when they concern the aspect of antisocial behavior that includes aggressive and violent behavior. Certain aspects of the concept of impulsivity might be of more crucial meaning for the associations found with biological measures (af Klinteberg et al., 1991), constituting vulnerability for developing this kind of disinhibitory symptoms and antisocial behaviors in the presence of environmental stress (see Figures 1-2). Results have been obtained, showing that a weak central serotonergic functioning was connected with earlier attempts of suicide in depressive patients and individuals with a personality disorder, and with impulsive aggressiveness, however only in individuals with a personality disorder (Coccaro, Kavoussi, Sheline, Lish, & Csernansky, 1996). This is of interest, in that a low central serotonergic turnover has shown to predict physical aggressiveness in a follow-up study of out-acting boys (Kruesi et al., 1992) and to characterize violent criminals (Garpenstrand et al., 2001).

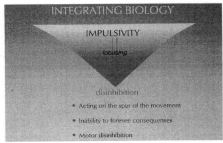

Figure 1

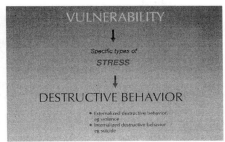

Figure 2

Relationships between early hyperactive behavior and later alcohol and violence problems. Results from the longitudinal studies, using prospective data, showed that hyperactive behavior in childhood was related to later alcohol problems. Moreover, it was shown that it was even related to violence against another individual. In studying whether these adjustment problems were appearing together in one and the same individual more often than could be expected by random, we used a configural frequency analysis (CFA) (von Eye, 1990). Two significant 'types' were obtained: One showing that a hyperactive behavior is closely connected to later alcohol problems and violent crime; and another that showed the common appropriate combination of low hyperactivity in childhood, no alcohol problems before 25 years of age and no violent crimes (af Klinteberg, Andersson, Magnusson, & Stattin, 1993). It is of interest to notify that in our study of normal adolescents, violence occurred together with alcohol problems and early hyperactivity in one and the same individual 10 times more often than could be expected by chance. It is even more interesting in the light of the finding that a problem sample characterized by violent crime(s) without alcohol problems and without hyperactive behavior occurred significantly more seldom (1/3) than could be expected according to the random model.

Since long, there is an ongoing discussion on alcohol abuse in terms of different forms (types) (Cloninger, Bohman, & Sigvardsson, 1981). One type is developing over the life span, starting later in life and more connected to stressful life events or conditions (Type I); the other type is developing early and is assumed to be under strong biological influence (Type II) (von Knorring, Bohman, von Knorring, & Oreland, 1985). Moreover, studies have repeatedly obtained connections between low MAO activity and early developed alcohol abuse; for an overview, see Oreland (1993). In our study, where the information on alcohol problems covered the development up to the age of 25 years, all those with alcohol problems were thus of this early type. This might indicate an underlying mechanism strongly connected to both hyperactive behavior and early alcohol problems. This form of early developed alcohol problems was reported to be associated with high impulsivity (von Knorring et al., 1987b) and with low platelet MAO activity (von Knorring, Hallman, von Knorring & Oreland, 1981). It is worth emphasizing that seven of ten individuals, characterized in our study by both alcohol problems and violence also had high ratings on hyperactivity related behaviors at the age of 13 years. This result gives empirical support to the earlier mentioned findings in the IDA project that boys rated high in hyperactivity-related behaviors by their teacher at the age of 13 yrs, displayed higher scores than the other boys on impulsivity at adult age (af Klinteberg et al., 1989), in turn shown to be related to alcohol abuse (Rydelius, 1983; von Knorring, Oreland, & von Knorring, 1987a), criminal behavior and psychopathy-related traits (af Klinteberg et al., 1992).

Motor disinhibition in individuals with low MAO activity. One of the results in our experimental studies on normal adolescents and pilot recruits (af Klinteberg et al., 1991; af Klinteberg, Levander, Oreland, Åsberg, & Schalling, 1987) was that male subjects with low MAO activity made pronounced more impulsive 'errors' in a two-choice visual reaction-time test with auditory signals for response inhibition (RTI) than intermediate and to some extent the high MAO activity groups. The subjects were asked not to press the button if an auditory signal was presented together with the visual stimulus. This test is assumed to reveal differences in ability to inhibit a motor response, irrespective of speed, and have some similarities with the passive avoidance test, in which psychopaths turned out not to be as good as non-psychopaths (Newman, Widom, & Nathan, 1985; Levander & Elithorn, 1986). Difficulties in holding back a learned motor response described here, is denoted 'motor disinhibition.' The relation between disinhibition in the RTI test and platelet MAO activity in our study is interesting because of the findings of low MAO activity associated with psychopathy in criminals, and with violence in mentally disordered offenders (Alm et al., 1996; Belfrage, Lidberg, & Oreland, 1992; Lidberg, Modin, Oreland, Tuck, & Gillner, 1985).

6. Psychopathic Personality and Violent Behavior

In a follow-up study of young criminals and controls (matched on age, social background and family-type, but without early criminal behavior) in the YLA project, we had the opportunity to studying personality in subgroups divided according to (1) evaluations on the Psychopathy Check List (PCL) (Hare, 1991) and (2) criminal activity during the age period 11-14 years (af Klinteberg et al., 1992). An important finding in this study was that individuals with high PCL scores (upper 25 percentage) manifested a personality pattern of higher scores than the low and intermediate PCL groups on psychopathy-related personality scales, according to the personality inventories, the Karolinska Scales of Personality (KSP) (Schalling, Åsberg, Edman, & Oreland, 1987), and the Eysenck Personality Questionnaire (EPQ) (Eysenck & Eysenck, 1975). The results were in line with theoretical assumptions of psychopathic personality; the high PCL group displayed markedly higher scores on impulsivity and sensation seeking scales, and lower on conformity, as indicated by low socialization and higher on somatic anxiety, suspicion and psychoticism scores. Psychoticism is a complex trait, characterized especially by non-conform conduct with a paucity of respect for rules and regulations in the society, and by hostile aggressiveness, closely connected to impulsivity and risktaking (Eysenck, Eysenck, & Barrett, 1985; Zuckerman, 1994). For the different criminal activity groups, the differences in personality were not that obvious; the criminal groups were higher on somatic anxiety and suspicion, and lower on socialization as compared to the controls. These results indicated some important empirical evidence for the construct of 'psychopathic personality'. The high impulsivity represents according to the theory, the specific type of vulnerability that might trigger the individual to act destructively in stressful situations, without foreseeing the consequences. It is also noteworthy that impulsivity in this study illustrated a quadratic increase with higher PCL scores (af Klinteberg et al., 1992). This indicate that a classification according to the PCL might be a powerful method when focusing on risk individuals for disinhibitory tendencies, especially violent offending.

In further support of this statement on the usefulness of PCL, the high PCL group in the present young criminals and controls were found, according to a random model (Bergman & El-Khouri, 1987), to be over-represented firstly among those who committed violent crime(s) after the age of 18 years, and secondly among those who were persistent violent offenders (violence both before and after the age of 18 yrs). The results are presented in Figure 3. It is noteworthy that the high PCL group subjects differed from the other groups in that they showed significantly higher scores on the self-reported personality trait psychoticism in addition to the higher scores on impulsivity, sensation seeking and nonconformity. This is a result of special interest, since psychoticism theoretically include not only non-conformity aspects of behavior but also cruelty and callous aggressivity.

PCL-scores

	Low	Int	High	
No Violent offending	e 80 o 95	e 42 o 43	e 40 o 24	162
Viol off early	e 5 o 3	e 3 o 4	e 3 o 4	11
Viol off late	e 7 o 0	e 4 o 2	e 4 o 13 p<.001	15
Viol off persistent	e 5 o 0	e 3 o 3	e 3 o 8 p<.01	11
	98	52	49	199

EXACON
e=expected freq.
o=observed freq.

Figure 3

7. Collaboration Between Clinicians and Researchers will make Possible Constructive Preventive Work with Children at Risk for Violent Behavior

In light of the results presented above, how are we to effectively prevent and treat these problem behaviors? How could we constructively work for an optimal prevention and contribute to an effective intervention of multiproblem violent youth? An impressive research have been reported, however, the issue is only partly resolved (Forth & Burke, 1998; Lipsey, Wilson, & Cothern, 1998; Lösel, 1998). It seems evident that a good collaborative work between clinical workers and researchers is needed. Recently different research aspects in the field 'the increasing criminality in society' have attempted to give a basis of knowledge for further discussions about attitudes to disinhibitory psychosocial disturbances such as alcohol and drug abuse, criminality and especially violence (Farrington & Loeber, 2000; Grisso, Appelbaum, Mulvey, & Fletcher, 1995; Raine, Brennan, Farrington, & Mednick, 1997; Selner-O'Hagan, Kindlon, Buka, Raudenbush, & Earls, 1998; Widom, 1997).

There is a need for a continuation of research studies of personality traits and neuropsychological and psychobiological factors connected to disinhibitory syndromes, as well as studies concerning early risk indicators for violent behavior and psychopathy. This direction of research to combine neuropsychological and psychobiological aspects is motivated by the research advances lately at the breakpoint between psychology and neurobiology. These advances have importantly contributed to our understanding of the developmental processes of different forms of disinhibitory psychosocial disturbances. Continued studies will focus more on assumptions about individual differences in the ability to interact effectively with the environment. It will specifically focus on inequalities in the experiencing of situations, as well as different ways of thinking, solving problems and memorizing. This is of special importance, when it concerns children who start early with antisocial behaviors and continue with these behaviors into adult age, since persistent antisocial individuals are found to be characterized by not only one or two risk factors but often a combination of multiple risk indicators. It is empirically evidenced that violent offending is connected to a great variation in criminality (criminal versatility) and do not represent any isolated behavior phenomena. Thus, persistent antisocial activity is specifically relevant when it concerns the development of abuse and violent behavior. Violence can be seen as the extreme behavior manifestation in a multifaceted picture of disinhibitory psychosocial disturbances.

* *Acknowledgements.* The research reviewed in this chapter includes studies which were made possible by access to data from the longitudinal research programs Individual Development and Adaptation (IDA) and Young Lawbreakers as Adults (YLA). Responsible for the planning, implementation, and financing of the collection of data was, for the IDA program, Professor D. Magnusson, and for the YLA program, Professors S. Ahnsjö, G. Carlson, and Associate Professor K. Humble. The research was supported by grants from the National Board of Institutional Care, the Swedish Council for Research in the Humanities and Social Sciences, the Swedish Council for Planning and Coordination of Research, the Swedish Medical Research Council, the National Institute of Public Health, and the National Council for Crime Prevention, Sweden. Special thanks are forwarded to Professor M. Carlson and Associate Professor M. Levander for valuable comments on an earlier version of the manuscript, and to L. Hagen for editing work.

References

af Klinteberg, B. (1988). Studies on sex-related psychological and biological indicators of psychosocial vulnerability: A developmental perspective. Stockholm University and the Karolinska Institute.

af Klinteberg, B. (1997). Hyperactive behaviour and aggressiveness as early risk indicators for violence: Variable and person approaches. *Studies on Crime and Crime Prevention, 6*, 21-34.

af Klinteberg, B. (2000). Psychobiological patterns at adult age: Relationships to personality and early behavior. In L. R. Bergman, R. B. Cairns, L.-G. Nilsson, & L. Nystedt (Eds.), *Developmental science and the holistic approach* (pp. 209-228). Mahwah, NJ: Erlbaum.

af Klinteberg, B., Humble, K., & Schalling, D. (1992). Personality and psychopathy of males with a history of early criminal behaviour. *European Journal of Personality, 6*, 245-266.

af Klinteberg, B., Andersson, T., Magnusson, D., & Stattin, H. (1993). Hyperactive behavior in childhood as related to subsequent alcohol problems and violent offending: A longitudinal study of male subjects. *Personality and Individual Differences, 15*, 381-388.

af Klinteberg, B., Levander, S., Oreland, L., Åsberg, M., & Schalling, D. (1987). Neuropsychological correlates of platelet monoamine oxidase (MAO) in female and male subjects. *Biological Psychology, 24*, 237-252.

af Klinteberg, B., & Magnusson, D. (1989). Aggressiveness and hyperactive behaviour as related to adrenaline excretion. *European Journal of Personality, 3*, 81-93.

af Klinteberg, B., Magnusson, D., & Schalling, D. (1989). Hyperactive behavior in childhood and adult impulsivity: A longitudinal study of male subjects. *Personality and Individual Differences, 10*, 43-50.

af Klinteberg, B., Oreland, L., Hallman, J., Wirsén, A., Levander, S., & Schalling, D. (1991). Exploring the connections between platelet monoamine oxidase activity and behavior: Relationships with performance in neuropsychological tasks. *Neuropsychobiology, 23*, 188-196.

Alm, P. O., af Klinteberg, B., Humble, K., Leppert, J., Sörensen, S., Thorell, L.-H., Lidberg, L., & Oreland, L. (1996). Psychopathy, platelet MAO activity and criminality among former juvenile delinquents grown up. *Acta Psychiatrica Scandinavica, 94*, 105-111.

American Psychiatric Association APA (1994). *Diagnostic and statistical manual of mental disorders* (DSM IV Ed.). Washington, DC: Author.

Barratt, E. S., & Patton, J. H. (1983). Impulsivity: Cognitive, behavioral, and psychophysiological correlates. In M. Zuckerman (Ed.), *Biological bases of sensation seeking, impulsivity and anxiety* (pp. 77-122). Hillsdale, NJ: Erlbaum.

Belfrage, H., Lidberg, L., & Oreland, L. (1992). Platelet monoamine oxidase activity in mentally disordered violent offenders. *Acta Psychiatrica Scandinavica, 85*, 218-221.

Bergman, L. R., & El-Khouri, B. (1987). Exacon: A Fortran 77 program for the exact analysis of single cells in a contingency table. *Educational and Psychological Measurement, 47*, 155-161.

Bongioanni, P. (1991). Platelet MAO activity and personality: An overview. *New Trends in Experimental and Clinical Psychiatry, 7*, 17-28.

Brown, G. L., Ebert, M. H., Goyer, P. F., Jimerson, D. C., Klein, W. J., Bunney, W. E., & Goodwin, F. K. (1982). Aggression, suicide, and serotonin: Relationship to cerebrospinal fluid amine metabolites. *American Journal of Psychiatry, 139*, 741-746.

Brown, T. E. (2000). *Attention deficit disorders and comorbidities in children, adolescents, and adults.* Washington, DC: American Psychiatric Press.

Carlson, M., & Earls, F. (2000). Social ecology and the development of stress regulation. In L. R. Bergman, R. B. Cairns, L.-G. Nilsson, & L. Nystedt (Eds.), *Developmental science and the holistic approach* (pp. 229-248). Mahwah, NJ: Erlbaum.

Cloninger, C. R., Bohman, M., & Sigvardsson, S. (1981). Inheritance of alcohol abuse. *Archives of General Psychiatry, 38*, 861-868.

Coccaro, E. F., Kavoussi, R. J., Sheline, Y. I., Lish, J. D., & Csernansky, J. G. (1996). Impulsive aggression in personality disorder correlates with tritiated paroxetine binding in the platelet. *Archives of General Psychiatry, 53*, 531-536.

Comings, D. E. (1997). Genetic aspects of childhood behavioral disorders. *Child Psychiatry and Human Development, 27*, 139-150.

Comings, D. E., Wu, S., Chiu, C., Ring, R. H., Gade, R., Ahn, C., MacMurray, J. P., Dietz, G., & Muhleman, D. (1996). Polygenic inheritance of Tourette syndrome, stuttering, attention deficit hyperactivity, conduct, and oppositional defiant disorder: The additive and subtractive effect of the three dopaminergic genes - DRD2, DβH, and DAT1. *American Journal of Medical Genetics (Neuropsychiatric Genetics), 67*, 264-288.

Ebstein, R., Novick, O., Umansky, R., Priel, B., Osher, Y., Blaine, D. et al. (1996). Dopamine D4 receptor (D4DR) exon III polymorphism associated with the human personality trait of Novelty Seeking. *Nature Genetics, 12*, 78-80.

Eysenck, S. B. G., & Eysenck, H. J. (1975). *Manual of the Eysenck Personality Questionnaire.* London: Hodder & Stoughton.

Eysenck, S. B. G., Eysenck, H. J., & Barrett, P. (1985). A revised version of the Psychoticism scale. *Personality and Individual Differences, 6*, 21-29.

Farrington, D. P., & Loeber, R. (2000). Epidemiology of juvenile violence. In D. O. Lewis & C. A. Yeager (Eds.), *Juvenile violence* (Vol. 9, pp. 733-748). Child and Adolescent Psychiatric Clinics of North America.

Farrington, D. P., Loeber, R., & Van Kammen, W. B. (1990). Long-term criminal outcomes of hyperactivity-impulsivity-attention deficit and conduct problems in childhood. In L. N. Robins & M. Rutter (Eds.), *Straight and devious pathways from childhood to adulthood* (pp. 62-81). Cambridge: Cambridge University Press.

Forth, A. E., & Burke, H. C. (1998). Psychopathy in adolescence: Assessment, violence and developmental precursors. In D. J. Cooke, A. E. Forth, & R. D. Hare (Eds.), *Psychopathy: Theory, research and implications for society* (pp. 205-229). Dordrecht, The Netherlands: Kluwer.

Garpenstrand, H., Longato-Stadler, E., af Klinteberg, B., Grigorenko, E., Damberg, M., Oreland, L., & Hallman, J. (2001). *Low platelet monoamine oxidase activity in Swedish imprisoned criminal offenders.* Submitted.

Gorenstein, E. E., & Newman, J. P. (1980). Disinhibitory psychopathology: A new perspective and a model for research. *Psychological Review, 87*, 301-315.

Grann, M., Långström, N., Tengström, A., & Kullgren, G. (1999). Psychopathy (PCL-R) predicts violent recidivism among criminal offenders with personality disorders in Sweden. *Law and Human Behavior, 23*, 205-217.

Grisso, T., Appelbaum, P. S., Mulvey, E. P., & Fletcher, K. (1995). The Mac Arthur treatment competence study. II. Measures of abilities related to competence to consent to treatment. *Law and Human Behavior, 19*, 127-148.

Hare, R. D. (1991). *The Hare Psychopathy Check List-Revised.* Toronto: Multi-Health Systems.

Hare, R. D. (1999). Psychopathy as a risk factor for violence. *Psychiatric Quarterly, 70*, 181-197.

Kruesi, M. J. P., Hibbs, E. D., Zahn, T. P., Keysor, C. S., Hamburger, S. D., Bartko, J. J., & Rapoport, J. L. (1992). A 2-year prospective follow-up study of children and adolescents with disruptive behavior disorders. *Archives of General Psychiatry, 49*, 429-435.

Lahey, B. B., & Loeber, R. (1997). Attention-Deficit/Hyperactivity Disorder, oppositional defiant disorder, conduct disorder and adult antisocial behavior: A life span perspective. In D. M. Stoff, J. Breiling, & J. C. Maser (Eds.), *Handbook of antisocial behavior* (pp. 51-59). New York: Wiley.

LaPierre, D., Braun, C. M. J., & Hodgins, S. (1995). Ventral frontal deficits in psychopathy: Neuropsychological test findings. *Neuropsychologia, 33*, 139-151.

Levander, S. E., & Elithorn, A. (1986). *An automated psychological test system.* (Manual). Institute of Psychiatry, University of Trondheim, Norway.

Lewis, C. E. (1991). Neurochemical mechanisms of chronic antisocial behavior (psychopathy): A literature review. *The Journal of Nervous and Mental Disease, 179*, 720-727.

Lidberg, L., Modin, I., Oreland, L., Tuck, J. R., & Gillner, A. (1985). Platelet monoamine oxidase activity and psychopathy. *Psychiatry Research, 16*, 339-343.

Linnoila, M., Virkkunen, M., Scheinin, M., Nuutila, A., Rimon, R., & Goodwin, F. K. (1983). Low cerebrospinal fluid 5- hydroxyindoleacetic acid concentration differentiates impulsive from nonimpulsive violent behavior. *Life Sciences, 33*, 2609-2614.

Lipsey, M. W., Wilson, D. B., & Cothern, L. (1998). Effective intervention for serious juvenile offenders: A synthesis of research. In R. Loeber & D. Farrington (Eds.), *Serious and violent juvenile offenders: Risk factors and successful interventions* (pp. 313-345). Thousand Oaks, CA: Sage.

Loeber, R., Wei, E., Stouthamer-Loeber, M., Huizinga, D., & Thornberry, T. P. (1999). Behavioral antecedents to serious and violent offending: Joint analyses from the Denver Youth Survey, Pittsburgh Youth Study and the Rochester Youth Development Study. *Studies on Crime and Crime Prevention, 8* (2).

Lynam, D. R. (1996). Early identification of chronic offenders: who is the fledgling psychopath? *Psychological Bulletin, 120*, 209-34.

Lösel, F. (1998). Treatment and management of psychopaths. In D. J. Cooke, A. E. Forth, & R. D. Hare (Eds.), *Psychopathy: Theory, research and implications for society* (pp. 303-354). Dordrecht, The Netherlands: Kluwer.

Magnusson, D. (1988). Individual development from an interactional perspective: A longitudinal study. In D. Magnusson (Ed.), *Paths through life* (Vol. 1). Hillsdale, NJ: Erlbaum.

Magnusson, D. (1999). Holistic interactionism: A perspective for research on personality development. In L. A. Pervin & O. P. John (Eds.), *Handbook of personality: Theory and research* (2nd ed., pp. 219-247). New York: Guilford.

Magnusson, D., af Klinteberg, B., & Stattin, H. (1994). Juvenile and persistent offenders: Behavioral and physiological characteristics. In R. D. Ketterlinus & M. E. Lamb (Eds.), *Adolescent problem behaviors - Issues and research* (pp. 81-91). Hillsdale, NJ: Erlbaum.

Mannuzza, S., Klein, R. G., Bessler, A., Malloy, P., & LaPadula, M. (1993). Adult outcome of hyperactive boys. Educational achievement, occupational rank, and psychiatric status. *Archives of General Psychiatry, 50*, 565-576.

Newman, J. P., Widom, C. S., & Nathan, S. (1985). Passive avoidance in syndromes of disinhibition: Psychopathy and extraversion. *Journal of Personality and Social Psychology, 48*, 1316-1327.

Oreland, L. (1993). Monoamine oxidase in neuro-psychiatric disorders. In H. Yasuhara, S. H. Parvez, K. Oguchi, M. Sandler, & T. Nagatsu (Eds.), *Monoamine oxidase: Basic and clinical aspects* (pp. 219-247). Utrecht: VSP Press.

Pine, D. S. et al. (1996). Platelet serotonin 2A (5-HT2A) receptor characteristics and parenting factors for boys at risk for delinquency: A preliminary report. *American Journal of Psychiatry, 153*, 538-544.

Raine, A., Brennan, P. A., Farrington, D. P., & Mednick, S. A. (Eds.). (1997). *Biosocial bases of violence.* New York, NY: Plenum.

Rydelius, P.-A. (1983). Alcohol-abusing teenage boys. *Acta Psychiatrica Scandinavica, 68*, 381-385.

Satterfield, J. (1978). The hyperactive child syndrome: A precursor of adult psychopathy? In R. D. Hare & D. Schalling (Eds.), *Psychopathic behavior: Approaches to research* (pp. 329-346). New York: Wiley.

Satterfield, J. H. (1987). Childhood diagnostic and neurophysiological predictors of teenage arrest rates: An eight-year prospective study. In S. A. Mednick, T. E. Moffitt, & S. A. Stack (Eds.), *The causes of crime* (pp. 199-207). Cambridge: Cambridge University Press.

Schalling, D. (1978). Psychopathy-related personality variables and the psychophysiology of socialization. In R. D. Hare & D. Schalling (Eds.), *Psychopathic behavior: Approaches to research* (pp. 85-106). Chichester: Wiley.

Schalling, D. (1993). Neurochemical correlates of personality, impulsivity and disinhibitory suicidality. In S. Hodgins (Ed.), *Mental disorder and crime* (pp. 208-226). Newbury Park, CA: Sage.

Schalling, D., Edman, G., Åsberg, M., & Oreland, L. (1988). Platelet MAO activity associated with impulsivity and aggressivity. *Personality and Individual Differences, 9*, 597-605.

Schalling, D., Åsberg, M., Edman, G., & Oreland, L. (1987). Markers for vulnerability to psychopathology: Temperament traits associated with platelet MAO activity. *Acta Psychiatrica Scandinavica, 76*, 172-182.

Selner-O'Hagan, M. B., Kindlon, D. J., Buka, S. L., Raudenbush, S. W., & Earls, F. J. (1998). Assessing exposure to violence in urban youth. *Journal of Child Psychology and Psychiatry, 39*, 215-224.

Shapiro, D. (1965). *Neurotic styles.* New York: Basic Books.

Soubrié, P. H. (1986). Reconciling the role of central serotonin neurons in human and animal behavior. *The Behavioral and Brain Sciences, 9*, 319-364.

Stattin, H., & Magnusson, D. (1996). Antisocial development: A holistic approach. *Development and Psychopathology, 8,* 617-645.

Stuss, D. T., Eskes, G. A., & Foster, J. K. (1994). Experimental neuropsychological studies of frontal lobe functions. In F. Boller & J. Grafman (Eds.), *Handbook of neuropsychology* (Vol. 9, pp. 149-185). Amsterdam: Elsevier Science.

Suomi, S. J. (in press). A biobehavioral perspective on developmental psychopathology: Excessive aggression and serotonergic dysfunction in monkeys. In A. J. Sameroff, M. Lewis, & S. Miller (Eds.), *Handbook of developmental psychopathology* (2nd ed.). New York: Plenum.

Tremblay, R. E. (2000). The development of aggressive behaviour during childhood: What have we learned in the past century? *International Journal of Behavioral Development, 24,* 129-141.

van der Molen, M. W., & Molenaar, P. C. M. (1994). Cognitive psychophysiology: A window to cognitive development and brain maturation. In G. Dawson & K. W. Fischer (Eds.), *Human behavior and the developing brain* (pp. 456-490). New York: Guilford.

Virkkunen, M., Kallio, E., Rawlings, R., Tokola, R., Poland, R. E., Guidotti, A., Nemeroff, C., Bissette, G., Kalogeras, K., Karonen, S.-L., & Linnoila, M. (1994). Personality profiles and state aggressiveness in Finnish alcoholic, violent offenders, fire setters, and healthy volunteers. *Archives of General Psychiatry, 51,* 28-33.

von Eye, A. (1990). *Introduction to configural frequency analysis: The search for types and antitypes in cross-classifications.* Cambridge: Cambridge University Press.

von Knorring, A.-L., Bohman, M., von Knorring, L., & Oreland, L. (1985). Platelet MAO activity as a biological marker in subgroups of alcoholism. *Acta Psychiatrica Scandinavica, 72,* 51-58.

von Knorring, A.-L., Hallman, J., von Knorring, L., & Oreland, L. (1991). Platelet monoamine oxidase activity in type I and type II alcoholism. *Alcohol and Alcoholism, 26,* 409-416.

von Knorring, L., Oreland, L., & von Knorring, A.-L. (1987a). Personality traits and platelet monoamine oxidase activity in alcohol abusing teenage boys. *Acta Psychiatrica Scandinavica, 75,* 307-314.

von Knorring, L., von Knorring, A.-L., Smigan, L., Lindberg, U., & Edholm, M. (1987b). Personality traits in subtypes of alcoholics. *Journal of Studies on Alcohol, 48,* 523-527.

Widom, C. (1997). Child abuse, neglect, and witnessing violence. In D. M. Stoff, J. Breiling, & J. C. Maser (Eds.), *Handbook of antisocial behavior* (pp. 159-170). New York: Wiley.

Williamson, S., Hare, R. D., & Wong, S. (1987). Violence: Criminal psychopaths and their victims. *Canadian Journal of Behavioural Science, 19,* 454-462.

Young, J. G., & Chiland, C. (1994). Cultivating and curing violence in children: A guide to methods. In C. Chiland & J. G. Young (Eds.), *Children and violence* (pp. 198-205). London: Aronson.

Zuckerman, M. (1994). Behavioral expressions and biosocial bases of sensation seeking. Cambridge: Cambridge University Press.

Multi-Problem Violent Youth
R.R. Corrado et al. (Eds.)
IOS Press, 2002

Psychopathy in Childhood and Adolescence: Implications for the Assessment and Management of Multi-Problem Youths

Gina M. Vincent and Stephen D. Hart

Personality plays an important role in organizing people's understanding of and interactions with their physical, psychological, and social environments. Although our understanding of its nature and development is far from complete, it is clear that personality has strong biological roots but also is shaped by life experiences. Personality influences us on a moment-to-moment basis, coloring our perceptions, motivations, thoughts, and feelings.

Many theoretical models of delinquent and antisocial behavior consider personality to be an important causal factor. For example, personality traits such as impulsivity, anxiety, and empathy are thought to influence the decisions that people make concerning whether or not to violate implicit and explicit social norms (e.g., Krueger et al., 1994; Loeber, 1990). When personality traits become extreme or rigid and cause persistent problems in psychosocial adjustment, they may be said to constitute a form of psychopathology known as personality disorder (see American Psychiatric Association, 1994). Not surprisingly, personality disorder — and in particular psychopathy, also known as psychopathic, antisocial, or dissocial personality disorder — also has been linked to criminality. Research in adults suggests that psychopathy plays a critical role in risk assessment and management (Hart, 1998; Hemphill, Hare, Wong, 1998; Salekin, Rogers, & Sewell, 1996). But for those interested in childhood and adolescence, it is unclear to what extent personality disorder exists or can be assessed reliably prior to the developmental stage of early adulthood.

Recently, the study of psychopathy and its relation to behavior problems and criminality in childhood and adolescence has become a focal point of much research. Proponents of young psychopathy research would argue that it may explain part of the heterogeneity in Conduct Disorder; may assist our understanding of the etiology of psychopathy which in turn would greatly enhance early prevention, intervention strategies and treatment; and may identify serious and violent offenders in need of future risk management. In this chapter, we start by summarizing what is known about psychopathy in adults. Next, we discuss some scientific and ethical concerns about studying psychopathy in childhood or adolescence and evaluate these in light of the existing research literature. Finally, we identify areas in need of further study and discuss how information about psychopathy can be used in the assessment and management of risks and needs in multi-problem youth.

1. Psychopathy in Adulthood

1.1 Assessment and Diagnosis

In major diagnostic systems such as the fourth edition of the *Diagnostic and*

Statistical Manual of Mental Disorders (DSM-IV; American Psychiatric Association, 1994) and the tenth edition of the *International Classification of Diseases and Causes of Death* (ICD-10; World Health Organization, 1990), personality disorder is defined as a disturbance in relating to one's self, others, and the environment that is chronic in nature, typically evident by childhood or adolescence and persisting into middle or late adulthood. Symptoms of personality disorder are rigid, inflexible, and maladaptive personality traits — tendencies to act, think, perceive, and feel in certain ways that are stable across time, across situations, and in interactions with different people. What distinguishes psychopathy from other personality disorders is its specific symptom pattern: Interpersonally, psychopathic individuals are arrogant, superficial, deceitful, and manipulative; affectively, their emotions are shallow and labile, they are unable to form strong emotional bonds with others, and they are lacking in empathy, anxiety, and guilt; behaviorally, they are irresponsible, impulsive, sensation seeking, and prone to delinquency and criminality (e.g., Cleckley, 1941; Hare, 1993; McCord & McCord, 1964).

What we now recognize as a personality disorder was first identified and described by alienists working with adults who were hospitalized or were in conflict with the law. The majority of patients suffered from "total insanity," illnesses that appeared to result in a general disintegration or deterioration of mental functions (Berrios, 1996). But some cases were characterized by rather specific disturbances of emotion and volition, demonstrating that mental disorder could exist even when reasoning was intact. The terms used to refer to such conditions included *manie sans délire, monomanie,* moral insanity, and *folie lucide* (Millon, 1981). The understanding of these conditions was refined over time, eventually leading to the modern clinical concept of personality disorder in the first half of the twentieth century.

There are different approaches to the assessment and diagnosis of psychopathy. Despite strong parallels at a conceptual level, at an operational level various diagnostic criteria sets for psychopathic, antisocial, dissocial, and sociopathic personality disorders are not equivalent. Perhaps the biggest difference is that diagnostic criteria for psychopathic or dissocial personality disorder typically include a broad range of interpersonal, affective, and behavioral symptoms. In contrast, diagnostic criteria for antisocial or sociopathic personality disorder tend to focus more narrowly on overt delinquent and criminal behavior. The differences between these two diagnostic traditions are discussed at length elsewhere (e.g., Dinges, Atlis, & Vincent, 1997; Hart & Hare, 1997). Perhaps the most important consequence of the focus on delinquent and criminal behavior in diagnostic criteria sets for antisocial or sociopathic personality disorder is that they lack specificity (i.e., misconduct can be a manifestation of other forms of disorders as well), and this can lead to over-diagnosis in forensic settings and under-diagnosis in other settings (see especially American Psychiatric Association, 1994; p. 647).

Much of the research on psychopathy in adulthood has used multi-item symptom construct rating scales such as the original or revised Psychopathy Checklist (PCL and PCL-R; Hare, 1980, 1991) or the Screening Version of the PCL-R (PCL:SV; Hart, Cox, & Hare, 1995). These tests require trained observers to rate the severity of personality disorder symptoms based on all available clinical data (e.g., interview with the respondent, review of case history information, interviews with collateral informants). The PCL and PCL-R were designed for use in adult male forensic populations, with some items being scored entirely or primarily on the basis of criminal records. The PCL-R comprises 20 items scored on a 3-point scale (0 = *item doesn't apply,* 1 = *item applies somewhat,* 2 = *item definitely applies*). Total scores can range from 0 to 40; scores of 30 or higher are considered diagnostic of psychopathy. The PCL:SV is a 12-item scale derived from the PCL-R. It was designed for use in adult populations, regardless of gender, psychiatric status, or criminal history. Scoring of the PCL:SV requires less information, and less detailed information, than does the PCL-R. Further, the PCL:SV can be scored even when the person does not have a criminal record or when the complete record is not available. Items are scored on the same 3-point scale used for the PCL-R. Total scores can range from 0 to 24; scores of 12 or higher indicate "possible psychopathy," and scores of 18 or higher indicate "definite psychopathy."

1.2 Major Research Findings

The PCL, PCL-R, and PCL:SV have been used extensively and successfully across a variety of adult settings and samples. A search of popular computerized databases indicates that the PCL scales have been used in more than two hundred empirical studies to date, including studies of adult male offenders and forensic psychiatric patients, female offenders, ethnic minority offenders in Canada and the United States (e.g., Kosson, Smith, & Newman, 1990; Loucks & Zamble, 1999; Wong, 1985), and nonoffender groups such as substance abusers, civil psychiatric patients, and university students (e.g, Forth, Brown, Hart, & Hare, 1996; Monahan et al., 2001; Rutherford, Cacciola, Alterman, & McKay, 1996). In this section, we summarize in general terms some of the major findings from research on psychopathy in adults. Detailed reviews are available elsewhere (e.g., Cooke, Forth, & Hare, 1998; Hart & Hare, 1997).

Psychopathy is a coherent and stable syndrome. Psychopathic symptoms co-vary in a reliable and meaningful manner. The structural reliability of the PCL scales, including internal consistency and item homogeneity, is high (e.g., Hare, 1991; Hart et. al, 1995). The interrater reliability of individual items, total scores, and categorical diagnoses as measured is good to excellent. The test-retest reliability of scores is moderate to high across periods ranging from one week to two years (Alterman, Cacciola, & Rutherford, 1993; Rutherford et al., 1999). According to Confirmatory Factor Analysis (CFA), symptoms of the disorder form a reliable hierarchical structure in which interpersonal, affective, and behavioral symptoms underlie a superordinate psychopathy factor (Cooke & Michie, in press). Finally, analyses based on Item Response Theory (IRT) indicate that the syndrome has substantial cross-cultural stability (Cooke, 1998; Cooke & Michie, 1999; Hare, Clark, Grann, & Thornton, 2000).

Individuals within and across settings differ with respect to psychopathic symptomatology. It is possible to differentiate among individuals on the basis of their psychopathic symptomatology, both within and across various settings. Even in correctional settings, only about 15% to 25% of serious offenders meet the diagnostic criteria for psychopathy on the PCL scales, and traits of the disorder are widely dispersed (Hare, 1991). Traits are generally higher in offenders housed in high-security versus low-security institutions (Wong, 1985). Prevalence rates are considerably lower in civil psychiatric patients (typically around 5%) and very low in community residents (about 1% or lower), although once again traits are widely dispersed even in these settings (e.g., Hart et al., 1995).

Psychopathy is associated with anomalies in neurocognitive processing. Neuropsychological, physiological, brain imaging, and biochemical studies indicate that psychopaths differ from other individuals in areas of neurocognitive and affective functioning. Learning and attentional irregularities are seen in their inability to learn from past mistakes when engaging in reward seeking behavior (see Newman, 1998, for a review). Differences in linguistic and emotional processing are expressed in a delayed or absent fear response (e.g., Patrick, Cuthbert, & Lang, 1994), and unusual brain activity and lateralization when processing emotional stimuli (Hare, 1998; Jutai, Hare, & Connolly, 1987). Finally, psychopathy may be associated with biochemical abnormalities, such as reduced monoamine oxidase activity (Belfrage, Lidberg, & Oreland, 1992; Lidberg, Modin, Oreland, Tuck, & Kristianson, 1985).

Psychopathy is associated with psychological morbidity. Psychopaths are at increased risk for certain mental disorders (see Hart & Hare, 1997). Specifically, psychopathy has a strong and positive association with substance use and Cluster B personality disorders. In contrast, psychopathy is negatively associated with psychotic, mood, anxiety, and Cluster A personality disorders.

Psychopathy is associated with criminal behavior. Many psychopaths, possibly as a result of symptoms such as impulsivity and unemotionality, engage in chronic criminal conduct. But psychopathy and criminality are distinct constructs: Only a minority (about 15% to 20%) of serious offenders meet PCL criteria for psychopathy.

Psychopaths who engage in crime generally have a specific pattern of offending in respects to the frequency and nature of their crimes. First, it has been discovered, both retrospectively and prospectively, that psychopaths commit offenses during their time at risk (time free from prison) more frequently than do nonpsychopaths (for a review, see Hart & Hare, 1997). Also, psychopathic offenders recidivate more quickly after release from institutions than do other offenders. For example, a recent meta-analysis by Hemphill et al. (1998) found that psychopathic offenders had a general recidivism rate that was three times higher, and a violent recidivism rate that was three to five times higher, than that of nonpsychopathic offenders.

2. Psychopathy in Childhood and Adolescence

2.1 Should We Study Psychopathy in Childhood and Adolescence?

The notion of identifying, and possibly ameliorating, psychopathy during childhood or adolescence is not new (see McCord & McCord, 1956). Presumably, the traits of a personality disorder do not have a sudden onset at the moment an individual turns 18 years of age. Some evidence for the existence of psychopathy in adolescence can be seen in the adult literature. Retrospective studies of adults have traced the onset of psychopathic symptoms back to childhood, as young as 6 to 10 years of age (e.g., Robins, 1978; Widiger et al., 1996). Also, a small proportion of people studied in past research on "adults" actually were younger than 18 years of age, and these youthful offenders and patients did not differ meaningfully from their older peers (e.g., Hare, McPherson, & Forth, 1988; Hare, Forth, & Strachan, 1992). For these reasons, some researchers recently have directed considerable attention to the investigation of psychopathy in children and adolescents.

But others are unconvinced by such evidence and have raised a number of concerns (e.g., Edens, Skeem, Cruise, & Kaufman, 2001). First, it has been argued that personality disorders in general, including psychopathy, may not or perhaps even cannot exist in childhood and adolescence. Second, even if psychopathy does exist in childhood and adolescence, it may be very difficult or even impossible to assess the disorder in a meaningful way. Third, even if psychopathy does exist in childhood and adolescence and can be assessed meaningfully, the concept is ethically problematic and raises the specter of inappropriate labeling and treatment of troubled youths. Let us evaluate each of the concerns separately.

Psychopathy may not or does not exist in childhood and adolescence. The first clinical descriptions of personality disorder were based on adults. Given the absence of systematic research by developmental psychopathologists, we cannot assume that personality disorder *per se* even exists prior to adulthood (e.g., Edens et al., 2001). Some mental disorders arise early in life and change greatly in nature or remit altogether as a result of developmental processes (both biological and social). Mental disorders may arise *de novo* in adulthood, without any manifest symptoms obvious earlier in life. It is possible, then, for a person who appears psychopathic in childhood to develop into an adult free from any sign of personality disorder, or for a person who appears normal in childhood or adolescence to be diagnosed with psychopathy in adulthood. If this is true then measures of psychopathic symptomatology in childhood and adolescence will have low test-retest reliability, especially across critical developmental stages, as well as poor long-term predictive validity with respect to problems in psychosocial adjustment.

Also, standard definitions of personality and personality disorder emphasize their stability. It is debatable whether personality, normal or abnormal, is stable or "crystallized" in children and young adolescents; even if it is, the limitations on life experience due to youth may make it impossible to draw inferences regarding the stability of traits across time or context. If this is true then measures of psychopathic symptomatology in childhood and adolescence will have low interrater reliability, especially across critical developmental stages, as well as poor short-term predictive validity with respect to problems in psychosocial adjustment.

Psychopathy is difficult or impossible to assess in childhood and adolescence. Even if personality disorder does exist in childhood or adolescence, it will not be manifested as it is in adulthood. It is not until late adolescence or early adulthood that people enter into important social roles and obligations, such as employment or marital relationships or parenthood, and have the opportunity to succeed or fail in them. The key problem is how to make accurate and reliable assessments of symptoms such as glibness and superficial charm, irresponsibility, and sexual promiscuity in very young people. Also, it is normal for adolescents to experience problems adjusting to their new roles and obligations; this has led some people to suggest (sometimes in all seriousness, sometimes with tongue in cheek) that all adolescents are — phenotypically and cross-sectionally — psychopathic. If this is true then measures of psychopathic symptomatology in childhood and adolescence will have low test-retest reliability, poor criterion-groups validity (i.e., psychopaths will not differ much from nonpsychopaths), and poor short-term predictive validity with respect to problems in psychosocial adjustment.

It is ethically problematic to study psychopathy in childhood and adolescence. Even if psychopathy exists in childhood and adolescence and can be assessed meaningfully, use of the concept raises a potential ethical problem. The diagnosis of psychopathy clearly carries extremely negative connotations. Affixing such a negative label to children and adolescents may cause decision-makers (courts, clinicians, caregivers) to think about them in very pessimistic terms, perhaps even failing to recommend and provide appropriate intervention. Use of the term psychopathy in relation to children and adolescents may also unwittingly foster and support the development of inaccurate negative stereotypes in the mind of the public. If this is true then measures of psychopathic symptomatology will be readily misinterpreted by mental health, social service, and criminal justice professionals, as well as by lay people.

2.2 Major Research Findings

We now turn to a summary of the existing research on psychopathy in children and adolescents. Our goal is to determine the extent to which the findings from children and adolescents parallel those from research on adults, as space limitations prohibit an exhaustive review. Even limiting our review to studies using rating scales whose development was influenced by the PCL-R (discussed below), since 1990 at least 50 articles in peer-reviewed journals have appeared, as well numerous book chapters, conference presentations, and graduate theses and dissertations (although the number of source studies is considerably smaller).

Most of the research reviewed below was relied on two assessment procedures. The *Psychopathy Checklist: Youth Version* (PCL:YV; Forth, Kosson, & Hare, in press) is a modified version of the PCL-R designed for use with adolescents, aged 13 to 18. It has the same format and administration and scoring procedures as the PCL-R. Item descriptions, and in some cases their labels, were modified to reflect adolescent experiences and the greater influence of family, peers, and school on their lives. For example, the PCL-R Item 17, *Many short-term marital relationships*, was changed to reflect more general *Unstable interpersonal relationships* in the PCL:YV. The *Psychopathy Screening Device* (PSD; Frick & Hare, in press) is a 20-item scale designed to measure psychopathy in children, aged 6 to 12. The authors attempted to adapt each PCL-R item into an analogous item that was more applicable to children. The result was 20 characteristics or statements rated on a 3-point rating scale (0 = *not at all true*, 1 = *sometimes true*, 2 = *definitely true*) by parents and teachers. The PSD also has been used in some research on adolescents; in these instances, it was administered in self-report format. A third measure, the *Childhood Psychopathy Scale* (CPS; Lynam, 1997), was designed to assess psychopathy in late childhood and early adolescence. It was constructed using archival data from a portion of the large community sample of boys, aged 12 or 13, involved in the Pittsburgh Youth Study. The CPS items were formed by combining behavioral ratings from mothers' reports on the Childhood Behavior Checklist (CBCL; Achenbach, 1991) and the Common Language Version of the California Child Q-set (CLQ; Caspi et al., 1992). Lynam was able to develop 13 CPS items that paralleled those from the PCL-R, but some PCL-R items (such as *Grandiose sense of self-worth* and *Proneness to boredom*) could not be assessed using the CBCL and CLQ ratings. To date, however, the CPS has been used only in its original construction sample.

Is psychopathy a stable syndrome in children and adolescents? With respect to structural reliability, studies have reported good internal consistency and item homogeneity for the PCL:YV in adolescents (e.g., Brandt et al., 1997; Forth, Hart, & Hare, 1990; Vincent, Corrado, Cohen, & Odgers, 1999) and for the PSD and CPS in children (Frick, Barry, & Bodin, 2000; Lynam, 1997).

Evidence concerning interrater reliability has been mixed. In adolescents, the interrater reliability of PCL:YV scores has been good to excellent, ranging from a low of .81 (Hume, Kennedy, Partick, & Partyka, 1996) to a high of .98 (Stafford & Cornell, 1997). In contrast, though, the interrater reliability between parents and teachers on the PSD for children has been weak, ranging from a low of .26 to a high of .40 (e.g., Frick, Lilienfeld, Ellis, Loney, & Silverthorn, 1999; Loney, Frick, Ellis, & McCoy, 1998). No interrater reliability data are available for the CPS.

There is no information available concerning the test-retest reliability of the PCL:YV in adolescents. In children, one study has reported acceptable test-retest reliability of teacher ratings on the CU scale of the PSD (McBurnett, Tamm, Nowell, Pfiffner, & Frick, 1994). No test-retest reliability data are available for the CPS.

The structure of psychopathic symptoms in children and adolescents is not clear. Early reports supported a two-factor structure for the PCL:YV in adolescents (Brandt et al., 1997; Forth, 1995) and, in children, for the PSD (Frick, O'Brien, Wootton, & McBurnett, 1994) and CPS (Lynam, 1997). But in these studies, the observed factor structures differed in important ways from those reported in adults (i.e., some items loaded on different factors). More recently, Frick et al. (2000) reported a three-factor solution for the PSD in community children, although once again its parallels to the 3-factor structure reported in adults are limited. A convincing answer to this question requires the application of CFA methods.

There has been no systematic research examining the stability of the PCL:YV, PSD, or CPS across gender, ethnicity, or nationality. Preliminary research using the PCL:YV in adolescents has found no association between total scores and age (e.g., Brandt et al., 1997; Forth et al., 1990; Vincent et al., 1999), but some association with gender (Myers, Burket, & Harris, 1995; Stanford, Ebner, Patton, & Williams, 1994) and ethnicity groups (Forth et al., 1990). In children, one study reported that PSD scores did not differ significantly across gender or ethnic groups (Frick et al., 1994), but in another it was significantly associated with ethnicity (Frick et al., 1999). Examination of group differences in the distribution of scores is not an adequate test of stability or bias, though; CFA and IRT methods are required to address this issue.

Do children and adolescents differ within and across settings with respect to psychopathic symptomatology? Studies consistently have found a good dispersion of psychopathy-related symptoms in various samples of children and adolescents. Studies of delinquent adolescents in high-security facilities using the PCL:YV have reported a distribution of scores very similar to that reported in incarcerated adult male offenders (e.g., Forth et al., 1990). Scores are lower in delinquents from open custody settings (Vincent et al., 1999), and lower still in civil psychiatric patients (e.g., Myers et al., 1995) and community residents (e.g., Chandler & Moran, 1990; Forth, 1995; Trevethan & Walker, 1989). Similarly, in children, PSD scores are higher among clinic referrals than among community residents (e.g., Frick et al., 2000).

Is psychopathy associated with anomalies in neurocognitive processing in children and adolescents? Systematic research on this issue is limited. Frick and colleagues (e.g., Christian, Frick, Hill, Tyler, & Frazer, 2000; O'Brien & Frick, 1996), as well as Lynam (1997), have examined the association between scores psychopathy-related traits, as measured by the PSD, and laboratory measures of behavioral disinhibition in children, reporting results consistent with those of Newman and colleagues in adults. Blair and colleagues, also using the PSD in children, have reported deficits related to both impulsivity and emotionality associated with psychopathy-related traits (e.g., Blair, 1999; Colledge & Blair, 2001; Stevens, Charman, & Blair, in press). Finally, various investigators have reported deficits in empathy and moral reasoning using the PCL:YV in adolescents (e.g., Chandler & Moran, 1990; Trevethan & Walker, 1989) and the PSD in children (e.g., Blair, Monson, & Frederickson, 2001).

Is psychopathy associated with psychological morbidity in children and adolescents? There have been few studies of other mental disorders. Consistent with the adult literature, there is some evidence that adolescents with high PCL:YV scores are more likely to have substance abuse problems than those with low scores (e.g., Forth, 1995; Mailloux, Forth, & Kroner, 1997; Vincent et al., 1998; but cf. Brandt et al., 1997). There is some evidence for an inverse relationship between trait-anxiety and psychopathy-related traits in children (Frick et al., 1999). There are no good studies looking at comorbidity with psychotic or affective disorders, or at comorbidity with other personality disorders.

There have been few studies of other mental disorders. Consistent with the adult literature, there is some evidence that adolescents with high PCL:YV scores are more likely to have substance abuse problems than those with low scores (e.g., Forth, 1995; Mailloux, Forth, & Kroner, 1997; Vincent et al., 1999; but cf. Brandt et al., 1997). There are no good studies looking at comorbidity with psychotic, affective, or anxiety disorders, or at comorbidity with other personality disorders.

Is psychopathy associated with criminal behavior in children and adolescents? A few studies have examined antisocial behavior in children (e.g., Lynam, 1997), but most of the research has focused on formal delinquency in adolescents. Consistent with the adult literature, adolescents with high scores on the PCL:YV engage in more antisocial behavior than do others, both in institutions (Forth et al., 1990; Hicks, Rogers, & Cashel, 2000; Rogers Johansen, Chang, & Salekin, 1997) and in the community (Brandt et al., 1997; Corrado, Vincent, Cohen, & Odgers, 2000; Forth et al., 1990; Toupin, Mercier, Déry, Côté, & Hodgins, 1996; Vincent et al., 1999). Adolescents with high PCL:YV scores also commit more serious antisocial behavior — specifically, more violence (Brandt et al., 1997; Stafford & Cornell, 1997). The predictive validity of the PCL:YV wit respect to serious antisocial behavior is evident over periods as long as 1 to 3 years (e.g., Corrado et al., 2000; Forth et al., 1990; Toupin et al., 1996).

3. Psychopathy and Multi-Problem Youth: Recommendations for Research and Practice

The research reviewed in the preceding section appears to undercut some of the arguments raised by those concerned with potential problems of applying the concept of psychopathy to children and adolescents. Specifically cross-sectional research suggests that it is possible to assess in adolescents, with good interrater reliability, traits that are phenotypically or superficially similar to those of psychopathy in adults. Also, these traits are predictive of problems in psychosocial adjustment, including institutional and community misbehavior, at least over time periods ranging from several months to several years. Put simply, we can reliably identify something in adolescents that is similar to psychopathy in adults and that is associated with future criminality, increased psychological morbidity, and some differences in neurocognitive processing. The problem is that we have no strong or direct evidence the thing we are measuring is actually psychopathy *per se*, a stable personality disorder that does not dissipate over time. This is particularly the case in children where the reliability of assessments across raters and contexts has yet to be established. With this in mind, we now turn to some recommendations for research and practice with respect to psychopathy and multi-problem youth.

3.1 Recommendations for Research

In our view, research on psychopathy in children and adolescence should focus on five major issues:

Temporal stability. We need to examine: (1) the temporal stability of symptoms, including the age onset, duration, and age offset, using methods such as event history analysis and test-retest reliability; and, (2) the temporal stability of syndrome/disorder across critical developmental stages using longitudinal designs. Until adequate studies of the temporal stability have been conducted, diagnoses of psychopathy in early developmental stages are entirely unwarranted. Though there is some evidence that traits are stable across contexts, at least in adolescents, it is possible that a person who appears psychopathic in childhood will develop into an adult free from any sign of personality disorder, or that a person who appears normal in childhood or adolescence will be diagnosed with psychopathy in adulthood.

Adequacy of measures. We need to examine: (1) concurrent validity of symptoms and syndrome/disorder across various measures at various ages using mono-trait/multi-method designs; and, (2) examination of possible measurement bias due to age, sex, and ethnicity/culture using confirmatory factor analysis (CFA) and item response theory (IRT). Though the results of current concurrent validity studies have been promising, multi-method designs have been used considerably more with children than with adolescents. There is some indication that measurement bias may be occurring given different factor structures of psychopathy measures from early childhood to adulthood. IRT methods would allow us to determine whether constructs such as this take on a different meaning (specifically, whether they are measuring the same latent trait) across different developmental stages. It is possible that many aspects of affect are not yet crystallized in youth.

Long-term predictive validity. We need to examine: (1) criminal careers, using longitudinal designs; and, (2) problems in psychosocial adjustment, including morbidity and mortality, also using longitudinal designs. Promising theories have suggested that psychopathy may explain the "chronic offender" subtype discussed in the juvenile delinquency literature (see Lynam, 1996). To validate these theories, we need no information as to the long-term predictive acumen of psychopathy from early developmental stages to later adjustment problems. This line of research would greatly enhance our understanding of early prevention, treatment, and risk assessment and management. If it is the case that the aging process alone ameliorates psychopathic symptoms for a significant proportion of youth, than psychopathy assessments would lack specificity for longer-term predictions.

Construct validity. We need to examine: (1) anomalies in neurcognitive processing, using experimental psychopathology methods; and, (2) comorbidity with other mental disorders in childhood/adolescence. Again, preliminary results in both of these areas are promising. However, information about the neurocognitive processing of young people with psychopathic characteristics is significantly lacking. This line of research would greatly enhance etiological theories of psychopathy. With respect to comorbidity, much work has examined psychopathy and disruptive behavior disorders, but little is known about other mental disorders such as psychotic, affective, and personality disorders.

Ethical issues. We need to examine: (1) stereotypes of psychopathy held by professionals and lay people; (2) differences between stereotypes for psychopathy and those for related mental disorders, such as CD and ODD; (3) extent to which negative stereotypes can be modified and corrected through education; and (4) extent to which information regarding assessment of psychopathy in childhood and adolescence can be communicated accurately. The significant and detrimental weight that psychopathy assessments carry in the adult courtroom continues to grow. It would be ethically irresponsible to assume that psychopathy research in childhood and adolescence may not lead to similar outcomes in juvenile courts, even when it is misused. Indeed, there are already examples where psychopathy assessments have been used in making decisions to transfer juveniles to adult court (see Lyon & Ogloff, 2000, for a review). Given the limitations noted in this chapter, we wish to stress that such uses of psychopathy measures are not appropriate at this time. Though no studies have been conducted per se, case law provides evidence that the psychopathy label in youth carries two assumptions to legal officials: first, the individual poses a high risk for criminality; and second, the individual is untreatable. In light of these concerns, it is important professionals subsume the role as educator. This is not to say that psychopathy is not a useful construct in childhood and adolescence. Here we will review the manners in which we believe the research supports and does not support its use.

3.2 Recommendations for Practice

In our view, it is possible to use the existing research on psychopathy in childhood and adolescence to assist the assessment and management of risks and needs in multi-problem youth. This must be done carefully, however, if one is to avoid the ethical problems discussed previously. At the present time, and until the outstanding issues discussed previously have been resolved, any clinical discussions of psychopathy in childhood and adolescence must take into account the following points:

First, *there is no general consensus in the scientific or professional literature that psychopathy — a form of personality disorder described and studied as a form of adult psychopathology — exists in childhood and adolescence.* According to basic principles of developmental psychopathology, one should not assume continuity of normal or abnormal behavior across stages of development. Convincing proof would require: (1) longitudinal studies demonstrating stability of (putative) psychopathic traits from childhood and adolescence through to early or middle adulthood; and, (2) experimental studies indicating that (putative) psychopathic traits observed in childhood and adolescence are associated with the same etiological factors as those observed in adulthood. An important corollary of this conclusion is that professionals should not use the diagnosis of psychopathy when evaluating children and adolescents.

Second, *it is possible to measure reliably personality features in childhood and adolescence that resemble symptoms of psychopathy in adulthood.* Tests such as the PCL:YV are potentially useful for making descriptive statements concerning the current or recent (i.e., cross-sectional) psychosocial adjustment of children and adolescents. The problem is we don't know exactly how to describe what these tests measure. At present, the most conservative approach is to describe them as tests of "personality features that may be associated with the development of personality disorder in adulthood"; a less conservative but perhaps still accurate description is "psychopathy-related personality features." Regardless, professionals should be careful not to claim that these personality features are symptoms of personality disorder *per se* — which, as discussed previously, may not even exist prior to adulthood. Also, professionals should be careful not to assume, infer, or claim that personality features observed in childhood or adolescence will persist over time. Maturation, changes in environment or living conditions, and — importantly — treatment all may lead to changes over time in these features.

Third, *personality features in childhood and adolescence that resemble symptoms of psychopathy in adulthood are associated with increased risk for institutional and community misbehavior.* Tests such as the PCL:YV are potentially useful for making prognostic statements concerning the psychosocial adjustment of children and adolescents over time periods ranging from several months to several years. But professionals should be careful not to overstate the extent to which these personality features are associated with misbehavior: Measures such as the PCL:YV were not designed to predict misbehavior, there is no evidence that they can be used to make specific predictions about the behavior of individuals with any reasonable degree of certainty, and they should not be used for this purpose. Administration of a single test (or a single battery of tests) does not constitute a comprehensive risk assessment.

Fourth, *personality features in childhood and adolescence that resemble symptoms of psychopathy in adulthood may be helpful in planning service delivery to multi-problem youth.* As has been discussed in relation to adults (e.g., Andrews & Bonta, 1998), personality features may be relevant to risk management in several ways. If one assumes that personality is causally related to misbehavior in a direct manner but temporally stable, then the presence of psychopathy-related personality features personality is a static or *fixed risk factor* that should signal the child or adolescent is a candidate for urgent or high-intensity services. If one assumes that personality is causally related to misbehavior in a direct manner but malleable, then the presence of psychopathy-related personality features is a criminogenic need or *variable risk factor* that should be considered an important target for intervention. Finally, if one assumes that personality is related to misbehavior only indirectly but is temporally stable, then the presence of psychopathy-related personality features is a co-factor, mediator, or *responsivity factor* that should influence decisions regarding how treatment is delivered to the child or adolescent. Of course, the only way to determine which of these approaches is most useful is to develop and evaluate treatment programs for multi-problem youth, examining the role of psychopathy-related personality features.

4. Conclusion

In adults, psychopathy is widely recognized as a mental disorder that has important implications for the assessment and management of risk for criminality and violence. In contrast, relatively little is known about psychopathy, or its developmental precursors, in childhood or adolescence. The existing research does, however, suggest that psychopathy-related personality features in childhood and adolescence may also be important to consider when developing, delivering, and evaluating services for multi-problem youth, although care must be taken to avoid potential ethical problems associated with its use. Given the disorder's prevalence and its impact on affected individuals and on society, research on developmental aspects of psychopathy should be a high priority.

References

Achenbach, T. M. (1991). *Manual for the Child Behavior Checklist and 1991 Profile*. Burlington, VT: University of Vermont, Department of Psychiatry.

Alterman, A. I., Cacciola, J. S., & Rutherford, M. J. (1993). Reliability of the Revised Psychopathy Checklist in substance abuse patients. *Psychological Assessment, 5,* 442-448.

American Psychiatric Association. (1994). *The diagnostic and statistical manual of mental disorders (4ᵗʰ ed.)*. Washington, DC: Author.

Andrews D. A., & Bonta J. (1998). *The psychology of criminal conduct*, 2nd ed. Cincinnati: Anderson.

Barry, C. T., Frick, P. J., DeShazo, T. M., McCoy, M. G., Ellis, M., & Loney, B. R. (2000). The importance of callous-unemotional traits for extending the concept of psychopathy to children. *Journal of Abnormal Psychology, 109,* 335-340.

Belfrage, H., Lidberg, L., Oreland, L. (1992). Platelet monoamine oxidase activity in mentally disordered violent offenders. *Acta Psychiatrica Scandinavica, 85,* 218-221.

Berrios, G. E. (1996). *The history of mental symptoms: Descriptive psychopathology since the nineteenth century*. Cambridge, UK: Cambridge University Press.

Blair, R. J. R. (1999). Responsiveness to distress cues in the child with psychopathic tendencies. *Personality and Individual Differences, 27,* 135-145.

Blair, R. J. R., Monson, J., & Frederickson, N. (2001). Moral reasoning and conduct problems in children with emotional and behavioural difficulties. *Personality and Individual Differences, 31,* 799-811.

Brandt, J. R., Wallace, A. K., Patrick, C. J., & Curtin, J. J. (1997). Assessment of psychopathy in a population of incarcerated adolescent offenders. *Psychological Assessment, 9,* 429-435.

Caspi, A., Block, J. H., Klopp, B., Lynam, D., Moffitt, T. E., & Stouthamer-Loeber, M. (1992). A "common language" version of the California Child Q-Set (CCQ) for personality assessment. *Psychological Assessment, 4,* 512-523.

Chandler, M., & Moran, T. (1990). Psychopathy and moral development: A comparative study of delinquent and nondelinquent youth. *Development and Psychopathology, 2,* 227-246.

Christian, R. E., Frick, P. J., Hill, N. L., Tyler, L., & Frazer, D. R. (1997). Psychopathy and conduct problems in children: II. Implications for subtyping children with conduct problems. *Journal of the American Academy of Child and Adolescent Psychiatry, 36,* 233-241.

Cleckley, H. (1941). *The mask of sanity*. St. Louis: The C. V. Mosbey Company.

Colledge, E., & Blair, R. J. R. (2001). The relationship between the inattention and impulsivity components of attention deficit and hyperactivity disorder and psychopathic tendencies. *Personality and Individual Differences, 30*, 1175-1187.

Cooke, D. J. (1995). Psychopathic disturbance in the Scottish prison population: The cross-cultural generalisability of the Hare Psychopathy Checklist. *Psychology, Crime and Law, 2*, 101-118.

Cooke, D. J. (1998). Psychopathy across cultures. In D. J. Cooke, A. E. Forth, & R. D. Hare (Eds.), Psychopathy: *Theory, research, and implications for society* (pp. 13-45). Dordrecht, The Netherlands: Kluwer.

Cooke, D. J., Forth, A. E., & Hare, R. D. (1998). *Psychopathy: Theory, research and implications for society.* Netherlands: Kluwer Academic Publishers.

Cooke, D. J., & Michie, C. (1999). Psychopathy across cultures: Scotland and North America compared. *Journal of Abnormal Psychology, 108*, 58-68.

Cooke, D. J., & Michie, C. (in press). Factor structure of the Hare Psychopathy Checklist. *Psychological Assessment.*

Corrado, R. R., Vincent, G. M., Cohen, I., & Odgers, C. (2000, July). *Predicting recidivism in young offenders: The utility of the PCL:YV.* Paper presented at the 25th Anniversary Congress on Law and Mental Health, International Academy of Law and Mental Health, Siena.

Dinges, N. G., Atlis, M. M., & Vincent, G. M. (1997). Cross-cultural perspectives on antisocial behavior. In D. M. Stoff, J. Breiling, & J. D. Maser (Eds.), *Handbook of antisocial behavior* (pp. 463-473), New York: John Wiley & Sons.

Edens, J. F., Skeem, J. L., Cruise, K. R., & Cauffman, E. (2001). Assessment of "juvenile psychopathy" and its association with violence: A critical review. *Behavioral Sciences and the Law, 19*, 53-80.

Forth, A. E. (1995). Psychopathy *and Young Offenders: Prevalence, family background, and violence.* Unpublished report, Carleton University, Ottawa, Ontario.

Forth, A. E., Brown, S. L., Hart, S. D., & Hare, R. D. (1996). The assessment of psychopathy in male and female noncriminals: Reliability and validity. *Personality and Individual Differences, 20*, 531-543.

Forth, A. E., & Burke, H. (1998). Psychopathy in adolescence: Assessment, violence, and developmental precursors. In D. J. Cooke, A. E. Forth, and R. D. Hare (Eds.), *Psychopathy: Theory, research and implications for society* (pp. 205-229), Dodrecht, The Netherlands: Kluwer Academic Publishers.

Forth, A. E., Hart, S. D., & Hare, R. D. (1990). Assessment of psychopathy in male young offenders. *Psychological Assessment, 2*, 342-344.

Forth, A. E., Kosson, D. S., & Hare, R. D. (in press). *Hare Psychopathy Checklist: Youth Version.* Unpublished test manual, Multi-Health Systems.

Forth, A. E., & Mailloux, D. L. (2000). Psychopathy in youth: What do we know? In C. B. Gacono (Ed.), *The clinical and forensic assessment of psychopathy: A practitioner's guide* (pp. 25-54), Mahwah, New Jersey: Lawrence Erlbaum Associates.

Frick, P. J., Barry, C. T., Bodin, S.D. (2000). Applying the concept of psychopathy to children: Implications for the assessment of antisocial youth. In C. B. Gacono (Ed.), *The clinical and forensic assessment of psychopathy: A practitioner's guide* (pp. 3-24), Mahwah, New Jersey: Lawrence Erlbaum Associates.

Frick, P. J., & Hare, R. D. (in press). *The Psychopathy Screening Device.* Unpublished test manual, Multi-Health Systems.

Frick, P. J., Lilienfeld, S. O., Ellis, M., Loney, B., & Silverthorn, P. (1999). The association between anxiety and psychopathy dimensions in children. *Journal of Abnormal Child Psychology, 27*, 383-392.

Frick, P. J., O'Brien, B. S., Wootton, J. M., & McBurnett, K. (1994). Psychopathy and conduct problems in children. *Journal of Abnormal Psychology, 4*, 700-707.

Hare, R. D. (1980). A research scale for the assessment of psychopathy in criminal populations. *Personality and Individual Differences, 1*, 111-119.

Hare, R. D. (1991). *The Hare Psychopathy Checklist — Revised.* Toronto, Ontario: Multi-Health Systems.

Hare, R. D. (1998). Psychopathy, affect and behavior. In D. J. Cooke, A. E. Forth, and R. D. Hare (Eds.), *Psychopathy: Theory, research and implications for society* (pp. 105-137), Dordrecht, The Netherlands: Kluwer Academic Publishers.

Hare, R. D., Clark, D., Grann, M., & Thornton, D. (2000). Psychopathy and the predictive validity of the PCL-R: An international perspective. *Behavioral Sciences and the Law, 18*, 623-645.

Hare, R. D., Forth, A. E., & Strachan, K. (1992). Psychopathy and crime across the lifespan. In R. Peters, R. McMahon, & V. Quinsey (Eds.). *Aggression and violence throughout the lifespan.* Newbury Park, CA: Sage

Hare, R. D., McPherson, L. E., & Forth, A. E. (1988). Male psychopaths and their criminal careers. *Journal of Consulting and Clinical Psychology, 56*, 710-714.

Hart, S. D. (1998). The role of psychopathy in assessing risk for violence: Conceptual and methodological issues. *Legal and Criminological Psychology, 3*, 123-140.

Hart, S. D., Cox, D. N., & Hare, R. D. (1995). *Manual for the Psychopathy Checklist: Screening Version (PCL:SV)*. Toronto: Multi-Health Systems.

Hart, S. D., & Hare, R. D. (1997). Psychopathy: Assessment and association with criminal conduct. In D. M. Stoff, J. Brieling, & J. Maser (Eds.), *Handbook of antisocial behavior* (pp. 22-35). New York: Wiley.

Hemphill, J. F., Hare, R. D., & Wong, S. (1998). Psychopathy and recidivism: A review. *Legal and Criminological Psychology, 3,* 141-172.

Hicks, M. M., Rogers, R., & Cashel, M. (2000). Predictions of violent and total infractions among institutionalized male juvenile offenders. *Journal of the American Academy on Psychiatry and Law, 28,* 183-190.

Hume, M. P., Kennedy, W. A., Patrick, C. J., & Partyka, D. J. (1996). Examination of the MMPI-A for the assessment of psychopathy in incarcerated adolescent male offenders. *International Journal of Offender Therapy and Comparative Criminology, 40,* 224-233.

Jutai, J. W., Hare, R. D., & Connolly, J. F. (1987). Psychopathy and event-related brain potentials (ERPs) associated with attention to speech stimuli. *Personality and Individual Differences, 8,* 175-184.

Kosson, D. S., Smith, S. S., & Newman, J. P. (1990). Evaluation of the construct validity of psychopathy in Black and White male inmates: Three preliminary studies. *Journal of Abnormal Psychology, 99,* 250-259.

Krueger, R. F., Schmutte, P. S., Caspi, A., Moffitt, T. E., Campbell, K., & Silva, P. A. (1994). Personality traits are linked to crime among men and women: Evidence from a birth cohort. *Journal of Abnormal Psychology, 103,* 328-338.

Lidberg, L., Modin, I., Oreland, L., Tuck, J. R., Kristianson, M. (1985). Platelet monoamine oxidase activity and psychopathy. *Psychiatry Research, 16,* 339-343.

Loeber, R. (1990). Development and risk factors of juvenile antisocial behavior and delinquency. *Clinical Psychology Review, 10,* 1-41.

Loney, B. R., Frick, P. J., Ellis, M., McCoy, M. G. (1998). Intelligence, callous-unemotional traits, and antisocial behavior. *Journal of Psychopathology and Behavioral Assessment, 20,* 231-247.

Loucks, A. D., & Zamble, E. (1999). Predictors of recidivism in female offenders: Canada searches for predictors common to both men and women. *Corrections, 61,* 26-32.

Lynam, D. R. (1996). Early identification of chronic offenders: Who is the fledgling psychopath? *Psychological Bulletin, 120,* 209-234.

Lynam, D. R. (1997). Pursuing the psychopath: Capturing the fledgling psychopath in a nomological net. *Journal of Abnormal Psychology, 106,* 425-438.

Lyon, D. R., & Ogloff, J. R. P. (2000). Legal and ethical issues in psychopathy assessment. In C. B. Gacono (Ed.), *The clinical and forensic assessment of psychopathy: A practitioner's guide* (pp. 139-173), Mahwah, New Jersey: Lawrence Erlbaum Associates.

Mailloux, D. L., Forth, A. E., & Kroner, D. G. (1997). Psychopathy and substance use in adolescent male offenders. *Psychological Reports, 80,* 529-530.

McBurnett, K., Tamm, L., Nowell, G., Pfiffner, L., & Frick, P. J. (1994, June). *Psychopathy dimensions in normal children*. Presented at the Scientific Meeting of the Society for Research in Child and Adolescent Psychopathology, London, England.

McCord, W., & McCord, J. (1956). *Psychopathy and delinquency*. New York: Grune and Stratton.

McCord, W., & McCord, J. (1964). *The psychopath: An essay on the criminal mind*. Princeton, NJ: Van Nostrand.

Millon, T. (1981). *Disorders of personality: DSM-III Axis II*. New York: Wiley.

Monahan, J., Steadman, H. J., Silver, E., Applebaum, P. S., Robbins, P. C., Mulvey, E. P., Roth, L. H., Grisso, T., & Banks, S. (2001). *Rethinking risk assessment: The MacArthur study of mental disorder and violence*. New York: Oxford University Press.

Myers, C., Burket, R. C., Harris, E. (1995). Adolescent psychopathy in relation to delinquent behaviors, conduct disorder, and personality disorders. *Journal of Forensic Sciences, 40,* 436-440.

Newman, J. P. (1998). Psychopathic behavior: An information processing perspective. In D. J. Cooke, A. E. Forth, and R. D. Hare (Eds.), *Psychopathy: Theory, research and implications for society* (pp. 81-104), Dordrecht, The Netherlands: Kluwer Academic Publishers.

O'Brien, B. S., & Frick, P. J. (1996). Reward dominance: Associations with anxiety, conduct problems, and psychopathy in children. *Journal of Abnormal Child Psychology, 24,* 223-239.

Patrick, C. J., Cuthbert, B. N., & Lang, P. J. (1994). Emotion in the criminal psychopath: Fear image processing. *Journal of Abnormal Psychology, 103,* 523-534.

Robins, L. N. (1978). Aetiological implications in studies of childhood histories relating to antisocial personality. In R. D. Hare & D. Schalling (Eds.), *Psychopathic behavior: Approaches to research* (pp. 255-271). Chichester, England: Wiley.

Rogers, R, Johansen, J., Chang, J. J., & Salekin, R. T. (1997). Predictors of adolescent psychopathy: Oppositional and conduct-disordered symptoms. *Journal of the American Academy of Psychiatry and Law, 25,* 261-271.

Rutherford, M. J., Cacciola, J. S., Alterman, A. I., & McKay, J. R. (1996). Reliability and validity of the Revised Psychopathy Checklist in women methadone patients. *Assessment, 3,* 43-54.

Salekin, R., Rogers, R., & Sewell, K. (1996). A review and meta-analysis of the psychopathy checklist and psychopathy checklist-revised: Predictive validity of dangerousness. *Clinical Psychology: Science and Practice, 3,* 203-215.

Stafford, E., & Cornell, D. G. (1997, August). *Psychopathy as a predictor of adolescents at risk for* inpatient *violence.* Poster presented at the annual meeting of the American Psychological Association, Chicago, Illinois.

Stanford, M., Ebner, D., Patton, J., & Williams, J. (1994). Multi-impulsivity within an adolescent psychiatric population. *Personality and Individual Differences, 16,* 395-402.

Stevens, D., Charman, T., & Blair, R. J. R. (in press). Recognition of emotion in facial expressions and vocal tones in children with psychopathic tendencies. *Journal of Genetic Psychology.*

Toupin, J., Mercier, H., Déry, M., Côté, G., & Hodgins, S. (1996). Validity of the PCL-R for adolescents. In D. J. Cooke, A. E. Forth, J. P. Newman, & R. D. Hare (Eds.), *Issues in Criminological and Legal Psychology: No. 24, International perspectives on psychopathy* (pp. 143-145). Leicester, UK: British Psychological Society.

Trevethan, S. D., & Walker, L. J. (1989). Hypothetical versus real-life mroal reasoning among psychopathic and delinquent youth. *Development and Psychopathology, 1,* 91-103.

Vincent, G. M., Corrado, R. R., Cohen, I., & Odgers, C. (1999, July). *Identification of psychopathy in* adolescents*: The utility of two measures.* Paper presented in R. R. Corrado (Chair) Psychological perspectives on young offenders: Identity, mental disorder, and violence. Symposium in the 1999 international conference of the European Psychology/Law Society and the American Psychology/Law Society, Dublin, Ireland.

Widiger, T. A., Cadoret, R., Hare, R. D., Robins, L., Rutherford, M., Zanarini, M., Alterman, A., Apple, M., Corbitt, E., Forth, A. E., Hart, S. D., Kultermann, J., Woody, G., & Frances, A. (1996). DSM-IV antisocial personality disorder field trial. *Journal of Abnormal Psychology, 105,* 3-16.

Wong, S. (1985). *Criminal and institutional* behaviors *of psychopaths.* Ottawa, Ontario: Programs Branch Users Report, Ministry of the Solicitor General of Canada.

World Health Organization. (1990). International *Classification of Diseases and Causes of Death (10ᵗʰ edition).* Geneva: Author.

Multi-Problem Violent Youth
R.R. Corrado et al. (Eds.)
IOS Press, 2002

164

Delinquency and Aggression of School Children in Relation to Maternal Age

Jan B. Deijen, Eric Blaauw and Frans Willem Winkel

Reports of shootings at high schools, killings and serious beatings by children, and other crimes involving very young offenders are frequent topics in the media. The age at which people display acts of violence and aggression appears to be decreasing. Numerous studies have revealed a relationship between crime, violence and aggression on the one hand and aggressive behaviors at an early age on the other hand (Hawkins et al., 1998). These findings indicate the importance of identifying problematic behavior at an early developmental stage. In addition, the findings demonstrate the importance of identifying the reasons for problematic behaviors among young children.

External factors play an important role in the development of child behavior. One of these external factors is the parental style of child rearing. Parents differ with respect to their cultural, social and economic background that may manifest itself in a diversity of child-rearing attitudes (Geronimus, Korenman, & Hillemeier, 1994). A certain style may influence school performance and relationships with other children. With respect to the maternal influence, a relationship has been demonstrated between the developmental status of school-aged children and maternal cognitive level (Camp, Swift, & Swift, 1982). Cross-sectional studies have shown differences between adolescent and adult mothers in authoritarian attitudes. A high authoritarian attitude was found in younger women, whereas this attitude decreased after the age of 21 (Ernhart & Loevinger, 1969). In addition, authoritarian attitudes in adult parents have been found to be negatively associated with performance in school-aged children regardless of socioeconomic status (Camp, 1996; Emmerich, 1969). These findings suggest that young maternal age may negatively influence child behavior.

In the United States, the potential consequences of teen childbearing for offspring is an important issue. The literature on child development in the United States has documented that, on average, children of young mothers perform more poorly on cognitive measures and show a poorer school achievement than children of older mothers (Roosa, Fitzgerald, & Carlson, 1982). In addition, more problem behavior has been found in children of teenage mothers (Furstenberg, Brooks-Gunn, & Morgan, 1987). Differences by maternal age in child development diminish, but nonetheless persist, when indicators of socioeconomic status are controlled (Broman, 1981). This evidence suggests that teen mothers are emotionally unprepared for motherhood and may have poorer parenting skills (Furstenberg, Brooks-Gunn, & Chase-Lansdale, 1989). Young mothers have indeed been found to be less sensitive and responsive to their infants (McAnarney, Lawrence, Ricciuti, Polley, & Szilagyi, 1986), to have negative child-rearing attitudes (Ernhart & Loevinger, 1969) and to have insufficient knowledge about child rearing (McAnarney et al., 1986). Although differences in parental behavior are generally considered to reflect different cultural, social or economic background circumstances rather than maternal age per se (Geronimus et al., 1994), the observed behavior problems in children of teenage mothers may be attributable to the biological immaturity of young mothers. Younger mothers have increased risk of adverse outcomes of pregnancy, even when the influence of socioeconomic status is controlled. These adverse outcomes may be low birth-weight, premature and small-for-gestational-age infants (Fraser, Brockert, & Ward, 1995). The association between maternal age and developmental problems in the offspring has mainly been studied in the context of teenage pregnancies by comparing the infants of (very young) teen mothers with those of older mothers. However, the study of differences between two quite separate age groups cannot shed light on the possibility that child behavior is associated with maternal age in a more continuous way in a broad age range. Age-related biological or cognitive changes in the mother may be associated with the degree of behavioral problems in their children. If age-related factors, other than those which are specific for teenage mothers, account for the observed behavioral problems in children, maternal age will be associated with problem behavior in offspring also in a non-teenage population. To investigate the possibility of a more continuous relationship between maternal age and problem behavior in offspring, we studied the problem behavior of school-aged children in a sample of mothers of a broad age range.

1. Method

1.1 Participants

The Child Behavior Checklist was completed by mothers on 90 children, 40 boys and 50 girls, aged between 8 and 12 (mean age 10 years, $SD = 0.9$). To check the validity of the mother's assessment each of eight teachers completed the Amsterdam Child Behavior Checklist. The maternal age at delivery ranged between 22 and 38 years (mean age, 29; $SD = 3$). The children were in school group 5, 6 or 7 in one of three different elementary schools.

A distinction was made between a "low maternal age" ($N= 49$) and a "high maternal age" group ($N = 41$), the first group comprising mothers below age 29 ("young") and the second group of mothers of 29 years and over ("old") at the time of birth of the child concerned. The age of 29 was the median of the total group of mothers. The mean age of the boys and girls in the group of young aged mothers was 10 ($SD = 0.72$) and 9.92 ($SD = 0.98$) respectively. In the group of the old aged mothers, the mean age of the boys and girls was 9.81 ($SD = 0.98$) and 9.88 ($SD = 0.97$), respectively. The two groups of boys and girls did not have statistically different age distributions.

1.2 Material

The Child Behavior Checklist (CBCL) (Achenbach, 1991), adapted for the Dutch language by Verhulst, Koot, Akkerhuis, and Veerman (1990) and the Amsterdam Child Behavior Checklist (ACBL) (De Jong, 1995) were used to obtain ratings of problem behavior. The CBCL has been used in a variety of settings and has been proven to be reliable (Achenbach, 1991). The questionnaire was designed to obtain parental ratings of problem-behaviors in 4-18 year old children. The list contains 113 items that can be scored with 0 (not true), 1 (partly true) or 2 (true). Scores are obtained for the problem categories *withdrawn* (score 0-18), *somatic complaints* (score 0-18), *anxious/depressed* (score 0-30), *social problems* (score 0-16), *thought problems* (score 0-14), *attention problems* (score 0-22), *delinquency* (score 0-26), *aggression* (score 0-40), and *sexual problems* (score 0-12). There are two second order factors, namely *externalizing* (score 0-100) and *internalizing* (score 0-100) behavioral problems. The sum of all problem categories gives one total problem score (range of scores 0-226).

The ACBL is an instrument to obtain teachers' ratings of behavior disorders within the classroom. The list consists of 35 items with 4 response alternatives, i.e., 1 (not true), 2 (partly true), 3 (quite true) and 4 (true). The scale is intended to assess the dimensions *inattention* (score 7-28), *restlessness* (score 4-16), *aggression* (score 6-24) and *anxiety* (score 4-16) in elementary school children. In the present study only the aggression dimension was evaluated to check the validity of the completion of the CBCL by the mothers.

1.3 Procedure

The study was performed in three different elementary schools. After having obtained the consent of the school management, the parents and teachers of the children concerned were asked for cooperation. Participation of the parents consisted of completing the CBCL at home. The teachers were asked to fill out the ACBL at work. As an indication of the social economic status of the family, the educational level of the fathers was assessed by means of scores ranging from 1 (elementary school) to 9 (university). In addition, the following parity categories were defined: 1 (only child), 2 (first child), 3 (second child) and, 4 (third or later child). The Pearson correlation coefficient was calculated to determine possible relationships between maternal age and behavior-problem scores. The father's educational level of the two different maternal age groups were analyzed by t-tests.

One-way multivariate analyses of covariance (Mancova) were performed with maternal age and sex as independent factors and parity as covariate. As the distributions of the dependent variables were highly skewed, a square root transformation was performed on all variables, except for father's educational level, to obtain a nearly normal distribution. In addition, the Pearson correlation was calculated. All statistical tests were two-tailed. In Table 1 the transformed variables are shown. The calculations were performed with the Statistical Package for the Social Sciences, 'SPSS 9.0 for Windows.

2. Results

The t-test indicated that the father's educational level of the low and the high maternal age group were not statistically different ($p = .53$). Therefore, educational level was not used as covariate in the analyses of variance. A Mancova on all CBCL scales, controlling for the effects of the confounding variable parity, indicated a multivariate difference between groups, $F(12, 75) = 1.87, p < .05$.

Table 1: Square root transformed means and SD's for all CBCL scales and the ACBL aggression scale

	maternal age	
	< 29 yr.	> 29 yr.
Aggression (CBCL) (rated by mothers)	2.08 ± 1.03	1.37 ± 1.17
Delinquency	0.71 ± 0.80	0.40 ± 0.63
Aggression (ACBL) (rated by teachers)	3.09 ± 0.70	2.88 ± 0.56
Withdrawal	1.08 ± 0.88	0.84 ± 0.90
Somatic complaints	0.82 ± 0.80	0.52 ± 0.68
Anxiety	1.36 ± 0.89	1.06 ± 1.07
Social problems	0.74 ± 0.83	0.80 ± 0.81
Thought problems	0.29 ± 0.56	0.30 ± 0.58
Attention problems	1.54 ± 0.87	1.20 ± 1.06
Internalizing	2.14 ± 1.12	1.62 ± 1.33
Externalizing	2.28 ± 1.15	1.50 ± 1.26

The results of the univariate *F*-tests indicated a significant higher *externalizing* score for the children in the low maternal age group, F (1,75) = 8.29, p < .005. No significant difference between groups was found for internalizing. In addition, with respect to the externalizing variables aggression and delinquency, tests of between subjects effects indicated significant differences between groups. A significant higher aggression score was found in the low maternal age group, F (1, 75) = 8.08, p < .01. In addition, compared to the high maternal age group, the children in the low maternal age group tended to have higher scores for delinquency, F (1,75) = 3.68, p = .06. The square root transformed means for aggression and delinquency are shown in Table 1. Correlation analysis also indicated that maternal age and CBCL scores were associated. Maternal age correlated negatively with the scores for *aggression*, r = -.34, p < .001, and *delinquency*, r = -.27, p < .01. The correlations indicate that a lower maternal age is associated with higher behavior problem scores.

Due to incomplete data, 70 ACBL-questionnaires, completed by the teachers, were evaluated (low maternal age: N = 41; high maternal age: N = 29). Regarding the ACBL a significant negative correlation was found between maternal age and *aggression*, r = -.28, p < .05. In addition, a significant positive correlation between *CBCL-aggression* (rated by parents) and *ACBL-aggression* (rated by teachers) was found, r = .27, p < .05; N = 70. A summary of the means and SD's is depicted in Table 1.

3. Discussion

The present study shows an inverse relationship between maternal age and behavior-problem ratings, in particular the externalizing behavioral categories aggression and delinquency. The younger the age of the mother at delivery, the higher the behavior problems rated by parents and teachers. The ratings of the parents indicate that children in the low maternal age group are judged as more externalizing, aggressive and delinquent. As the scales were completed by the mothers, one could argue that this effect can be attributed to a difference between young and older mothers in the way they complete this type of question-naire. From such a point of view younger mothers would be expected to be more insecure than older ones about their role as a mother and consequently more worried about the behavior and cognitive functioning of their child. However, the children whose mothers were young at their birth were rated as more aggressive by *both* their mothers and teachers. Both the significant negative correlation between maternal age and child aggression rated by the teachers and the positive correlation between aggression scores from parents (CBCL) and teachers (ACBL) support the idea that parents and teachers give similar ratings for aggressive child behavior. Thus, the higher ratings for aggressive and delinquent behavior of children whose mothers were young at their birth cannot be explained by the response strategy of their parents. Also, as father's educational level was not different between the two maternal age groups and the influence of parity was controlled for, the difference in ratings for aggression and delinquency cannot be explained by educational level or parity. Furthermore, it is unlikely that the maternal age effect is associated with the difference between teenage versus non-teenage mothers, because our sample consisted of mothers aged between 22 and 38 years at the time of delivery. As authoritarian attitudes in adult parents may negatively influence child behavior, a decrease in authoritarian attitude with increasing age may account for the present results. In addition, the reduced problem behavior in children of older mothers may be accounted for by the notion that older mothers are emotionally prepared for motherhood and, as a consequence, exhibit more parenting skills.

The results of the present study suggest the existence of continuously decreasing problem behavior with increasing maternal age. However, it is important to note that the scores of all children on the CBCL were below the clinical problem range and that the problem scores of the children in the young maternal age group do not indicate the existence of clinically relevant problem behavior. Nonetheless, the finding that aggression and delinquency are inversely associated with maternal age, possibly mediated by an authoritarian attitude or a lack of parenting skills of the mother, may have important implications for the prevention of the development of criminal behavior. Early detection of aggressive tendencies may prevent the occurrence of the consistent (Hawkins et al., 1998) relationship between clinical aggressiveness in childhood and later violent or criminal behavior. Clearly, it is important to prevent the development of clinically relevant problem behavior that eventually may turn into criminal behavior.

References

Achenbach, T.M. (1991). *Manual for the Child Behavior Checklist and 1991 Profile*. Burlington, VT: University of Vermont.

Broman, S.H. (1981). Longterm development of children born to teenagers. In K.G. Scott, T. Field, & E.G. Robertson (Eds.), *Teenage parents and their offspring* (pp.195-226). New York: Grune & Stratton.

Camp, B.W., Swift, W.J., & Swift, E.W. (1982). Authoritarian parental attitudes and cognitive functioning in preschool children. *Psychological Reports, 50*, 1023-1026.

De Jong, P.F. (1995). Validity of the Amsterdam Child Behavior Checklist: A short rating scale for children. *Psychological Reports, 77*, 1139-1144.

Emmerich, W. (1969). The parental role: A functional-cognitive approach. *Monographs of the Society for Research in Child Development, 34*, 1-71.

Ernhart, C.B., & Loevinger, J. (1969). Authoritarian family ideology: A measure, its correlates and its robustness. *Multivariate Behavioral Research Monographs, 74*, 1-82, 1969.

Fraser, A.M, Brockert, J.E, & Ward, R.H. (1995). Association of young maternal age with adverse reproductive outcomes. *New England Journal of Medicine, 332*, 1113-1117.

Furstenberg, F.F., Brooks-Gunn, J., & Chase-Lansdale, L. (1989). Teenage pregnancy and childbearing. *American Psychologist, 44*, 313-320.

Furstenberg, F.F., Brooks-Gunn, J., & Morgan, P. (1987). Adolescent mothers and their children in later life. *Family Planning Perspectives, 19*, 142-151.

Geronimus, A.T, Korenman, S., & Hillemeier, M.M. (1994). Does young maternal age adversely affect child development? Evidence from cousin comparisons in the United States. *Population and Development Review, 20*, 585-609.

Hawkins, J.D, Herrenkohl, T., Farrington, D.P., Brewer, D., Catalano, R.F., & Harachi, T. (1998). A review of predictors of youth violence. In R. Loeber & D.P. Farrington (Eds.), *Serious & violent juvenile offenders* (pp. 106-146). London: Sage.

McAnarney, E.R., Lawrence, R.A., Ricciuti, H.N, Polley, J., & Szilagyi, M. (1986). Interactions of adolescent mothers and their 1-year-old children. *Pediatrics, 78*, 585-590.

Roosa, M.W., Fitzgerald, H.E., & Carlson, N.A. (1982). Teenage and older mothers and their infants: A descriptive comparison. *Adolescence, 17*, 1-17.

Verhulst, F.C, Koot, J.M., Akkerhuis, G.W., & Veerman, J.W. (1990). *Praktische handleiding voor de CBCL*. Assen: van Gorcum.

II

Assessment Issues

Multi-Problem Violent Youth
R.R. Corrado et al. (Eds.)
IOS Press, 2002

Review of Clinical Assessment Strategies and Instruments for Adolescent Offenders

Marc Le Blanc

This chapter reviews the strategies for the clinical assessment of juvenile offenders: process-based, classification-based, and risk/need based. A number of instruments are describe and analyzed using seven criteria: preliminary evaluation, ad hoc instruments, case study, rating of risks/needs, multiaxial assessment, and adolescent based. Finally, some issues are discussed relative to the clinical assessment of multiproblem adolescents and serious and violent juvenile offenders.

In this chapter, we will discuss the following question: Can criminology help policy, decision-makers and clinicians select appropriate assessment strategies and construct accurate instruments for adolescent offenders? We answer this question through responses to subquestions. What does criminology know and suggest about clinical assessment? What are the available assessment strategies in the juvenile justice system? Can a clinical instrument be accurate and valid? What behavioral, social, biological, and psychological factors should be consider while assessing? Do we need different assessment instruments for different age, gender, and ethnic or racial groups, types of programs, decision points in the juvenile justice system? We will answer these questions along a review of the main strategies and instruments for the clinical assessment of juvenile offenders. We will also present the instrument that we constructed over the last 25 years. Finally, we will comment on various issues.

Definition and Purpose of Assessment

Assessment occupies a central role in behavioral sciences. In criminology, classifications of offenders have been used in corrections since the middle of the nineteenth century (Barnes & Teeters, 1945) and they are routinely employed nowadays (Brennan, 1987; Glaser, 1987; Sechrest, 1987). The screening for potential juvenile delinquents is also a long-standing subject since the initial efforts of Binet and Simon (1907) in France. Decision-making devices for the juvenile justice system have been developed more recently (LeBlanc, 1998; Wiebush, Baird, Krisberg, & Onek, 1995). Finally, Pinatel (1963) indicated that Lombroso (1890), at the Congrès Pénitenciaire International de Saint-Petersburg in 1890, advocated the clinical assessment of delinquents. At that time it was defined as medical and psychological, while Garofalo, in 1891, suggested the necessity of the social inquiry.

Types of Assessment

In criminology, decision-making, according to Gottfredson (1987), is customary in criminal justice practice. In the juvenile justice system, there are numerous decisions, such as to arrest, prosecute, adjudicate, grant parole, and select a level of security or a treatment program. Some of these decision points regulate the flow, such as prosecution, while others involve programming decisions, such as a correctional classification, either for reasons of security or type of treatment. For most of the decision points, the factors that influence decisions are well documented in the literature according to Gottfredson and Gottfredson (1985). When intervention programming is in question, our knowledge about which program is most efficient for which type of offenders is insufficient according to Palmer (1992). Le Blanc (1998) review shows that what is important in discussing decision-making is the accuracy of the decision or the appropriateness of the treatment program or the security level selected. The particular target of serious offenders may impose certain constraints to the construction of a decision-making instrument. In addition, decision-making implies some prediction of future behavior. In consequence, a target such as serious offenders imposes certain conditions to prediction according to Le Blanc (1998). Particularly important is the question of the very low base rate of these offenders in the population and of their relatively low base rate at the entry point in the juvenile justice system. What we must keep in mind is that the identification of offenders does automatically imply the prediction of possible future offenses. This is an assessment of risks.

Clinical assessment of offenders was lengthily discussed in Europe according to Pinatel (1963). This type of assessment of offenders was defined as a medical, psychological, and social assessment. It was progressively introduced in correction, courts and, in non common law countries, it became, and still is, part of criminal procedure codes. This assessment involves direct observation and the use of instruments. It has three components, a diagnostic of dangerousness, a social prognostic, and a treatment plan. The diagnostic of dangerousness involves an investigation in two directions, the criminal personality and the social adaptation of the offender. The social prognostic is composed of an evaluation of the offender potential of recidivism and possibility of social integration. The treatment plan supposes the choice of a measure or program and an individual treatment plan.

There is confusion about the difference between decision-making and clinical assessments. These two categories of assessment are sometimes confounded because there is a fashion, in criminology, that consists in calling these two kinds of evaluation risk assessment. For example, Hoge and Andrews (1996a) classify under the heading of broad risk instruments the Wisconsin juvenile probation and aftercare assessment risk scale, a decision-making device, and the Youth Level of Service Inventory, an in-depth clinical instrument to assess risk/needs. In this chapter, we will accept that the distinction between these types of assessments is a question of depth, this independently of the question of their specific purpose. In the case risk and need devices for decision-making purposes, the assessment is short in terms of time demanded of the assessor, limited in terms of amount of information and number of informants used, and specialized to a decision point in the juvenile justice system in terms of the nature of the indicators employed. In the case of clinical assessments, the evaluation is detailed in terms of the quantity of *information* gathered, the time devoted to this operation, the number of sources of information, the contexts and informants considered, and, most often, because it is followed by a case conference.

From a comparative point of view, we should note some differences of interest between European, western as well as eastern, and North-American countries. Early on in US criminal justice, risk and need assessment, starting with the Burgess prediction table in 1928, became a method to structure and standardize decisions in the juvenile justice system, as in some other common law countries. There were no central institutions specialized in the training of police officers, judges, social workers, or correctional officers or even an institution, like in France, responsible for the ongoing training of all juvenile justice personnel. In consequence, instruments, rather then the laws or rules, became an alternative way to structure decision-making. In continental Europe, there is another reason why there was less interest for empirical decision-making assessment, procedural codes exist in these countries and they greatly limit the discretion of the agents of the juvenile justice system. Concerning clinical assessment, European criminology has a long tradition of evaluation of offenders going back to Lombroso and Garofalo. These medical, psychological, and social assessment procedures have been integrated in the criminal code procedural laws and in juvenile delinquency laws for many decades. This difference between European and North-American countries is very well illustrated in Ferracuti and Wolfgang (1983) book on criminological diagnosis in various countries.

Criteria of the Evaluation of Assessment Instruments

The fields of prediction and classification have attained a certain level of maturity in criminology according to the Farrington and Tarling (1985) and Gottfredson and Tonry (1987) books. Since the expertise of this literature is not always used in the domain of clinical assessment, it appears essential to identify criteria that should be used to evaluate instruments. The following seven criteria have been retained for the evaluation of clinical devices: comprehensivity and parsimony; a theoretically and empirically pertinent set of risk factors; an adequate base rate; the estimation of the reliability of risk factors and outcome; the selection of an appropriate method for combining risk factors; the measurement of the predictive accuracy; and the validation of the device. They can constitute guidelines for the development of new instruments. Since we discussed the last five more technical criteria in a recent publication (LeBlanc, 1998), we will limit ourselves to the first two in this chapter.

Comprehensivity and parsimony.   Almost every behavioral scientist would agree that the causes of juvenile offending are multidimensional. For example, they can be biological, social (family, school, peers, and so on), and psychological. For each dimension, the potential causes are numerous. For example, the family factors that can encourage juvenile offending compose a complex system: social and economic disadvantage of the family; parental conflict and deviance; attachment and involvement of family members; social constraints such as rules, discipline, and supervision. The complexity of the causes of juvenile offending imposes that instruments consider a variety of factors. However, the assessment has to be parsimonious. A clinical evaluation cannot consider the whole range of causal factors. Such an assessment would be too costly, in terms of time, money, and psychological energy, for the juvenile delinquent and the clinician alike. We have to recognize that there is no evident solution between comprehensivity and parsimony. However, we can propose three guidelines. First, decision-making devices should be shorter and less detailed then clinical instruments. Second, all instruments should include factors representing different categories of causes of juvenile offending such as past behavior, family, school, peers, and others dimensions according to the target and the nature of the instrument. Third, an instrument should limit itself to the most active factors according to empirical research.

A theoretically and empirically pertinent set of indicators. In the field of criminology, the choice of risk factors was traditionally determined by the availability of the information in case records and, consequently, only one source of information was used. Instead, Farrington and Tarling (1985) and Gottfredson (1987) concur in recommending the selection of risk factors on theoretical grounds. There is another method of selection of indicators. This strategy involves the use of a meta-analysis to identify the most significant risk factors; Lipsey and Derzon (1998) propose a list of such factors. According to Le Blanc

(1998), there are many advantages and disadvantages to each of these strategies for the selection of risk factors. Independently of the adoption of either a strictly empirical strategy or a combined theoretical and empirical strategy, the question of the nature and the sources of information will have to be addressed. Risk studies, as shown by Lorion, Tolan, and Wahler (1987), are in two categories. First, there is the behavioral pattern strategy that focuses on aggressive and antisocial behaviors that preceded offending. This research tradition identifies strong behavioral risk factors such as early problem behavior and age of onset. Second, there is the psychosocial strategy that seeks a large set of individual and social indicators. In this research tradition, there is no consensus as to the appropriate risk factors because studies always focus on a specific set of indicators rather than on a comprehensive set. As for the sources of information, Farrington and Tarling (1985) cite studies that indicate that the use of case record risk factors is improved by home background data and personality tests. In the domain of the identification of potential offenders for prevention, studies from Loeber, Dishion and Patterson (1984) to Charlebois, Le Blanc, Gagnon, Larivée (1994), show that the use of multiple informants from multiple settings is a more efficient strategy than the selection of only one source of information. In sum, the selection of active risk factors should involve multiple informants from multiple settings and it should rest on solid empirical evidence, as well as theoretical significance. Decision-making instruments require a very careful selection of risk factors because they use a very limited number of them, generally less then twenty. This process is less important for clinical instrument because they rest on a large number of risk factors that are confirming each other.

 Validation of the assessment device. We should distinguish three categories of validation: psychometric, accuracy, and practical. Psychometric validation takes the form of the concurrent, discriminant, and predictive methods to assess the validity of scales. Validation is also an empirical procedure used to obtain an estimate of the accuracy of the prediction or classification produce by an instrument. This form of validation requires a representative sample and involves cross-validation, while bootstrapping may increase the utility of prediction (LeBlanc, 1998). Practical validation is also recommended because we can expect that the relative importance of risk factors may vary from sample to sample and over time. For example, Wolfgang et al.'s operational definition of a chronic offender born in Philadelphia in 1945 may not be representative of national or other state or city samples for later decades. The same argument can be stated for risk factors. In fact, Farrington and Loeber's paper (1995) compares risk factors in London in the 1960s and in Pittsburgh in the 1980's and there are similarities and differences. The best solution to this problem of validation may be the result of a meta-analysis that has enough studies. With such a data set, different combinations of risk factors could be tested. The replication of the risk factors from one study to the other may be the strongest form of validation that we could expect. Such replication, as advocated by Farrington, Ohlin, and Wilson (1986), should include different geographical areas and times. Le Blanc (1998) concludes that this type of validation is essential for decision-making devices because of the particularities of juvenile justice systems. Such validation may be unnecessary for clinical instruments, but the demonstration has to be made.

1. Review of Clinical Assessments

Clinical assessment has been a controversial issue in criminology for more then a century. We can subdivide this period in two phases. From the 1890s to 1960s, mainly in Europe, the most important question was the strategy and the methodology of assessment, irrespective of the age of the offender. From 1940s on, essentially in the United States, the main question became the development of instruments to assess and classify juvenile offenders. In this section, we will discuss the strategy of the clinical evaluation and we will review some particularly interesting instruments using the criteria outlined earlier on.

1.1 The Strategies for the Clinical Evaluation

Our reading of the development of applied criminology concerning clinical assessment suggest that three strategies are used by agencies in the juvenile justice system: process-based, classification-based, and risk/need based. These strategies are not in conflict, as we will see, they are complementary because they emphasize one or more aspects of the clinical assessment—the method, the diagnostic, the prognostic, or the treatment plan.

The process-based strategy. In 1890, Lombroso (1890) argued for the necessity of a medical and psychological investigation of the criminal for the management of correctional institutions and his treatment. Garofalo (1891) added that a social inquiry was also essential. Some years later Kinberg (1911) added that it should be compulsory in the justice system. Clinical criminology had received its basic parameters that are still debated today. Pinatel (1963) shows how clinical assessment was gradually implemented in courts and corrections in various countries, and how it was integrated in the codes of criminal procedure of many European countries. Pinatel then defines the process of the clinical assessment in the following way.

The clinical assessment should involve four basic methods with their particular instruments: the social inquiry, the medical examination, the psychiatric, and the psychological investigations. It could also involve some direct observations and the use of complementary data gatherings, for example biological tests for sexual offenders or, now, urine tests for drug abusers. Then a multidisciplinary case conference should arrive at a diagnostic, a prognostic, and a treatment plan. There are very few data on the nature and content of the clinical assessment of juvenile offenders. Based on the situation in Quebec and our impressionistic information about other provinces in Canada and other countries, I would tentatively state that contemporary applied criminology relies nearly exclusively on various forms of the social inquiry; the psychological examination is most often optional and reserved for some specific cases; the medical and psychiatric investigations are much less frequent then the psychological evaluations. To our knowledge, there is only one data set that is pertinent to this question. Cronin (1996) surveyed 300 juvenile justice agencies in the United States—only 20 agencies in six states declared an assessment program and 9 of these were operational and 3 in a planning phase. Only six sites had an in-depth clinical assessment. Some had psychological measures and urinalysis for drug use. Some provided a case management system, by there were no indication of a case conference in his report. This situation of the clinical assessment is probably not different in other country; it is very similar in Quebec concerning the presentence report and other assessments in the juvenile justice system (Jasmin, 1995). Compared with the expectations of early criminologists, the contemporary situation of the clinical assessment of juvenile offenders is disappointing. The methodology of such assessment has not progressed much over the last century.

Pinatel (1963) was even more idealistic; he recommended a multidisciplinary case conference to arrive at a diagnostic, a prognostic, and a treatment plan. The diagnostic should involve an evaluation of the offender criminal capacity, through the assessment of his personality, and social adaptation utilizing a social inquiry about his criminal conduct and his integration into social institutions such as school. The diagnostic should be formulated in terms of levels of criminal capacity and social adaptation and then synthesized in terms of degrees of dangerousness. The diagnostic becomes a prognostic when it is associated with a probability of recidivism. With the diagnostic and the prognostic, the case conference would select a program or type of placement, a treatment method and schedule that are appropriate for each offender. Favard (1991) proposed a computer-assisted method to arrive at a synthesis of the case conference, particularly when it is multidisciplinary.

The assessment process describe by Pinatel is compatible with recent advances in that domain in the United States. The Office of Juvenile Justice and Delinquency Prevention developed the concept of Community Assessment Center (1995). This center should offer: a single point of entry in the juvenile justice system; provide an immediate and comprehensive assessment in a community-based setting; follow the requirements of the Juvenile Justice and Delinquency Prevention Act; maintain a management information system; use an integrated case management system, and provide input to the policy making process.

The classification-based strategy. Classifications of juvenile offenders were developed mainly after the Second World War in the United States. Three main psychological classifications of juvenile offenders exist. They have a theoretical rationale and some empirical base, at least for their construction, but sometimes also for their validation, and two multidimensional typologies have been proposed. Leaving aside the Diagnostic and Statistical Manual of the American Psychological Association (see Hoge & Andrews, 1996a for critique in relation to juvenile delinquency), the psychological classifications are derived from the psychodynamic, the developmental, and the psychometric perspectives. The psychodynamic point of view is represented by Hewitt and Jenkins' (1946) proposal of three categories of juvenile offenders (overinhibited, underinhibited and pseudosocial). Quay's (1971, 1987), using a questionnaire and multivariate statistical analyses, concluded that there is a fourth type that extended Hewitt and Jenkins' classification. The developmental perspective adopts the ego development paradigm and proposes the interpersonal maturity classification (Sullivan, Grant, & Grant, 1957, Warren, 1966, 1983) that is operationalized psychometrically with the Jesness Inventory (1983) (see Hoge & Andrews, 1996a review of problems). Finally, the psychometric perspective is represented by the MMPI route with the Megargee and Bohn (1979) categorization of youthful offenders (see Hoge & Andrews, 1996 for a review of deficiencies). There are many types in each of the major classifications that have similar characteristics according to Blackburn's (1993) comparison.

Two multidimensional classifications are relevant to delinquency, but also to serious offending. They are the Elliott (Dunford & Elliott, 1984, Elliott, 1994) and Le Blanc (Fréchette & LeBlanc, 1987; LeBlanc & Fréchette, 1989; LeBlanc & Kaspy, 1998) classifications. These classifications use self-reported offending as a starting point and propose a serious offender category. They draw, for each type of offenders, the social profile, in Elliott's typology with a national sample of American adolescents, and the social and psychological profile, in Le Blanc's typology using an adjudicated urban sample of adolescents. In addition, they validate their typology using official delinquency. Le Blanc (1995a) shows that the distribution of adolescents across the three main types of juvenile offending careers (common, temporary, and chronic delinquency) is comparable and the characteristics of the offenders are generally replicated from one study to the other. The Fréchette and Le Blanc (1987) classification is pertinent for the diagnostic because it is used to propose differential treatment strategies and programs (LeBlanc, Dionne, Proulx, Grégoire, & Trudeau-LeBlanc, 1998) and targeted secondary prevention programs (LeBlanc, 1995a).

In sum, the classification-based strategy targets the diagnostic and the treatment plan and some components of the clinical assessment process. The question of the prognostic is not directly addressed by these typologies. This strategy is more parsimonious than the process-based strategy that favors four domains of investigation and a multidisciplinary case conference. However, this method lacks comprehensivity because it is generally limited to only one domain of variables, personality. We can also characterize this strategy as statistical because standardized measures are used to assess juvenile offenders, and they are sometimes classified with the help of some statistical function. In addition, these classifications are theoretically and empirically based which includes some evaluation of reliability and validity even if the estimation of the predictive accuracy is left out. Purely judgmental systems, like the I-Level Interview (Warren, 1966), are increasingly left aside because of their cost in terms of manpower and their low reliability. In contrast, psychometric forms of classifications are used by some agencies of the juvenile justice system, particularly treatment agencies.

The risk/need-based strategy. This in-depth clinical assessment strategy was initially developed for adult offenders and required 10 years of research and clinical validation (Andrews & Bonta, 1995). An adolescent version of the Level of Service Inventory and Case Management Inventory has also been proposed (Hoge & Andrews, 1994, 1999) and applied in Ontario (MacLeod, 1995). A similar strategy is proposed for boys under 12 with the EARL-20B (Augemeri, Webster, Koegl, & Levene, 1998). We will present the content of these instruments in the next section. This strategy involves four sources of data: an interview with the offender, the review of the file, test scores, and other information about the offender. The offender is then rated on 42 risk/need items regrouped under 8 subscales and a global score is derived. Mitigating circumstances are considered. Finally, case management is supported through indications of contact level and the request for stating goals and means of the treatment plan. This strategy incorporates many components of the ideal clinical assessment process. The many data sources used could involve professionals from various disciplines. This strategy is comprehensive by the range of variables considered (including offense history, family circumstances, education, peer relations, substance abuse, leisure and recreation, personality and behavior, and attitudes). However, it remains parsimonious because the indicators are limited in number, and their selection is research based and theoretically grounded. This assessment strategy is helpful at the level of the diagnostic, the prognosis, and the treatment plan.

In this section, we have reviewed three strategies of clinical assessment used by agencies in the juvenile justice system: process-based, classification-based, and risk/need based. These strategies are not in conflict; they are complementary because they emphasize one or more aspects of the in-depth assessment, the method, the diagnostic, the prognostic, or the treatment plan. We will now turn to the most interesting instruments for the in-depth assessment of juvenile offenders. The distinction between a strategy and an instrument is the following. Strategies incorporate many data sources and many aspects of the assessment process, while instruments are appropriate for one data source even if they can help for the diagnostic, the prognostic, and the treatment plan.

1.2 Instruments for Clinical Assessment

Cronin (1996) reviewed US assessment centers in juvenile justice systems and found that only three of seven states used a validated instrument. In all other agencies of the juvenile justice systems, we can imagine that there must be some clinical assessment of juvenile offenders that take one of the following forms: preliminary evaluation, ad hoc instruments, case study, rating of risks/needs, multiaxial assessment, and adolescent-based. These forms of in-depth assessment of juvenile offenders are used in Canada and, probably, also in most countries in Europe. From our experience, we can state that the case study is the dominant form of clinical assessment everywhere, while the use of ad hoc instruments comes next. Preliminary evaluation, ratings of risk/need, and adolescent-based instruments are employed in some systems of juvenile of justice. For example, Cronin (1996) found that POSIT, a preliminary assessment instrument, is used in three states. In-depth ratings of risk/need are widely used in Ontario, a Canadian province. Adolescent-based instruments are in used in Philadelphia (ProDES) and Quebec (MASPAQ). More often, all these methods of clinical assessment are employed only in a particular agency in a juvenile justice system. What is most interesting is that while of risk and need decision-making devices were developed during the 1980s, the clinical assessment instruments were constructed during the 1990s as if the first category of instruments were unsatisfactory for clinicians.

Preliminary assessments are increasingly used. One of these, POSIT (Problem Oriented Screening Instruments for Teenagers), is a self-report questionnaire comprised of 139 yes-no items. They were designed to flag potential problems in ten areas: substance use/misuse, physical and mental health status, leisure and recreation, peer relations, educational and vocational status, social skills, family relations, and aggressive behavior/delinquency. This instrument was developed in the domain of drug use (Rahdert, 1991); it is now employed with juvenile offenders. Some preliminary data are published with samples of juvenile offenders and some psychometric properties are calculated for that population (Dembo & Brown, 1994; Dembo, Schmeidler, Borden, & Turner, 1996; Dembo, 1996). These instruments could easily replace the risks and needs decision-making devices that are based on practitioners judgements on the same problem areas. However, they do not qualify as an in-depth clinical assessment. They can be used as a case management instrument at the entry in the juvenile justice system or to decide which offender will be the object of an in-depth assessment with various other instruments.

The ad hoc clinical assessment takes the following form. An agency or a practitioner selects some instruments according to the type of cases he encounters, particular individual situations, a professional training, the philosophy of an agency, and many other criteria. As showed by Cronin (1996), there is a variety of such mix of instruments in his review of assessment centers in seven states. Hoge and Andrews (1996a) book lends credibility to that type of clinical assessment. Their book is such a tool kit. They review numerous instruments concerning the assessment of aptitudes and achievements, personality, attitudes, and behaviors, and the environment of the adolescent (family functioning, school performance and adjustment, and peer group association). For example, they identify eleven instruments concerning the assessment of the family, six about family functioning and five in relation to parental management. Then they recommend two instruments, one for each of these dimensions, because of their psychometric qualities and their pertinence. The difficulty with this tool kit method of clinical assessment is the lack of systematization. The clinical assessment of a juvenile offender may be conducted with different instruments in various agencies, may vary between practitioners in the same agency, and may change for particular juvenile offenders even if the same practitioner in a particular agency evaluates them. In addition, this type of in-depth assessment is far from parsimonious. Each of the instruments reviewed by Hoge and Andrews (1996a) can require from fifteen minutes to an hour or more. In consequence, to do an assessment of various aspects of an adolescent offender may represent a long process of data gathering, without calculating the time to compile, analyze, and interpret the data from many instruments. In addition, such tool kit is not very practical because it does not favor the communication between partners in a criminal justice system. For example, if the presentence report is not standardize and if it does not always use the same instruments, the legal personnel of the court or the treatment personnel in correctional institutions will not be able to understand the content of the data from every instrument.

The case study is the dominant form of clinical assessment in all juvenile justice systems. It has been mark out, particularly for presentence investigation reports (Clear, Clear, & Burrell, 1989). It is impossible to report on the various forms of case study because, to our knowledge, there is no inventory of its different operationalizations. We can reasonably say that there is no consensus on the content of the case study in the professional literature, even if the dominant headings are standard: offense, family, social, and developmental history, and attitudes/personality. Juvenile delinquency laws or the rules of practice in some agencies sometimes define the content of these case studies. For example, in Canada the Young Offender Law (1995) specifies the table of content of the presentence report and in Quebec the practitioners have a reference manual with the detailed content that is expected from the case studies at various points in the juvenile justice system under that law (1993). An example of a case study method can be taken from MST (Multisystemic Treatment of Antisocial Behavior of Children and Adolescents; Henggeler, Schoenwald, Borduin, Rowland, & Cunningham, 1998). They use a social-ecological model of behavior to determine how each factor, individually or in combination, increase or decrease the problem behaviors of the adolescent. These factors are assessed in the different subsystems, family, school, peers, and individual. This evaluation targets specific problems in real-world context. The information comes from different sources in each subsystems and from observations and interviews. The assessment is continuous but the initial phase serves to formulate hypothesis to be tested about the proximal and concrete factors that are responsible for the behavior problems. This stage is done in consultation with the supervisor and the treatment team. The assessment of the family functioning takes the form of a genogram for the description of the systemic structure and interaction patterns and the parenting style, the marital functioning, and the subsystem changes are evaluated by direct questioning of family members, observing family interactions, and asking family members to monitor and record particular behavior. The same means are used for the evaluation of the relations of the adolescents with peers. Concerning academic performance, intelligence, achievement, and learning disabilities testing, if necessary, complement these means of assessment in addition of contacts with the school of the adolescent. Psychological tests are rarely used, but the practitioner performs a cognitive-behavioral analysis. For each subsystem, there is a comprehensive list of elements to be assessed. When the practitioner suspects some psychiatric or other disorders in family members, he can ask for a specific evaluation. From a process-based strategy, this type of case study has all the main components. The MST assessment is also theoretically and empirically grounded. However, it lacks standardization, like the ad hoc clinical assessment, because it is individualize. In addition, there is no preoccupation with the reliability, validity (other then face), and the predictive accuracy of the assessment. The structure and standardization of the assessment process can be attained only through the careful training and monitoring of the practitioners. This is not the case in most agencies in the juvenile justice system.

Multiaxial assessment has been proposed by Achenbach and McConaughy (1987) for adolescent boys and girls with conduct problems, including delinquency. The evaluation involves five axes. The first axis is composed of parents reports based on the CBCL and a parent interview on the developmental history of the adolescent. The second axis involves the school, the CBCL teacher's report form and school records. The third axis refers to a cognitive assessment, that is intelligence, achievement, and speech and language tests. The fourth axis comprise the physical exam, height and weight, medical and neurological exams. Finally, the fifth axis involves a direct assessment of the adolescent with the CBCL, direct observations, a semi structured clinical interview, and self-concept and personality measures. This form of assessment is partly parsimonious particularly because of axes three and four. It would be much more if it was only based on the CBCL gathered from various informants. The reliability and validity of the CBCL is well established in many countries, but not for the other components of the assessment. To our knowledge, the multiaxial CBCL assessment is not use in the criminal justice system.

Clinical assessment by an in-depth rating of risks/needs is a recent innovation for juvenile offenders. The Youth Level of Service/Case Management Inventory developed by Hoge and Andrews (1994, 1999) is, to our knowledge, to most sophisticated and psychometrically sound such instrument and the case management inventory has been gradually implemented in Ontario juvenile justice system, the largest province in Canada (MacLeod, 1995; Loza, & Simourd, 1994). This clinical assessment was originally developed for adults' (Andrews, & Bonta, 1995) with psychometric properties well documented (Loza & Simourd, 1994). It has been adapted to juvenile offenders after reviews of the literature (Andrews, Hoge, & Leschied, 1992; Leschied, Andrews, & Hoge, 1992) and a series of empirical studies on the risks and needs for Ontario's delinquents (Hoge, Andrews, & Leschied, 1994, Hoge, Andrews, & Leschied, 1995, Hoge, Andrews, & Leschied, 1996). The psychometric properties are estimated such as internal consistency and inter-rater reliability and predictive, concurrent, and construct forms of validity (Andrews, Robinson, & Balla, 1986; Costigan, 1999; Jung, & Rawana, 1999; Schnidt & Hoge, 1999; Simourd, Hoge, & Andrews, 1991). The ethnicity (Native or not) and the gender are inconsequential concerning the validity of the instrument (Jung & Rawana, 1999).

This assessment kit is designed for frontline workers in the juvenile justice system and involves six steps. The first step is an assessment of risk/needs. This evaluation starts with a review of file data, then an interview with the adolescent is conducted, and test scores are consulted. There are eight categories of risk/needs composed of 42 items to be rated: offense history (prior and current offenses/dispositions), family circumstances/parenting, education/employment, peer relations, substance abuse, leisure and recreation, personality and behavior, and attitudes/orientation. The assessor rates each risk/need item and there is a section to indicate strengths. The second step is the calculation of each subscale risk/need scores and a summary score that can be compared to a normative sample of adjudicated adolescents (the normative sample is rather small compared to the size and diversity of the Ontario juvenile justice system and there is no possibility to compare juvenile offenders to the population of adolescents). An empirically based typology has been developed according to the levels of risk/needs (Simourd, Hoge, Andrews, & Leschied, 1994). The third step is the consideration of a variety of particular factors that are relevant for the adolescent, factors that may not be directly linked to his criminal activity but that may be important for the case planning. The fourth step is the confirmation of the risk/needs, particularly current risk/needs, based on the estimate of the assessor or a reassessment if the adolescent is at another decision point in the juvenile justice system. The fifth step concerns the choice of the placement or the definition of the contact level; this is the case management assessment. This component of the instrument is composed of four categories of risk and 12 items: offenses, problematic behaviors, personal characteristics, and administrative concerns. There are preliminary data on the distribution of these risk factors for probation and open and secure custody (Wormith, 1995). Finally, the sixth step supports the case management plan, the statement of the goals of the intervention and the identification of the means of achieving those goals. There is a comparable instrument, in nature and psychometric properties, to assess the risk/needs of children under 12, the EARL-20B for boy's (Augemeri, Webster, Koegl, & Levene, 1998) and one in development for girls.

If we turn to the criteria that we selected for the evaluation of assessment strategies and instruments, we can comment in the following way on this instrument. The YLS/CMI is comprehensive and parsimonious. This instrument is comprehensive because it considers a wide range of factors that are empirically associated with juvenile offending. The only missing set of factors is biological. This instrument is parsimonious because it assessed a limited number of risks and needs and because the six steps of the procedure can be climbed in a few hours (this is our estimation since we did not read any information about the duration of the assessment process in the documents that we consulted). The rating of risk/needs satisfies another criterion, as it reflects the current theoretical and empirical literature. In addition, this instrument rests on original research to identify risks/needs in the population of juvenile offenders. An additional criterion is respected by this rating of risks/needs; there are preliminary data, to use Hoge and Andrews (1996a) own qualification, on reliability and validity. The YLS/CMI uses the simplest method of combining the indicators, the additive method without weights. Over and above these strengths of this in-depth rating of risk/needs, we have to recognize that this instrument displays practical advantages. The rating provides direct information on juvenile offenders and intervention. Frontline workers can use it easily when trained. This instrument standardizes the data collection. It provides a common language in the juvenile justice system. The YLS/CMI coordinates the components of the juvenile justice system by the possibility of reassessment at each of his decision points.

The last form of clinical assessment is the *adolescent-based* instrument. In our inventory of instruments, we identified two such instruments, the ProDES (the Program Development and Evaluation System) and the MASPAQ (Measuring Adolescent Social and Personal Adaptation in Quebec). ProDES, developed by Harris and Jones (1996a, 1996b) is a data system for juveniles entering programs in Philadelphia that start with the court disposition and covers intake, discharge, and follow-up. The intake evaluation involves socio-demographic information, prior and current offenses, risks/needs assessment, educational and vocational data, and self-reported scales from various instruments on self-esteem, values, school and family bonding, and drug and alcohol use. There are only descriptive data and no psychometric information on this intake assessment kit. In consequence, we will review an instrument that provides such data.

The MASPAQ (Measures of Adolescent Social and Personal Adaptation; LeBlanc, 1996) was developed during 25 years of longitudinal and evaluation research. Initially, the MASPAQ was a nine hours structured and semi-structured interview that included deviant behavior, social experience, and half a dozen psychological tests with adjudicated adolescents. In addition, a 2.5 hour self-administered questionnaire was given to a representative sample of a few thousand adolescents. Subsamples of these subjects have been reinterviewed until 40 years of age. During the 1980s and the 1990s, other similar samples have been recruited for longitudinal or treatment evaluation studies. The recent version of the MASPAQ requires a two hour personal interview. This instrument applies to adolescents, necessarily to juvenile offenders, between 10 and 18, whatever their gender and ethnic group. The use of the MASPAQ rest on a fundamental assumption, that the most important information for decision-making and intervention is the adolescent's own perception of his life and situation. Even if this source of data involves distortions of the reality, memory deficiencies, personal interpretations of these experiences, confabulations, and so on, it is the most basic and necessary material for a clinical intervention when adolescents are concerned. During the intervention process, after an initial assessment with the MASPAQ, all the data are to be corroborated, and the adolescent may have to develop, if necessary, a more adequate perception of certain realities.

The interview kit is composed of a card sorting phase concerning current and past offending and problem behaviors (drug use, sexual promiscuity, family rebellion, school maladjustment, and other risk behaviors), and a structured interview about family and school experiences, peer relations, routine activities, attitudes and values toward deviance and justice. In addition, psychological tests are administered (Jesness, Eysenck, Beck). A computer program compiles these data, calculates various information, for example, probabilities of dropping out of school, type of school drop-out, being part of a particular type of family or personality type (Interpersonal maturity), and produces a list of results and graphs. It is also possible to compare not only the results of an offender with normative data from the general population of adolescents, but also with data from other populations such as adjudicated youths, violent adolescents, and drug abusers. The scales were constructed and validated with more then 8,000 adolescents, mostly representative of the general youth population and some adjudicated youth. The samples were recruited during the 1970s, 1980s and 1990s in Canada. This instrument is also being validated in Spain, France, and North Africa. Le Blanc (1996) reports the reliability for each scale globally, by sexes, ages, decades, and subgroups of the construction sample. Most of the scales display a 12-item standardized alpha above 0.80. There is a validated lie scale and there are good indications of validity when comparing the parent and the adolescent descriptions of the family functioning. Le Blanc (1996) also reports validation data in terms of concurrent (higher correlation's between scales of the same domain, for example family, than with scales from another domain, for example, school), discriminant (known groups of offenders and protected youths), and predictive validity of the scales (official and self-reported juvenile and official adult offending).

This instrument displays several features that we expect from assessment instruments. The large number of domains of the adolescent life that are covered reflects the comprehensiveness of the MASPAQ. The depth of the instrument can be illustrated by the evaluation of the family; this part of the interview uses 113 questions that covers the socio-economic status, the family status, marital relations, parental deviance and characteristics, bonds (attachment, involvement), and constraints (rules and their legitimacy, supervision, methods of discipline). There are 29 scales that characterize family functioning, including 16 scales that are subdivided for the mother and the father. This depth and comprehensiveness are not contradictory, and there is a reasonable level of parsimony. In fact, the interview is two hours in duration, and the analysis does not require much more time. In contrast, case studies of problem youth in Quebec are far longer, averaging 28 hours (Lebon, 1999). In addition, the psychometric qualities of this instrument have been validated far more than all the instruments that we reviewed above. What is more important is the fact that both the theoretical and empirical backgrounds of this instrument are not remote or indirect like most other instruments. The section on offending and problem behavior operationalizes the developmental paradigm (Le Blanc & Loeber, 1998; Loeber & Le Blanc, 1990) and reflects our longitudinal data (Fréchette & Le Blanc, 1987; Le Blanc & Fréchette, 1989; Le Blanc & Kaspy, 1998). The other sections of the interview refer to an integrative social and self control theory that has been published (Le Blanc, 1997a) and was based on a formalization of control theory (Le Blanc & Caplan, 1993). This theoretical model, as well, was validated with data from different decades and samples of representative and adjudicated adolescents (Le Blanc & Biron, 1980; Le Blanc, 1983, 1986, 1997b; Le Blanc, Ouimet, & Tremblay, 1998). Middle range theories about family, school, peers, and constraints also have been developed and tested cross-sectionally and longitudinally (Le Blanc, 1994, 1995b; Le Blanc, McDuff, & Kaspy, 1998; Le Blanc, Vallières, & McDuff, 1992; Le Blanc, Vallières, & McDuff, 1993). Fréchette and Le Blanc (1987) additionally have developed a screening device to distinguish juvenile offender who should be diverted, and a multidimensional classification of offenders that distinguish common, temporary, and persistent delinquents. Accordingly, differential prevention and treatment programs have been proposed (respectively Le Blanc, 1995a and Le Blanc,

Dionne, Proulx, Grégoire, & Trudeau-Le Blanc, 1998).

In this section on instruments for the clinical assessment of juvenile offenders, we reviewed its five forms. The instruments of preliminary assessment were not reviewed because their depth is insufficient. The choice of an instrument for the clinical assessment depends on the main purpose of this operation in the juvenile justice system. If we accept that the objectives of clinical assessment should be to facilitate communication between practitioners in the juvenile justice system, then the ad hoc instruments and the case study are not the most useful methods. They neither foster the systematization or the standardization of the clinical assessment, nor facilitate the interchange of information between agencies of the juvenile justice system. In addition, the psychometric properties of the ad hoc instruments and the case study are very rarely established. Finally, we are left with the in-depth ratings of risks/needs and adolescent-based instruments.

The YLS/CMI is more directly linked to the juvenile offender laws and the operations of the juvenile justice system. The MASPAQ is more comprehensive and, probably, more useful in the application of the individualized treatment plan. These forms of assessment are equally parsimonious in terms of evaluation time. The psychometric properties of the YLS/CMI are less well documented, particularly for population samples of adolescents, than for the MASPAQ. However, both instruments have to be validated in other jurisdictions (this is less the case for the MASPAQ since many of its scales have been widely used in research, for example, the parental supervision scale is adapted from Hirschi (1969). The theoretical and empirical backgrounds of the MASPAQ are more direct and thorough, while it is more remote and less developed for the YLS/MCI. Overall, both instruments have advantages and disadvantages. They reflect a fundamental Canadian philosophical dilemma in juvenile justice, i.e. the priority of the protection of society versus the needs of adolescents. Quebec, historically, is more treatment oriented and more often uses diversion alternatives and less institutional measures. Other provinces in Canada are more justice oriented, and therefore, the offense and prior offending then become the starting point of the administration of juvenile justice. These two instruments reflect these two basic points of view, as the YLS/CMI focuses on risks and needs while the MASPAQ focuses on the treatment plan.

2. Issues

We have reviewed strategies and instruments for the clinical assessment of juvenile offenders. We observed that during the last two decades several instruments have been developed. However, it is impossible to unqualifiedly recommend a particular strategy and instrument for clinical assessment. All the available instruments suffer from deficiencies, particularly in terms of reliability, validity, or predictive accuracy. However, some strategies and instruments constitute important steps toward the development of useful empirical assessment instruments, particularly the YLS/MCI and the MASPAQ. Ultimately, the use of an instrument is dependent on the particular legal system. This context of legal, ethical, and policy restraints have been discussed in more depth elsewhere (Le Blanc, 1998).

In this chapter, we focused on the assessment of adolescent offenders. Adolescence usually includes the period between 12 and 18 years of age, although in some countries it overlaps latency, from 10 to 12, or young adulthood, from 18 to 21. The instruments that we discuss can be applied without restrictions to these age groups. Some of them, like the MASPAQ have been validated for these ages. Since we are targeting juvenile offenders for clinical assessment, our task is simplified, but still not easy. In fact, juvenile offenders are neither necessarily violent offenders nor chronic offenders and there are many juvenile offenders that are drug abusers or multi-problem adolescents but not all of them. In consequence, there is a need to have a common understanding of these categories of adolescents. The adjectives serious and multi-problem refer to many specific legal and research definitions. Legal definitions of seriousness revolve around habitual offender laws. However, juvenile delinquency statutes generally do not provide habitual offender definitions. Because of the absence of common operational definitions across states, we cannot find in the legal definitions of seriousness, violence, or chronicity any clear advice for targeting these types of offenders for the design of a clinical assessment. Research definitions of serious offending vary enormously. However, the analysis of offending is guided by a theoretical paradigm, the criminal career view (Wolfgang, Figlio, & Sellin, 1992) or the developmental criminology perspective (Le Blanc, & Loeber, 1998). In this paradigm, seriousness, chronicity, and violence are complementary descriptors of the nature of offending. The degree of involvement in a criminal career is assessed through the seriousness of the offenses, the frequency and continuity in offending, and the use of violence. In addition, from a measurement point of view, in this paradigm there is no equivalence between definitions of the same construct when researchers use official or self-reported data. Even if all operational definitions are arbitrary statistical decisions in particular universes, some definitions of these constructs are selected more often then others.

Concerning official data, the Wolfgang, Figlio and Sellin (1992) definition of juvenile chronic offenders, those with five arrests or more, is widely referred to in studies of serious offenders. This operational definition of a chronic juvenile offender has been criticized because it does not take into account the exposure time of the offender (Wolfgang, Figlio, & Sellin, 1992). In addition, there have been very few systematic tests of this cut-off point as compared to others. In addition, the data from their 1958 Philadelphia cohort indicates an increase in the proportion of chronic juvenile offenders from 6.3% to 7.5% and in the proportion of crimes they commit, from 52% to 61%. The five or more cutoff point may also be an underestimation. Cohen (1986) estimates, irrespective of age and samples, that active violent offenders commit an average of two to four serious assaults a year and active property offenders perpetrate an average of five to ten crimes for each of the property crimes they are arrested for. In addition, Blumstein, Farrington and Moitra (1985), in the only test of the Wolfgang cutoff point, show that in four cohorts the best dividing point between chronic and other offenders is six arrests rather than five. Self-reported studies, for their part, tend to use only one construct to define serious, violent or chronic juvenile offenders. In general, they prefer a scale of variety of crimes or of frequency of offenses, or sometimes subscales of violent or serious delinquency. In consequence, the category chronic juvenile offenders is rarely present in the self-reported literature, while the constructs of chronicity, violence, or seriousness, when they are used, are operationalized as in the official data studies. For example, the proportion of chronic juvenile offenders defined by self-reports does not differ from its official measurement. Dunford and Elliott (1984) classify 6% of the population as serious career offenders, while Fréchette and Le Blanc (1987) identify 5% of their representative sample as continuous, frequent and serious offenders.

There is a consensus between criminology and criminal justice on the list of violent crimes. They are the assaults, rapes, robberies, and homicides. These violent crimes are also the most serious crimes. Some other crimes are defined as serious, such as burglary, arson, drug trafficking, motor vehicle theft, and theft over $100. However, the sum of these crimes and the violent crimes represent only around 40% of juvenile arrests if we look at crime statistics (Hamparian, Schuster, Dinitz, & Conrad, 1978). When we look at serious violent offenders, Elliott (1994) reports that the cumulative ever prevalence, to age 18, is nearly 40% of the black males compared to 30% of the white males. Most of them became serious violent offenders before age 16. Without a thorough review of the literature, we can estimate approximately the proportion of chronic juvenile offenders that are violent. Based on official data, violent juvenile delinquents represent between 53% (an estimate based on the Wolfgang et al., 1992 study) and 71% of the chronic delinquents.

These definitions tend to propose two categories of offenders, chronic and others, while chronic and violent offenders overlap. It may be better to define three categories of juvenile offenders. Le Blanc and Fréchette (1989) proposed a developmental typology of adolescent criminal activity using a representative sample of adolescents. The typology was constructed with the frequency, the variety and the seriousness of 28 self-reported offenses across a few waves of data. Before the end of adolescence, 3% of the representative sample of adolescents did not report a criminal offense. Transient delinquency accounted for 69% of the sample with 13% occasional delinquents reporting less than three offenses during a particular time until the end of adolescence, such as vandalism, minor theft or public mischiefs. In that group of transient delinquents, there was 32% of them that reported intermittent offending, a few offenses at different periods during adolescence, mainly minor offenses. There were also 24% of them that admitted less than 20 minor offenses during their whole adolescence. Persistent delinquency accounted for 28% of the representative sample of adolescents; it is composed of 15% of continuous and frequent delinquency and 13% of continuous, frequent and serious offending. Dunford and Elliott (36) developed a typology of career offenders with a national sample of American adolescents that validates this typology. The member of their non-offender's group, who represented 49% of their sample, are the abstainers, the occasional delinquents and the intermittent delinquents, representing 48% of Le Blanc and Fréchette sample. Their group of persistent delinquents represented 51% of their sample, whereas Le Blanc and Fréchette group accounts for 52% of the sample. Finally, their non-career group corresponds to Le Blanc and Fréchette delinquents who show medium and very stable activity, 38% in their sample against 41% in Dunford and Elliott sample. This classification of transient and persistent offenders was renamed adolescent-limited and life-course persistent offenders by Moffitt (Moffitt, 1993, 1997). Since then, this classification was empirically supported, for example by Nagin et al. (1995).

If we turn to the notion of multi-problem adolescents, it is probably unspecific in the adolescent population. Even if we remain in the limited field of deviant behavior, we have to conclude that juvenile offenders display the deviant behavior syndrome. The existence of this syndrome has been often demonstrated with various types of samples and we have showed that it particularly characterizes samples of juvenile offenders over the last decade (Le Blanc & Bouthillier, 2001; Le Blanc & Girard, 1997).

The focus on serious offenders as opposed to other juvenile offenders may have important implications for the assessment strategy and the choice of an instrument. In that context, the assessment strategy will probably need to be multistage because the potential offenders will have to be distinguished from the non-offenders and then the serious offenders will have to be detected among offenders.

3. Conclusion

In this chapter, the most important characteristics of clinical assessment strategies and instruments were identified. They should involve multiple sources of data from multiple domains. In addition, specific instruments should be evaluated in terms of the balance between comprehensivity and parsimony, the reliability and validity of the criterion and risk factors, the appropriateness of the method of combining the risk factors, the predictive accuracy, and the practical utility. We concluded that all instruments display some deficiencies. The most adequate instruments for clinical assessment were the YLS/CMI and the MASPAQ. Other clinical assessment devices display a high face validity, but they lack other important features. Complete instruments still have to be constructed, therefore, there is considerable work still to be done before criminology and psychology can propose definitive clinical instruments for adolescent offenders. Some instruments are promising and they can be recommended for immediate use to policy-makers and practitioners. The more adequate clinical assessment instruments, such as the YLS/CMI and the MASPAQ, still have to be validated in other jurisdictions,. The state of the art of clinical assessment of juvenile offenders is such that the research community can indicate how to develop good instruments, what deficiencies have to be corrected, and which are instruments before are the most useful. Clinical assessment is more challenging when the clients are violent or have multiple problems. No clinical instrument has specific capacities to distinguish these types of adolescents from others, with the exception of chronic delinquents Finally, we must remember that assessment is not always appropriate or necessary. At some points in the juvenile justice system, assessment of offenders may not be pertinent or useful because the flow of adolescents is very low. Similarly, for some juvenile justice measures, a clinical assessment is not needed; for example, imposing 50 hours of community work does not need a comprehensive evaluation of the adolescent. In such a case, the cost of the assessment is higher then the cost of the measure.

References

Achenbach, T.M., & McConaughy, S. H. (1987). *Empirically based assessment of child and adolescent psychopathology.* Newbury Park: Sage.

Andrews D.A., Robinson D., & Balla, M. (1986) Risk principle of case classification and the prevention of residential placements: An outcome evaluation of the Share the Parenting Program. *Journal of Consulting and Clinical Psychology, 54,* 203-207.

Andrews, D.A., & Bonta, J. (1995). *Level of Service Inventory-Revised.* Toronto: Multi-Health Systems.

Andrews, D.A., Hoge, R.D., & Leschied, A. (1992). *A review of the profile, classification, and treatment literature with young offenders: A social and psychological analysis.* Toronto: Ontario Ministry of Community and Social Services.

Augimeri, L.K., Webster, C.D., Koegl, C.J., & Levene, K.S. (1998). *EARL-20B: Early assessment risk list for boys.* Toronto: Earlscourt Child and Family Center.

Barnes, H.E., & Teeters, N.K. (1945). *New horizons in criminology.* New York: Prentice Hall.

Binet, A., & Simon, T. (1907). *Les enfants anormaux : Guide pour l'admission des enfants anormaux dans les classes de perfectionnement.* Paris: Armand Colin.

Blackburn, R. (1993). *The psychology of criminal conduct: Theory, research and practice.* Toronto: Wiley.

Blumstein A., Cohen J., Roth, J.A., & Visher, C.A. (1986). *Criminal career and "career criminals."* Washington, D.C.: National Academy Press.

Blumstein A., Farrington, D.P., & Moitra, S. (1985). Delinquency careers: Innocents, desisters, and persisters. *Crime and Justice: An Annual Review, 7,* 187-219.

Brennan, T. (1987). Classification: An overview of selected methodological issues. *Crime and Justice: An Annual Review, 9,* 201-248.

Charlebois, P., Le Blanc, M., Gagnon, C., & Larivée, S. (1994). Methodological issues in multiple-gating screening procedures for antisocial behaviors in elementary students. *Remedial and Special Education, 15,* 44-55.

Clear, T.R., Clear, V.B., & Burrell, W.D. (1989). *Offender assessment and evaluation: The presentence report.* Cincinnati: Anderson Publishing.

Cohen, J. (1986). Research on criminal careers: Individual frequency rates and offense seriousness. In A. Blumstein, J. Cohen, J.A.Roth, & C.A.Visher (Eds.), *Criminal career and "career criminals"* (Volume I). Washington: National Academy Press.

Costigan, S. (1999). *Critical evaluation of the long-term validity of the risk/need assessment and its young offender typology.* Thunder Bay: Unpublished M.A. thesis, Lakehead University.

Cronin, R.C. (1996). *OJJDP community assessment center fact finding project: Final report.* Washington, D.C., Office of Juvenile Justice and Delinquency Prevention, US Department of Justice.

Dembo, R. (1996). Problems among youths entering the juvenile justice system, their service needs and innovative approaches to address them. *Substance Use and Misuse, 31,* 81-94.

Dembo, R., & Brown, R. (1994). The Hillsborough County Juvenile Assessment Center. *Journal of Child and Adolescent Substance Abuse, 3,* 25-43.

Dembo, R., Schmeidler, J., Borden, P., Turner, G., Sue, C.C., & Manning, D. (1996). Estimation of the reliability of the problem oriented screening instrument for teenagers (POSIT) among arrested youths entering a juvenile assessment center. *Substance Use and Misuse, 31,* 785-824.

Dunford, F.W., & Elliott, D.S. (1984). Identifying career offenders using self-reported data. *Journal of Research in Crime and Delinquency, 21,* 57-86.

Elliott, D.S. (1994). Serious violent offenders: Onset, developmental course, and termination, The American Society of Criminology 1993 Presidential Address. *Criminology, 32,* 1-21.

Farrington, D.P., & Loeber, R. (1995). *Transatlantic replicability of risk factors in the development of delinquency.* Paper given at the Meeting of the Society for Life History Research in Psychopathology in Catham, Massachusetts.

Farrington, D.P., & Tarling, R. (1985). (Eds.). *Prediction in criminology.* NY: State University of New York Press.

Farrington, D.P., Ohlin, L.E., & Wilson J.Q. (1986) *Understanding and controlling crime: Toward a new research strategy.* N.Y.: Springer Verlag.

Favard, A.- M., (1991) *L'évaluation clinique en action sociale.* Toulouse, Érès.

Ferracuti, F., & Wolfgang, M.E. (1983). *Criminological diagnosis: An International perspective* (vol. I, & II). Lexington: Lexington Books.

Fréchette, M., & Le Blanc, M. (1987). *Délinquances et délinquants.* Boucherville, Québec: Gaëtan Morin Éditeur.

Garofalo, R. (1891). *Criminologia.* Turin, Bocca.

Glaser, D. (1987). Classification for risk. In D.M. Gottfredson & M. Tonry (Eds.), Prediction and classification: Criminal justice decision-making. *Crime and Justice: An Annual Review, 9,* 249-292. Chicago, IL: University of Chicago Press.

Gottfredson, D.M. (1987). Prediction and classification in criminal justice decision-making. *Crime and Justice: An Annual Review, 9,* 1-20. Chicago, IL: University of Chicago Press.

Gottfredson, D.M., & Tonry, M. (1987). (Eds.), Prediction and classification: Criminal justice decision-making. *Crime and Justice: An Annual Review.* Chicago, IL: University of Chicago Press.

Gottfredson, S.D., & Gottfredson, D.M. (1985). Screening for risk among parolees: Policy, practice, and method. In D.P. Farrington & R. Tarling (Eds.), *Prediction in criminology.* NY: State University of New York Press.

Gouvernement du Québec. (1993) Loi sur les jeunes contrevenants, manuel de référence. Québec, Ministère de la santé et des services sociaux, Direction de l'adaptation sociale.

Hamparian, D.M., Schuster, R., Dinitz, S., & Conrad, J.P. *The violent few: A study of dangerous juvenile offenders.* Lexington: Lexington Books, 1978.

Harris, P.W., & Jones, P.R. (1996a). *ProDES: Second annual report.* Philadelphia, Crime and justice research institute.

Harris, P.W., & Jones, P.R. (1996b). *ProDES: The program development and evaluation system.* Philadelphia, Crime and justice research institute.

Henggeler, S.W., Schoenwald, S.K., Borduin, C.M., Rowland, M.D., & Cunningham, P.B. (1998). *Multisystemic treatment of antisocial behavior in children and adolescents: Treatment manuals for practitioners.* New York: Guilford Press.

Hewitt, L.E., Jenkins, R.L. (1946). *Fundamental patterns of maladjustment: The dynamic of their origin.* Springfield: State University of Illinois.

Hirschi, T. (1969). *Causes of delinquency.* Berkeley: University of California Press.

Hoge R.D., Andrews D.A., & Leschied, A.W. (1995). Investigation of variables associated with probation and custody dispositions in a sample of juveniles. *Journal of Clinical Child Psychology, 24,* 279-286.

Hoge R.D., Andrews, D.A., & Leschied A.W. (1996). An investigation of risk and protective factors in a sample of youthful offenders. *Journal of Child Psychology and Psychiatry, 37,* 419-424.

Hoge R.D., Andrews, D.A., & Leschied, A.W. (1994). Tests of three hypotheses regarding the predictors of delinquency. *Journal of Abnormal Child Psychology, 22,* 547-559.

Hoge, R.D., & Andrews, D.A. (1994). *The youth level of service/case management inventory and manual.* Ottawa: Department of Psychology, Carleton University.

Hoge, R.D., & Andrews, D.A. (1996a). *Assessing the youthful offender: Issues and techniques*. New York: Plenum Press.

Hoge, R.D., & Andrews, D.A. (1996b). *Assessing risk and need factors in the youthful offender*. Toronto: Annual Conference of the American Psychological Association.

Hoge, R.D., & Andrews, D.A. (1999). *The youth level of service/case management inventory and manual (revised)*. Ottawa: Department of Psychology, Carleton University.

Jasmin, M. (1995). Les jeunes contrevenants: au nom et au-delà de la loi. Québec, Rapport du groupe de travail chargé d'étudier l'application de la Loi sur les jeunes contrevenants au Québec, Ministère de la justice et Ministère de la santé et des services sociaux.

Jesness, C.F., & Wedge, R.F. (1983) *Classifying offenders: The Jesness Inventory Classification System Technical Manual*. Sacramento: Department of Youth Authority.

Jung, S., & Rawana, E.P. (1999). Risk and need assessment of juvenile offenders. *Criminal Justice and Behavior, 26*, 69-89.

Kinberg, O. (1911). L'examen medico-psychologique. Cologne, VI Congrès d'Anthropologie Criminelle.

Le Blanc, M. (1983). Vers une théorie intégrative de la régulation de la conduite délinquante. *Annales de Vaucresson, 20*, 1-34.

Le Blanc, M. (1986). Pour une approche intégrative de la conduite délinquante des adolescents. *Criminologie XX* 73-96.

Le Blanc, M. (1992). Family Dynamics, Adolescent Delinquency and Adult Criminality. *Psychiatry, 55*, 236-253.

Le Blanc, M. (1994). Family, school, delinquency and criminality: The predictive power of an elaborated social control theory for male. *Mental Health and Criminal Behavior, 4*, 101-117.

Le Blanc, M. (1995a). Common, temporary, and chronic delinquencies: Prevention strategies during compulsory school. In P.O. Wikström, J. McCord, & R.W. Clarke (Eds.), *Integrating crime prevention strategies: Motivation and opportunity* (169-205). Stockholm: The National Council for Crime Prevention.

Le Blanc, M. (1995b). The relative importance of internal and external constraints in the explanation of late adolescence delinquency and adult criminality. In J. McCord (Ed.), *Coercion and punishment in long-term perspectives*. London: Cambridge University Press.

Le Blanc, M. (1996). MASPAQ: Mesures de l'adaptation sociale et individualnelle pour les adolescents québécois. Montréal, Groupe de recherche sur l'inadaptation psychosociale chez l'enfant, Université de Montréal.

Le Blanc, M. (1997a). A generic control theory of the criminal phenomenon, the structural and the dynamical statements of an integrative multilayered control theory. In T.P. Thornberry (Ed.), *Developmental theories of crime and delinquency. Advances in theoretical criminology, 7*, 215-286. New Brunswick: Transaction.

Le Blanc, M. (1997b). Socialization or propensity: A test of an integrative control theory with adjudicated boys. *Studies in Crime and Crime Prevention, 6*, 200-224.

Le Blanc, M. (1998). Screening of serious and violent juvenile offenders: Identification, classification, and prediction. In R. Loeber & D.P. Farrington (Eds.), *Serious and violent juvenile offenders: Risk factors and successful interventions* (167-193). Thousand Oaks, CA: Sage.

Le Blanc, M., & Biron, L. (1980). Vers une théorie intégrative de la régulation de la conduite délinquante des garçons. Montréal, Groupe de recherche sur l'inadaptation juvénile, Université de Montréal.

Le Blanc, M., & Bouthillier, C. (2001). A developmental test of the general adolescent deviance syndrome with adjudicated girls and boys and confirmatory factor analysis. *Criminal Behaviour and Mental Health*

Le Blanc, M., & Caplan, M. (1993). Theoretical formalization, a necessity: The example of Hirschi's social control theory. *Advances in Criminological Theory, 4*, 329-431.

Le Blanc, M., & Fréchette, M. (1989). *Male criminal activity, from childhood through youth: Multilevel and developmental perspectives*. N.Y.: Springer-Verlag.

Le Blanc, M., & Girard, S. (1997). The generality of deviance: Replication over several decades with a Canadian sample of adjudicated boys. *Canadian Journal of Criminology, 39*, 171-183.

Le Blanc, M., & Kaspy, N. (1998). Trajectories of delinquency and problem behavior: Comparison of synchronous and non synchronous paths on social and personal control characteristics of adolescent. *Journal of Quantitative Criminology, 14*, 181-214.

Le Blanc, M., & Loeber, R. (1998). Developmental criminology updated. *Crime and Justice: A Review of Research, 23*, 115-198.

Le Blanc, M., Dionne, J., Proulx, J., Grégoire, J., & Trudeau-Le Blanc, P. (1998). *Intervenir autrement: le modèle différentiel et les adolescents en difficulté*. Montréal: Presses de l'Université de Montréal.

Le Blanc, M., McDuff, P., & Kaspy, N. (1998). Family and preadolescence delinquency: A comprehensive sequential family control model. *Early Child Development and Care, 142,* 63-91.

Le Blanc, M., Ouimet, M., & Tremblay, R.E. (1988). An integrative control theory of delinquent behavior: a validation 1976-1985. *Psychiatry, 51,* 164-176.

Le Blanc, M., Vallières, E., & McDuff, P. (1992). Adolescents' school experience and self-reported offending: A longitudinal test of a social control theory. *International Journal of Youth and Adolescence, 8,* 197-247.

Le Blanc, M., Vallières, E., & McDuff, P. (1993). The prediction of males adolescent and adult offending from school experience. *Canadian Journal of Criminology, 35,* 459-478.

Lebon A. (1999). État de la situation en regard des listes d'attente en protection de la jeunesse et de l'accessibilité aux services à la jeunesse. Québec: Gouvernement du Québec, Ministère de la santé et des services sociaux.

Leschied, A.W., Andrews, D.A., & Hoge, R.D. (1992). *Youth at risk: A review of Ontario young offenders, programs, and literature that support effective intervention.* Toronto, Ministry of community and social services.

Lipsey, M.W., & Derzon, J.H. (1998). Predictors of violent and serious delinquency in adolescence and early adulthood : A synthesis of longitudinal research. In R. Loeber, & D.P. Farrington (Eds.), *Serious and violent juvenile offenders: Risk factors and successful interventions.* Thousand Oaks, CA: Sage. 86-105.

Loeber, R., & Le Blanc, M. (1990). Toward a developmental criminology. In M. Tonry, & N. Morris (Eds.), *Crime and Justice: An Annual Review, 13,* 1-98. Chicago: University of Chicago Press.

Loeber, R., Dishion, T.J., & Patterson, G.R. (1984). Multiple gating: A multistage assessment procedure for identifying youths at risk for delinquency. *Journal of Research in Crime and Delinquency, 21,* 7-32.

Lombroso, C. (1890). L'examen medico-psychologique. Saint-Pétersbourg, Actes du Congrès pénitentiaire international. Tome II: 440-48.

Lorion, R.P. Tolan, P.H., & Wahler, R.G. (1987). Prevention. In H.C. Quay (Ed.), *Handbook of juvenile delinquency.* NY: John Wiley, & Sons. 383-416.

Loza, W., & Simourd, D.J. (1994). Psychometric evaluation of the level of supervision inventory (LSI) among male Canadian federal offenders. *Criminal Justice and Behavior, 21,* 68-80.

MacLeod, S. (1995). *Youth management assessment guide.* Toronto, Ontario Ministry of the solicitor general and correctional services.

Megargee, E.I., & Bohn, M.J. (1979). *Classifying criminal offenders: A new system based on the MMPI.* Beverly Hills: Sage.

Moffitt, T.E. (1993). "Life-course-persistent" and "Adolescent-limited" antisocial behavior: A developmental taxonomy. *Psychological Review, 100,* 674-701.

Moffitt, T.E. (1997). Adolescence-limited and life-course-persistent offending: A complementary pair of developmental theories. *Advances in Theoretical Criminology, 7,* 11-54.

Nagin, D., Farrington, D.P., & Moffitt, T. (1995). Life-course trajectories of different types of offenders. *Criminology, 33,* 111-139.

Office of Juvenile Justice and delinquency Prevention. (1995). Community Assessment Centers: A discussion of the concept's efficiency. Washington, U.S. Department of Justice, Office of Juvenile Justice and Delinquency Prevention.

Palmer, T. (1963). *The re-emergence of correctional intervention.* Newbury Park: Sage, 1992.

Pinatel, J. *Criminologie.* Paris: Cujas.

Quay, H.C. (1987). Patterns of delinquent behavior. In H.C.Quay (Ed.), *Handbook of juvenile delinquency.* New York: Wiley.

Quay, H.C., & Parsons, L.B. (1971). *The differential classification of juvenile offender.* Washington, U.S. Bureau of Prisons.

Rahdert, E.R. (1991). Manual. *The adolescent assessment/referral system.* Rockville, National institute of drug abuse.

Schnidt F., & Hoge, R.D. (1999). *An investigation of a sample of mentally disordered young offenders.* Thunder Bay: Unpublished report, Lakehead University.

Sechrest, L. (1987). Classification for treatment. In D.M. Gottfredson, & M. Tonry (Eds.), Prediction and Classification: Criminal Justice Decision-making. *Crime and Justice: An Annual Review, 9,* 293-322.

Simourd D.J., Hoge D.R., Andrews, D.A., & Leschied, A.W. (1994). An empirically-based typology of male young offenders. *Canadian Journal of Criminology, 36,* 447-461.

Simourd, D.J., Hoge, D.R., & Andrews, D.A. (1991). *The Youth level of service inventory: An examination of the development of a risk/needs instrument.* Calgary: Annual Conference of the Canadian Psychological Association.

Snyder, H.N., Sickmund, M., & Poe-Yamagata, E. (1996). *Juvenile offenders and victims: 1996 update on violence.* Washington: Office of Juvenile Justice and delinquency Prevention.

Sullivan, C., Grant, J.D., & Grant, M.Q. (1957). The development of interpersonal maturity: Application to delinquency. *Psychiatry, 20,* 373-385.

Tracy, P.E., Wolfgang, M.E., & Figlio, R.M. (1990). *Delinquency career in two birth cohorts.* New York: Plenum Press.

Warren, M.Q. (1966). *Interpersonal maturity level classification: Juvenile diagnosis and treatment for low, middle, and high maturity delinquents.* Sacramento, CA: Youth Authority.

Warren, M.Q. (1983). Applications of interpersonal maturity theory to offender populations. In W.S. Laufer & J.M. Day (Eds.), *Personality theory, moral development, and criminal behavior.* Lexington: Lexington Books.

Wiebush, R.G., Baird, C., Krisberg, B., & Onek, D. (1995). Risk assessment and classification for serious, violent, and chronic juvenile offenders. In J.C. Howell, B. Krisberg, J.D. Hawkins, & J.J. Wilson (Eds.), *Serious, violent, and chronic juvenile offenders.* Thousand Oaks: Sage.

Wolfgang, M.E., Figlio, R.M., & Sellin T. (1992). *Delinquency in a birth cohort.* Chicago, IL: University of Chicago Press.

Wormith, J.S. (1995). The Youth Management Assessment: Assessment of young offenders at risk of serious reoffending. *Forum of Correctional Research, 7,* 23-27.

Young Offenders Act. Ottawa, Ministry of Justice, Government of Canada, Act, 1984, revised 1986, 1992, 1995.

Multi-Problem Violent Youth
R.R. Corrado et al. (Eds.)
IOS Press, 2002

191

The Risk Factor Profile Instrument: Identifying Children at Risk for Serious and Violent Delinquency[*]

Trisha Beuhring

Longitudinal studies in England, New Zealand, and the United States indicate that the vast majority of serious and violent juvenile crime is attributable to a relatively small group of chronic offenders. Up to 70% of such crimes are committed by no more than 8% of the males in a normative sample (Farrington & West, 1993; Moffitt, 1993; Snyder, 1998). Even in the inner cities of the United States, only 15-18% of adolescents are responsible for the vast majority of serious and violent juvenile crimes (Howell, 1995; Kelley, Huizinga, Thornberry, & Loeber, 1997). The cost to society is substantial. Chronic delinquents typically become career criminals who cost society an estimated US$1.7 to US$2.3 million *each* over the course of their lifetime (Cohen, 1998). A study in England and Wales estimated that the cost to government alone totals in the billions (Bagley & Pritchard, 1998).

The disproportionate impact of chronic offenders has major implications for policy makers worldwide. Although programs aimed at the remaining 82-92% of youth are worthwhile, they are unlikely to reduce serious and violent crime by more than one-quarter, even if highly successful. Programs that prevent chronic delinquency will therefore be the linchpin of any effort to permanently reduce the cost to society of serious violent and nonviolent crime (Howell, 1995; Loeber & Farrington, 1998).

In order to prevent serious and violent delinquency (SVJ) a program must be able to identify youth who are at risk of going down that path. Few instruments exist, however, for identifying children's risk of future delinquency (Howell, 2001a; LeBlanc, this book; Webster, Augimeri, & Koegl, this book). This chapter describes the development and preliminary validation of a risk assessment instrument for use with children under the age of 10 who exhibit early signs of serious problem behavior, including delinquency. Known as the Risk Factor Profile, the instrument is grounded in research on risk factors for SVJ. It is being used in a community intervention program to match unusually young offenders—children under 10 who have committed chargeable offenses—with interventions that are appropriate to their level of risk.[1]

1. Challenges in Early Identification

Accurately identifying children who are at risk of becoming serious and violent juvenile delinquents (SVJ) presents both practical and ethical challenges. Practically, self-report studies indicate that many delinquent youth go undetected by the legal system (Loeber & Farrington, 1998). Yet identifying children who are genuinely at risk is the key to cost-effectiveness, especially for expensive interventions that begin in childhood (Aos, Barnoski, & Lieb, 1998 Cohen, 1998). Ethically, the case for screening during early childhood is strengthened by evidence that parents seldom seek professional help until long after delinquent behavior is entrenched, if help is sought at all (Stouthamer-Loeber, Loeber, VanKammen, & Zhang, 1995). The benefits of early prevention, however, must be weighed against the risk of negative effects due to labeling (Horwitz & Wasserman, 1979) or unanticipated consequences of the intervention, such as those that occur when antisocial youth are treated in groups (Dishion, McCord, & Poulin, 1999; Peters, Thomas, & Zambrelan, 1997).). Early identification is complicated by the fact that only 2%-5% of young children are at detectable risk of becoming SVJ, even using sensitive criteria (Burrell & Warboys, 2000; Keenan, 2001; Loeber, Wung, Keenan et al., 1993; Snyder, 2001; Wasserman & Seracini 2001). Baserates this low make it difficult to accurately identify the target population without including large numbers of children who are not really at risk (Kraemer, 1992; Kraemer, Kazdin, Offord et al., 1999; Offord, Lipman, & Duku, 2001).

The solution is a "multiple gate" identification strategy (Kelley et al., 1997). In the ACE program, the first gate is commission of a chargeable offense before age 10: this identifies a group of children who have a 50% or greater baserate of risk for becoming chronic SVJ (Blumstein, Farrington, & Moitra, 1985; Snyder, 2001). The second gate is a review of risk factors for SVJ in the child, family, school and neighborhood: this helps distinguish between child delinquents who are likely to desist from offending and those who are likely to escalate. The final gate is a review of special circumstances that may enhance or reduce risk, including services already in place for the child and the severity and diversity of past delinquent behavior (see Beuhring & Melton, in press). The remainder of this chapter focuses on the development and validation of the risk assessment instrument that is used as the second gate in the identification process.

2. Risk Factor Profile Instrument

Instruments used in community programs must meet practical as well as scientific standards. The ACE program needed a risk assessment that was relatively short, easy to administer, and able to handle missing information. The assessment also had to be completed within one to two weeks of referral to be considered timely. The solution was an instrument that was grounded in research but which relied on members of the program's Screening Team to make judgments based on information gathered from various databases and interviews that are conducted at a confidential screening meeting. The computerized rating instrument takes 15 minutes to complete at the end of the meeting. A score that estimates the child's overall level of risk for chronic SVJ is used a guide for matching children to interventions that differ in scope and intensity.

2.1 Development of the Instrument

The Risk Factor Profile was grounded in a review of the literature regarding risk factors for serious and violent juvenile delinquency (see Hawkins, Herrenkohl, Farrington et al., 2000; Howell, 1995; Loeber & Farrington, 1998; 2001). Risk factors may be either causal factors or indicators of risk (see Kraemer, Kazdin, Offord et al., 1999). Both types are useful for prediction, but only the manipulation of causal factors will lead to changes in the outcome. The Risk Factor Profile instrument therefore focused on risk factors that were likely to be causal or that were proxies for causal processes. In general, this meant focusing on risk factors that were proximal to the child (parents, siblings, peers, school and neighborhood) rather than distal (race/ethnicity, gender, family structure, and poverty). Proximal risk factors are better predictors of individual differences in problem behavior than distal risk factors (Blum, Beuhring, & Mann-Rinehart, 2000). Proximal risk factors are more amenable to intervention (Wasserman, Miller, & Cothern, 2000). Excluding distal risk factors also avoided ethical concerns about perpetuating stereotypes based on group membership.

Risk factors are intercorrelated in complex ways, which can be a strength for risk assessments. Increasing the number of intercorrelated variables in a test improves its reliability and validity (Nunnally, 1978). It is also consistent with theory and research which indicates that serious and violent delinquency is the result of an accumulation of risks in multiple domains (Hawkins et al., 2000; Reid & Eddy, 1997; Thornberry, 1996)). This psychometric approach contrasts sharply with an empirical multiple regression strategy for test development. Multiple regression is meant to identify the minimum number of variables that will efficiently predict an outcome, and may therefore reduce reliability and construct validity. It also has serious statistical limitations as a method for selecting and weighting predictors when the candidate variables are intercorrelated (Pedhazur, 1997). Consequently, the author relied on the interdisciplinary ACE Screening Team to help decide which risk factors were most important to retain in the instrument, and how those risk factors should be weighted. The team included a county attorney, corrections worker, financial services worker, psychologist, public health representative, and social worker, all of whom had extensive experience with multi-problem children and families.

2.2 Content

A variety of studies indicate that taking both the child's history of problem behavior and psychosocial risk factors into account improves the ability to predict future delinquency compared to relying on past delinquent or problem behavior alone (Barnoski, this book; Day & Hunt, 1996; Earls, 1998; LeBlanc, this book; Moffitt, 1993; Sampson & Laub, 1997; Tolan & Thomas, 1995; Webster, Augimeri, & Koegl, this book). Consistent with this, the Risk Factor Profile includes scales that measure problem behaviors as well as the accumulation of risk factors in multiple domains. The instrument and its scoring are summarized in an Appendix.

The Risk Factor Profile includes 30 risk constructs (such as family disorganization and history of neglect or abuse) that are organized into 8 subscales: Referring Offense; Behavior History; Child Risk Factors; Child Temperament; Parent Risk Factors; Sibling Risk Factors; Peer Risk Factors; Community Risk Factors. The subscales are grouped into two scales: Child Scale and Contextual Scale. A third scale was developed to capture the interaction between a child's temperament and the context in which s/he was being raised (see Appendix). The total risk score is the average of the Child, Contextual, and Interaction scale scores. The total score currently determines how children are matched to interventions, although the profile of Child, Contextual, and Interaction scale scores eventually will be used to improve the match.

Although the instrument covers most of the key risk factors for SVJ, a few (such as antisocial attitudes and harsh or neglectful parenting practices) were omitted because it was unlikely that informants at the screening meeting could provide sufficient information. These influences are measured indirectly, however, through other risk factors such as parent or sibling criminality and child abuse or neglect. Ratings of school and neighborhood risk also serve as a proxies for peer influences when that information is not directly available.

2.3 Administration

Members of the inter-disciplinary ACE Screening Team make independent ratings of the severity of each risk factor following a discussion of objective indicators of risk (child protection case openings, parent criminal records, restraining orders for domestic violence, etc.) and interviews with school staff and other professionals who are able to attend a confidential screening meeting. The confidential screening meeting—which includes a review of background information, interviews, completion of the computerized risk instrument, and making an intervention decision—takes one hour per case. Collecting background information prior to the meeting takes an additional two to four hours per case. The near impossibility of contacting parents during the one to weeks between referral and screening made it necessary to exclude parent and family interviews as a source of information.

2.4 Scoring

The reliability and validity of the total score was improved by averaging ratings across Screening Team members, items, subscales, and scales. At each step, random error in judgment and individual differences in rating biases tend to cancel each other out, giving a better estimate of the child's "true" risk score (Nunnally, 1978). The weighting scheme for risk factors within each subscale was devised by pilot testing options on the first 30 cases screened by the ACE program. The benchmark was a global assessment of risk for SVJ made by the Screening Team immediately after each case was reviewed. Feedback from the Screening Team regarding their decision-making process also guided scoring and weighting decisions.

SVJ-risk scores, which take into account both the accumulation and severity of risk factors, varied widely among the young offenders referred to ACE over a period of 20 months (Table 1). Scores ranged from 0.8 to 6.2 on a scale of 0 (no risk for SVJ) to 7 (extreme risk for SVJ). The scores were normally distributed with a mean of 3.4 (SD 1.26) and a median of 3.1. The older the child at the time of screening, the more variation there was in estimates of risk for future delinquency. The Screening Team's global ratings of risk, rather than statistical guidelines, were used to establish the cut-point score that defines high risk of SVJ (Rutter, 1997). Generally, young offenders who scored 3.0 or more were considered high risk and assigned to a comprehensive, long-term intervention; those who scored 2.9 or less were referred to shorter and less intensive interventions in the community.

Table 1: Distribution of Risk Scores Among Young Offenders

Scoring Range (N=8)	Extremely Young 4 to 5 yrs (N=7)	6 yrs (N =26)	7 yrs (N =33)	Unusually Young 8 yrs (N =80)	9 yrs (N=154)	Total	
0.0 to 0.9	1	--	--	--	1	1%	(2)
1.0 to 1.4	1	--	3	2	7	8%	(13)
1.5 to 1.9	1	1	2	4	6	9%	(14)
2.0 to 2.4	2	2	2	4	7	11%	(17)
2.5 to 2.9	1	--	3	3	14	14%	(21)
3.0 to 3.4	1	1	6	5	9	14%	(22)
3.5 to 3.9	1	1	5	7	13	18%	(27)
4.0 to 4.4	--	1	3	3	4	7%	(11)
4.5 to 4.9	--	1	1	--	8	7%	(10)
5.0 to 5.4	--	--	1	3	7	7%	(11)
5.5 to 5.9	--	--	--	1	4	3%	(5)
6.0 to 7.0	--	--	--	1	--	1%	(1)

Scores for children referred to the ACE program between 9/1/99 – 4/30/01.

2.5 Interaction between Temperament and Context

Children who were clearly at risk for violent offending, in the view of the Screening Team, had two things in common: volatile temperaments that led to rejection by peers and chaotic environments that exacerbated this predisposition. A key feature of the screening instrument was therefore a score that captured the interaction between the child's predisposing temperament and the context in which s/he is being raised. Although there are debates in the research literature regarding the nature of interaction effects in practice, theoretical models emphasize the importance of interactions in the dynamic course of development (Loeber & Stouthamer-Loeber, 1998; Pilkonis & Klein, 1997; Reid & Eddy, 1997; Roosa, 2000; Rutter, 1985; Zimmerman, 1994). Their relevance for identifying potentially violent juvenile delinquents is apparent in the actual case histories summarized in Table 2. The interaction scale score also helps distinguish different degrees of risk for SVJ among siblings and friends who co-offended.

Table 2: Different Risk Scores for Similar Offenses

Child	Male, 8 years old	Child	Male, 9 years old
Offense	Broke windows at school. One co-offender (age 8)	Offense	Broke windows in a home. Two co-offenders (ages 9 and 10)
SVJ-risk score[a] 1.2		SVJ-risk score[a] 5.8	
History	Child shows little evidence of aggression or other problem behaviors. Single mother has no criminal or drug history, and is employed. Father is a drug addict and schizophrenic, but there is no evidence of domestic violence or child abuse. Siblings show little evidence of behavior problems. Child's neighborhood and school are not ideal but they are not high-risk. Child's residence and school have been stable for three years.	History	Child is ADHD but does not get medication regularly. Child was expelled from two schools for threatening to kill teachers. Single mother is a recovering cocaine addict with a criminal history. Father has an extensive criminal record, including assault on a police officer and domestic violence. Siblings are also aggressive and violent. Family lives in a high-crime neighborhood and moves frequently. Child changes schools frequently and has poor marks.

[a] Risk of becoming a serious and violent juvenile delinquent on a scale of 0 (no risk of SVJ) to 7 (extreme risk of SVJ).

2.6 Sample Characteristics

A total of 190 children were referred to the Ramsey County, Minnesota ACE program over a period of 21 months (9/1/99—5/31/01). Of these, 161 (85%) met the criteria for further screening (under age 10, a county resident, and sufficient evidence to charge the offense in juvenile court if the child were older). Although 15% of the referrals came from the suburbs, most suburban child delinquents lived in apartment buildings and trailer parks where there was high crime and poverty and low social cohesion (see Sampson, 1995; Sampson & Radenbusch, 2001). Ratings of neighborhood risk were therefore based on the characteristics of these "micro neighborhoods" rather than on census statistics.

Half (52%) of the children referred to the ACE program had committed a chargeable aggressive or violent offense (Table 3). Criminal damage to property was the most common property crime. These patterns are quite different from referrals to the Earlscourt program, which were primarily for shoplifting and other thefts (Webster et al., this book). One in ten children referred to ACE for a chargeable offense was under age 7. This was consistent with research that indicates that an unusually early onset of problem behavior is associated with greater seriousness and persistence of delinquency (Henry, Caspi, Moffitt, & Silva, 1996; Loeber & Farrington, 1998; Moffitt, 1993; Rutter, 1997). In fact, many of the highest risk children referred to ACE had a history of serious violent behavior that crossed the threshold into being a chargeable offense before the age of 6.

Although African American children were over-represented in referrals, there was little indication this was due to racial profiling: referral statistics for Anglo and African American children were nearly identical in every category except age and gender. The disproportionate referral rate was consistent with self-report and victim data for national samples of adolescents (Hawkins, Laub, Lauritsen, & Cothern, 2000). The higher proportion of females among minority child delinquents was consistent with racial/ethnic differences in violence reported by Anglo, Hispanic, and African American adolescent females in a nationally representative US sample (Blum, Beuhring, & Mann-Rinehart, 2000). Overall, the ratio of male to female referrals (85% vs. 15%) was similar to the 9:1 ratio of males to females among serious and violent juvenile delinquents. The under-representation of other sizeable racial/ethnic groups in Ramsey County (Hispanic, Hmong Asian, and Native Americans) was largely due to cultural and language factors.

Table 3: Child Delinquents Referred to ACE (9/1/99 – 5/31/01)

	Ethnicity			SVJ-risk Score at Screening[b]				
	Anglo Amer. (N=48)	African Amer. (N=89)	Other (N=26)	Low 0.8 – 1.4 (N=16)	Moderate 1.5 – 2.9 (N=56)	High 3.0 – 4.4 (N=63)	Very high 4.5 – 6.2 (N=28)	TOTAL (N=163)
Referral Source								
% Police	85	81	42	75	91	73	79	80
% Schools	10	11	46	0	4	25	14	14
% County Attorney	4	8	12	25	5	2	7	6
Offense								
% Person	56	52	38	31	46	54	64	52
% Property	40	44	39	44	43	43	36	41
% Other	4	4	23	25	11	3	0	7
Prior police contact (%)	27	30	12	6	23	24	50	26
Age								
% 4 – 6 years	8	8	19	12	12	10	4	10
% 7 – 8 years	38	53	27	38	38	56	36	44
% 9 years	54	39	54	50	50	35	60	46
Gender[a]								
% male	94	82	81	69	89	87	85	85
% female	6	18	19	31	11	13	15	15

[a] Race/ethnicity and gender are *not* included as risk factors in the screening instrument. "Other" includes Hispanic, Asian, and Native Americans.
[b] Risk for serious and violent juvenile delinquency on a scale of 0 (no risk of SVJ) to 7 (extreme risk of SVJ).

2.7 Reliability

Three types of reliability analyses are in progress: internal consistency analyses, inter-rater agreement, and test-retest. Internal consistency reliability is expected to be high because of the intercorrelation among risk factors (Cronbach, 1971).). Inter-rater reliability is also expected to be high because of the process of averaging ratings described in the section on scoring (Nunnally, 1978). Test-retest reliability, however, is likely to be modest.

Current estimates are based on repeat screenings that occurred because a child was referred for another offense; intervals varied from three weeks to 12 months. Some children who were referred for another offense had no notable change in their risk profile; they were screening "misses". Others, however, had genuinely escalated in both levels of risk and antisocial behavior. This was to be expected, given that the ages of 7 to 9 appear to be a transition period for the emergence of delinquency (Loeber & Farrington, 2001; Loeber et al., 1993).

2.8 Concurrent Validity

Several lines of evidence indicate the Risk Factor Profile score ia a valid measure of a very young offender's risk for escalating into serious and violent juvenile delinquency. Risk scores were strongly correlated with global assessments of risk for SVJ that had been made by the Screening Team before the Risk Factor Profile score was known ($r = .83$, $N = 153$). Children who scored as being at very high risk of SVJ were nearly twice as likely to have been referred for a person crime as a property crime, and more than half (53%) had prior contact with the police (Table 3). Risk scores consistently differentiated among siblings and children who co-offended: risk scores for nine pairs of siblings referred to ACE differed by an average of 1.3 points (on a scale of 0-7) regardless of whether they co-offended or committed different offenses months apart. Differences in risk scores for unrelated children who co-offended were comparable. Half (56%) of the 161 child delinquents referred to ACE scored above the independently determined cut-point that defines high risk of SVJ; this figure is very similar to the proportion of early onset offenders who would be expected to become SVJ in the absence of special intervention (Snyder, 2001). Finally, independent indicators of risk in the family and child were strongly related to risk scores, as was an independent measure of the severity and diversity of the child's history of delinquency. These data are summarized below.

Family indicators of risk. The proportion of children with independent indicators of risk in the family—such as a father who was a violent career criminal—was strongly related to the SVJ-risk score as can be seen in Table 4. Note that 33% to 53% of the fathers of the highest SVJ-risk children were career criminals with an arrest for murder and/or multiple serious violent crimes.

Table 4: Range of Family Risk Factors Among 6- to 9-year old Offenders

Family Characteristics and Indicators of Risk	SVJ-risk Score at Screening[a]			
	Lowest Risk		Highest Risk	
	0.8 – 1.4 (N=15)	1.5 – 1.9 (N=14)	4.5 – 4.9 (N=10)	5.0 – 6.2 (N=17)
CHARACTERISTICS				
Family Structure[b]				
% Two parents or partners	40	36	20	35
% One parent	60	50	60	53
% Relative or out-of-home placement	0	14	20	12
Children[b]				
Mother <18 at first child (%)	33	42	70	35
Number of children (mean)	2.0	3.8	3.5	3.8
Number of partners who fathered children (mean)	1.5	1.8	2.3	2.1
Financial Assistance (%)[b]	56	65	80	82
FAMILY CRIMINALITY				
Father (Mother)[c]				
% Chronic SV+	0 (0)	10(0)	33 (0)	53 (0)
% Chronic SV or Chronic S	14 (14)	50(17)	22 (50)	18 (12)
% Less severity or frequency	26 (26)	30(33)	11 (30)	23 (41)
% None	60 (60)	10(50)	33 (20)	6 (47)
Siblings (age 7 – 20+)[c]				
% SJV or SVJ-risk	33	65	68	90
OTHER INDICATORS OF RISK				
History of Domestic Violence				
% Any	67	57	70	88
Severity where present[d]	2.5	3.0	3.3	4.4
Substance Use and/or Mental Illness				
% One or both parents	47	79	90	88
Severity where present[d]	4.0	4.6	4.8	4.4
Abuse, Neglect, Poor Supervision[d]	1.5	2.7	4.6	4.9
Child Protection Services (% ever)	27	71	80	82
Home Environment Severity of disorganization[d]	2.7	2.9	5.0	5.0
Number of moves per year (mean)	1.1	1.3	1.7	1.3

[a] Risk of becoming a serious and violent juvenile delinquent on a scale of 0 (no risk) to 7 (extreme risk). Data are for the lowest
and highest 17-18% of the scoring distribution for N=161 child delinquents under age 10.
[b] Race/ethnicity, gender, family structure and poverty are *not* included as risk factors.
[c] Chronic: 5+ arrests. SV+: murder or 2+ serious violent crimes. SV: 1-2 serious violent crimes. S: 1-2 serious non-violent crimes. SVJ: serious and violent juvenile delinquency. Identical percentages for low-risk fathers and mothers
occurred by chance and represent different pairings (e.g., chronic S fathers and chronic S mothers were *not* partners).
[d] Average rating by the Screening Team on a scale of 0 (none) to severe (6) taking all sources of information into account.

Child indicators of risk. The proportion of children with independent indicators of risk – such as a mental health diagnosis – was also strongly related to the SVJ-risk score as can be seen in Table 5. The only notable exception from the literature was ability. Ability was normally distributed in the sample as a whole, except for the very highest SVJ-risk children where 35% (6 of 17) had tested as very bright or gifted (the finding for the next highest risk group was an anomoly that may be due to the much smaller sample).

Table 5: Range of Child Risk Factors Among 6- to 9-year-old Offenders

| | SVJ-risk Score at Screening[a] | | | |
| | Lowest Risk | | Highest Risk | |
Child characteristics and indicators of risk	0.8 – 1.4 (N=15)	1.5 – 1.9 (N=14)	4.5 – 4.9 (N=10)	5.0 – 6.2 (N=17)
CHARACTERISTICS				
Race/ethnicity[a]				
% Anglo American (non-Hispanic)	33	21	20	24
% African American	33	43	70	70
% Other	33	36	10	6
Gender[a]				
% male	67	79	70	94
% female	33	21	30	6
Mean age (years)	7.9	7.9	8.5	8.6
INDICATORS OF RISK				
Mental health diagnoses				
% Conduct/oppos. defiant disorders	7	0	20	29
% ADHD	29	13	70	65
% Depression	7	13	20	47
% Other (anxiety, bipolar, etc.)	0	13	20	47
Residential treatment center				
(% ever)	0	0	10	18
Volatile temperament[b]	1.2	.9	5.4	5.7
Isolation from peers[b]	0.6	1.4	4.9	5.5
Emotional or behavioral disability				
% EBD 5 (student:teacher ratio is 3:1)	0	0	50	63
% EBD 2 – 4	0	0	30	19
% mainstream classroom	100	100	20	18
Ability[b]				
% Learning disabled, dev. delayed	14	14	40	12
% Very bright or gifted	0	0	0	35

[a] Risk of becoming a serious and violent juvenile delinquent on a scale of 0 (no risk of SVJ) to 7 (extreme risk of SVJ).
Race/ethnicity, gender, family structure and poverty are *not* included as risk factors in the screening instrument.
[b] Average rating of severity made by the Screening Team on a scale of 0 (none) to 6 (extreme) taking all sources of information into account. Ability was estimated from IQ tests, school interviews or other records provided at screening.

Child's history of delinquency. Research indicates that children who become SVJ gradually escalate in both the severity and diversity of their problem behavior (Loeber et al., 1993; Kelley et al., 1997). Diversification occurs across three pathways: authority conflicts (defiance and status offenses); covert acts (property crimes); and overt acts (person crimes). Children's status in these pathways at the time of screening was strongly related to their SVJ-risk score, as can be seen in Table 6. (A child's standing on this measure were not discussed until after the global rating of risk and the Risk Factor Profile ratings had been made so that it would not bias the Screening Team's judgments.) Behavior was counted as a chargeable offense if it resulted in a sanction that was the developmental equivalent of an arrest or conviction (expelled from preschool; suspended from elementary school, denied school bus privileges, expelled from a community center, denied entry to a store, referred to a treatment center, etc.). This improved the sensitivity of the measure (police reports underestimate rates of offending) without having to obtain self-reports from children or their parents (which would have been subject to mandatory reporting laws in Minnesota). Requiring that the behavior resulted in a sanction ensured that it crossed an objective threshold of severity, and that it was negatively impacting the child's life, much as an arrest or conviction would do later on (Richters & Cicchetti, 1993).

Table 6: History of Delinquency Among 6- to 9-year old Offenders

| | SVJ-risk Score at Screening | | | |
	Lowest Risk		Very High Risk	
Indicators of diversity, severity and frequency	0.8 – 1.4 (N=15)	1.5 – 1.9 (N=14)	4.5 – 4.9 (N=10)	5.0 – 6.2 (N=17)
History of delinquency[a]				
% Person, property and status offenses	0	0	50	24
% Person and property (stage 3 or 4)	0	8	30	24
% Person only (stage 4, or 3 if adult victim)	20	14	10	47
% Property only (stage 4)	13	14	0	0
% Lesser severity	67	64	10	5
Felony-level offense (% ever)	27	29	80	71
Suspended from school (% ever)	7	11	100	100
Frequency				
% Chronic (5+)	7	14	100	82
% 2-4 offenses	13	14	0	18
% Referring offense only	80	72	0	0

[a] Includes all chargeable offenses that resulted in a sanction. Stage 4 offenses are felonies in Minnesota.

    Serious violent offenses committed by the high SVJ-risk children at some point prior to screening included setting fire to an occupied crib (two children), choking a brother in his sleep, choking a classmate while threatening to kill her, and a group rape of a young girl who was lured into an empty building. Moderately serious violent offenses included sexual predation of classmates in kindergarten, kicking female classmates in the head after knocking them down stairs, knocking the braces off a child's teeth, hitting a teacher from behind with a chair, punching a teacher in the face, and kicking a pregnant teacher in the stomach. Serious non-violent offenses included burglary of an occupied home, constructing and using fire bombs, car theft, and doing more than $10,000 in damage to cars with a forklift (one child) and bricks or bats (two children).
    Studies of normative populations provide additional perspective, as shown in Figure 1. Note that children with very high scores on the Risk Factor Profile instrument have a much more violent offense profile than either children who scored low on the instrument or normative samples of youth. Many of the highest risk 6- to 9-year-old children are already serious and violent delinquents.

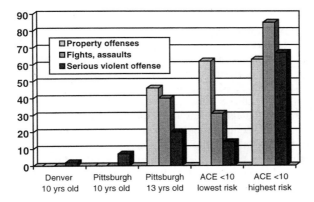

Comparison groups are samples of 10 year old boys in Denver and Pittsburgh (Keenan et al., 1997) and 13 year old boys in Pittsburgh (Loeber et al., 1993). Data on property offenses and fights were not available for the first two samples. ACE samples combine low with moderately low risk children (N=29) and very high with extremely high risk children (N=27).

Figure 1: ACE Young Offenders Compared to Normative Samples

2.9 Estimates of Predictive Validity

The predictive validity of the Risk Factor Profile score is still to be determined (Kraemer et al., 1999; Kraemer, 1992). In the meantime, the very young age of children referred to ACE raises a question about whether the identification process has improved detection of children at risk of SVJ, or merely identified children who would otherwise be considered ordinary pranksters (Richters & Cicchetti, 1993).

Two studies by other programs for very young offenders indicate that risk scores predict later delinquency in community-referred samples. In one study, 69% of child delinquents under age 10 who scored in the highest quartile on a risk assessment were re-arrested by age 13 or 14, compared to only 26% of those in the lowest quartile of risk (Stevens, Owen, & Lahti-Johnson, 1999). In the second study, 56% of child delinquents under 12 who had a high risk score were re-arrested by the end of adolescence, compared to 24% of those with a low risk score (Webster et al., this book).

2.10 Moderating Risk and Protective Processes

Child development is a dynamic process and high-risk children are likely to attract numerous interventions, both formal and informal, by the age of 18 (Loeber & Stouthamer-Loeber, 1998; Pilkonis & Klein, 1997; Reid & Eddy, 1997; Roosa, 2000; Rutter, 1985; Thornberry, 1996). How, then, can an assessment of risk made during childhood be expected to predict outcomes a decade later? Experience with the 161 very young offenders referred to ACE provides some insights into moderating risk and protective processes that may help explain the predictive validity of early problem behavior and psychosocial risks.

Parent opposition to intervention. The most unexpected finding to date has been the strong relationship between a child's assessed risk for SVJ and parent opposition to services that would reduce that risk: A total of 6%, 26%, 30%, and 47% of children with moderately high (3.0-3.4), high (3.5-4.4), very high (4.5-4.9), and extremely high (5.0-6.2) risk scores have parents who actively or passively refuse to cooperate with assessments, medications, counseling or other services that might benefit them or their child. Resistance is not limited to the ACE program; teachers, counselors and child protection workers repeatedly express frustration over their inability to provide services to the children who need it most. The reasons are complex , ranging from parent distrust of the system to parent dysfunction and active criminality (Beuhring & Melton, in press). Parent opposition remains the single biggest barrier to effective intervention with these children (also see Augimeri, Koegl, & Goldberg 2001; Harrell, Cavanagh, & Sridharan, 1999; Stevens et al., 1999).

Explicit socialization. Young offenders referred to ACE may not simply be modeling antisocial values and behavior (Hawkins & Farrington, 1999). Many are being explicitly taught antisocial behavior or reinforced for exhibiting it. Examples include a parent who drives her children to shopping malls with instructions on how to shoplift; fathers who watch pornographic movies with their 8- and 9-year-old sons; a mother who calls her prostitution clients during a conference at school about her son's aggressive sexual behavior; parents who encourage their child to be destructive while in foster care so they will be returned home; and parents or relatives who praise the child for delinquent behavior ("Look, the school gave you a vacation day for hitting that girl!"). The higher the risk, the more likely the child is being explicitly socialized into antisocial behavior.

Child "likability." Case managers, counselors, teachers, and others who have contact with child delinquents almost inevitably refer to the child's "likability" or lack thereof. Children who are perceived as sweet, cute, funny or talented despite their challenging behavior receive more positive attention from these adults; these adults also indicate a greater willingness to persist in providing services to oppositional children who are likeable than to oppositional children who are perceived as strange or bizarre. A similar phenomenon has been observed by Augimeri, who believed it had sufficient clinical significance to include as a factor in the EARL20-B risk assessment instrument (Augimeri, Koegl & Goldberg, 2001). Interestingly, likability shows no relationship to risk scores among the very young offenders referred to ACE.

The author will be examining the role of these and other moderating risk and protective processes as part of the longitudinal evaluation study for the ACE program.

2.11 Next Steps

The Risk Factor Profile screening instrument is only one of several instruments that will be needed over time in a community intervention program for very young offenders. Moreover, other agencies that interact with the child over the course of development will have their own assessment needs. Ideally, instruments would be:

> Linked across developmental stages—the ability to predict behavior, and track changes in risk, will depend on instruments being linked to a common developmental framework that specifies how risk factors change in importance over time (such as peer influences) as well as how their manifestation changes (such the nature of connections to others).

> Linked across purposes—screening, needs assessments, and evaluation tools serve different purposes but should tap into similar constructs in order to provide an integrated picture of the individual at each stage. They should also complement each

other in identifying subgroups of children who may need, or benefit from, different intervention strategies.

➢ Linked across systems—schools, corrections, and mental health systems, to name just a few, have different reasons for assessing risk and different expectations of change. The fact that many of these systems are involved in a youth's life at the same time underscores the need to link screening, needs assessment, and evaluation tools to a common framework so that the efforts of each system will complement each other rather than work at cross purposes.

The integration of services and interventions across systems and developmental stages is the direction that theory and research are taking community programs (Farrington, Loeber, & Kalb, 2001; Garry, 1997; International Centre for the Prevention of Crime, 2001; Schorr, 1999). An integrated framework of constructs, instruments, and interventions would provide the foundation for more effective and cost-effective intervention to prevent serious and violent delinquency.

3. Appendix. Risk Factor Profile Instrument

Risk Factors	Indicator(s)	
Child Scale (0 – 6)	Referring Offense	
- Severity for age	School and/or police report of offense.	
- Intention	School and/or police report of offense.	
Behavior History		
- Aggression, torture animals		Prior police contact, school suspensions,
- Arson, theft, vandalism		interviews with teachers, principals,
- Other (drugs, suicide risk,		school counselors and/or case workers
sexual behavior, lying, gangs)		who attended the confidential screening.
Child Risk Factors		
- History of abuse, neglect		Child Protection records and interviews.
- Likelihood of academic failure		School records and interviews.
Child's Temperament		
- Volatile, impulsive		Documented diagnoses and interviews.
- Social isolation	Interviews.	
Context Scale (0 – 6)	Parent Risk Factors[a]	
- Family disorganization		Financial assistance records.
- Criminal history	County, state, federal police records.	
- Drugs, mental health		Arrests for drugs, treatment by county.
- Domestic violence		Police records, court orders.
- Abuse, neglect of siblings		Child Protection Services records.
Sibling Risk Factors		
- Delinquency	Police and court records.	
- Problem behaviors		Interviews (if known to school, others).
Peer Risk Factors		
- Delinquency	Generally unavailable (unless peer	
- Problem behaviors		was a co-offender and screened).
Community Risk Factors		
- Neighborhood disorganization		Local crime and poverty statistics.
- Neighborhood resources		Team knowledge of neighborhood.
- School academic quality		School statistics; team knowledge.
- Student climate	Team knowledge of school.	

Interaction Scale (0 – 9)   Statistical interaction between the temperament score and the score for the highest-risk context, with correction to 0-9 scale.

Risk Profile Score (0 – 7)   Average of child, context, and interaction scale scores.

[a] Unstable family composition, residential history, and financial history are rated under family disorganization. Where possible, information is rated for both biological parents as well as any partner or other adult who is residing in the household.

Notes

* Members of the ACE Screening Team provided the clinical insights and critical feedback on which the validity and reliability of the Risk Factor Profile instrument rests: Hope Melton (ACE Program Director, Department of Public Health); Ed Frickson (Psychologist, Children's Mental Health Services); Kelly Higgins (Probation Officer, Corrections); Jack Jones (Sr. Social Worker, Child Protection Services); Connie McKee (Financial Services Worker); and Leslie Norsted (Assistant County Attorney). Their unflagging commitment and patience with months of pilot testing and revisions exemplified the value of close research-community partnerships.

1 All Children Excel (ACE) in Ramsey County (St. Paul) Minnesota, is one of only a handful of programs worldwide designed for very early onset offenders (Howell, 2001b; Webster, Augimeri, & Koegl, this book). Sponsored by local government, ACE offers a comprehensive, long term intervention to the highest risk children that is grounded in research, supports the integration of services across government units, and promotes collaboration among police, schools, and community non-profit organizations toward a common goal—preventing delinquency through promoting healthy development (Beuhring & Melton, in press).

References

Aos, S., Barnoski, R., & Lieb, R. (January, 1998). *Watching the bottom line: Cost-effective interventions for reducing crime in Washington.* Olympia, Washington: Washington State Institute for Public Policy. [http: //www.wa.gov/wsipp]

Augimeri, L.K., Koegl, C.J., & Goldberg, K. (2001). Children under age 12 years who commit offenses: Canadian legal and treatment approaches. In R. Loeber & D.P. Farrington (Eds.), *Child delinquents: Development, intervention, and service needs.* Thousand Oaks, CA: Sage.

Bagley C., & Pritchard C. (1998). The billion dollar costs of troubled youth: Prospects for cost-effective prevention and treatment. *International Journal of Adolescence and Youth, 7,* 211-225.

Barnoski, R. (1998). Juvenile rehabilitation administration assessments: Validity review and recommendations. Olympia, Washington: Washington State Institute for Public Policy. [http: //www.wa.gov/wsipp]

Beuhring, T., & Melton, H. (in press). *Identifying and intervening with unusually young offenders: The ACE program.* [Juvenile Justice Bulletin.] Washington, DC: U.S. Department of Justice, Office of Juvenile Justice and Delinquency Prevention.

Blum, R.W., Beuhring, T., Rinehart, P.M. (2000). *Protecting teens: Beyond race, income and family structure.* Center for Adolescent Health. Minneapolis, MN: University of Minnesota.

Blumstein, A., Farrington, D.P., & Moitra, S.D. (1985). Delinquency careers: Innocents, desisters, and persisters. In M. Tonry & N. Morris (Eds.), *Crime and justice: An annual review of research.* Chicago: University of Chicago Press.

Burrell, S., & Warboys, L. (2000, July). *Special education and the juvenile justice system.* [Juvenile Justice Bulletin] Washington, DC: U.S. Department of Justice, Office of Juvenile Justice and Delinquency Prevention.

Cohen, M.A. (1998). The monetary value of saving a high-risk youth. *Journal of Quantitative Criminology, 14,* 5-33.

Cronbach, L.J. (1971). Test validation. In R.L. Thorndike (Ed.), *Educational measurement* (2nd ed.). Washington, DC: American Council on Education.

Day, D.M., & Hunt, A.C. (1996). A multivariate assessment of a risk model for juvenile delinquency with an 'under 12 offender" sample. *Journal of Emotional and Behavioral Disorders, 4,* 66-72.

Dishion, T.J., McCord, J., & Poulin, F. (1999). When interventions harm: Peer groups and problem behavior. *American Psychologist, 54,* 755-764.

Earls, F. (1998, September). *Linking community factors and individual development.* [Research Preview] Washington, D.C.: U.S. Department of Justice, National Institute of Justice.

Farrington, D.P., & West, D.J. (1993). Criminal, penal and life histories of chronic offenders: Risk and protective factors and early identification. *Criminal Behaviour and Mental Health, 3(4),* 492-523.

Farrington, D.P., Loeber, R., & Kalb, L.M. (2001). Key research and policy issues. In R. Loeber & D.P. Farrington (Eds.), *Child delinquents: Development, intervention, and service needs.* Thousand Oaks, CA: Sage.

Garry, E.M. (1997, December). *Performance measures: What works?* [Fact Sheet] Washington, DC: U.S. Department of Justice, Office of Juvenile Justice and Delinquency Prevention.

Harrell, A., Cavanagh, S., & Sridharan, S. (1999, November). *Evaluation of the Children at Risk program: Results 1 year after the end of the program.* [Research in Brief] Washington, D.C.: U.S. Department of Justice, National Institute of Justice.

Hawkins, J.D., Herrenkohl, T.I., Farrington, D.B., Catalano, R.F., Harachi, T.W., & Cothern, L. (2000, April). *Predictors of violence.* [Juvenile Justice Bulletin] Washington, DC: U.S. Department of Justice, Office of Juvenile Justice and Delinquency Prevention.

Hawkins, J.D., & Farrington, D.P. (Eds.). (1999). Mechanisms of transmission and change in antisocial behavior. [Special issue] *Criminal Behavior and Mental Health, 9,* 3-117.

Hawkins, D.F., Laub, J.H., Lauritsen, J.L., & Cothern, L. (2000, June). *Race, ethnicity, and serious and violent juvenile offending.* [Juvenile Justice Bulletin]Washington, DC: U.S. Department of Justice, Office of Juvenile Justice and Delinquency Prevention.

Henry, B., Caspi, A., Moffitt, T.E., & Silva, P.A. (1996). Temperamental and familial predictors of violent and nonviolent criminal convictions: Age 3 to age 18. *Developmental Psychology, 32,* 614-623.

Horwitz, A., & Wasserman, M. (1979). The effect of social control on delinquent behavior: A longitudinal test. *Sociological Focus, 12(1),* 53-70.

Howell, J.C. (Ed.). (1995). *Guide for implementing the comprehensive strategy for serious, violent and chronic juvenile offenders.* Washington, DC: US Dept. of Justice, Office of Juvenile Justice and Delinquency Prevention.

Howell, J.C. (2001a). Risk/needs assessment and screening devices. In R. Loeber & D.P. Farrington (Eds.), *Child delinquents: Development, intervention, and service needs.* Thousand Oaks, CA: Sage.

Howell, J.C. (2001b). Juvenile justice programs and strategies. In R. Loeber & D.P. Farrington (Eds.), *Child delinquents: Development, intervention, and service needs.* Thousand Oaks, CA: Sage.

International Centre for the Prevention of Crime (2001 April). *The role of local government in community safety.* [Monograph] Washington, D.C.: U.S. Department of Justice, Bureau of Justice Assistance.

Keenan, K. (2001). Uncovering pre-school precursors to problem behavior. In R. Loeber and D.P. Farrington (Eds.), *Child delinquents: Development, intervention, and service needs.* Thousand Oaks, CA: Sage.

Kelley, B.T., Loeber, R., Keenan, K., & DeLamatre, M. (1997 December). *Developmental pathways in boys' disruptive and delinquent behavior.* [Juvenile Justice Bulletin] Washington, DC: US Dept. of Justice, Office of Juvenile Justice and Delinquency Prevention.

Kelley, B.T., Huizinga, D., Thornberry, T.P., & Loeber, R. (1997 June). *Epidemiology of serious violence.* [Juvenile Justice Bulletin] Washington, DC: U.S. Department of Justice, Office of Juvenile Justice and Delinquency Prevention.

Kraemer, H.C. (1992). *Evaluating medical tests: Objective and quantitative guidelines.* Thousand Oaks, CA: Sage.

Kraemer, H.C., Kazdin, A.E., Offord, D.R., Kessler, R.C., Jensen, P.S., & Kupfer, D.J. (1999). Measuring the potency of risk factors for clinical or policy significance. *Psychological Methods, 4,* 257-271.

Loeber, R., & Farrington, D.P. (Ed.). (1998). *Serious and violent juvenile offenders: Risk factors and successful interventions.* Thousand Oaks, CA: Sage.

Loeber, R., & Farrington, D.P. (Ed.). (2001). *Child delinquents: Development, intervention, and service needs.* Thousand Oaks, CA: Sage.

Loeber, R., & Stouthamer-Loeber, M. (1998). Development of juvenile aggression and violence: some common misconceptions and controversies. *American Psychologist,* 242-259.

Loeber, R., Wung, P., Keenan, K., Giroux, B., Stouthamer-Loeber, M., VanKammen, W.B., & Maughan, B. (1993). Developmental pathways in disruptive child behavior. *Development and Psychopathology,* 103-133.

Moffitt, T.E. (1993). Adolescence-limited and life-course-persistent antisocial behavior: A developmental taxonomy. *Psychological Review, 100,* 674-701.

Nunnally, J. (1978). *Psychometric theory* (2nd edition). NY: McGraw-Hill.

Offord, D., Lipman, E., & Duku, E. (2001). Epidemiology of problem behavior up to age 12 years. In R. Loeber and D.P. Farrington (Eds.), *Child delinquents: Development, intervention, and service needs.* Thousand Oaks, CA: Sage.

Pedhazur, E.J. (1997). *Multiple regression in behavioral research* (3rd Edition). NY: Harcourt Brace. (Pp. 294-318)

Peters, M., Thomas, D., & Zamberlan, A. (1997 September). *Boot camps for juvenile offenders.* Washington, DC: U.S. Department of Justice, Office of Juvenile Justice and Delinquency Prevention.

Pilkonis, P.A., & Klein, K.R. (1997). Commentary on the assessment and diagnosis of antisocial behavior and personality. In D.M. Stoff, J. Breiling, & J.D. Maser (Eds.), *Handbook of antisocial behavior.* NY:

Wiley.

Reid, J.B., & Eddy, J.M. (1997). The prevention of antisocial behavior: Some considerations in the search for effective interventions. In D.M. Stoff, J. Breiling, & J.D. Maser (Eds.), *Handbook of antisocial behavior.* NY: Wiley.

Richters, J.E., & Cicchetti, D. (1993). Mark Twain meets DSM-III—R: Conduct disorder, development, and the concept of harmful dysfunction. *Development and Psychopathology, 5(1-2)*, 5-29.

Roosa, M.W. (2000). Some thoughts about resilience versus positive development, main effects versus interactions, and the value of resilience. *Child Development, 71*, 567-569.

Rutter, M. (1985). Resilience in the face of adversity: Protective factors and resistance to psychiatric disorder. *British Journal of Psychiatry, 147*, 598-611.

Rutter, M. (1997). Antisocial behavior: Developmental psychopathology perspectives. In D.M. Stoff, J. Breiling, & J.D. Maser (Eds.), *Handbook of antisocial behavior.* NY: Wiley.

Sampson, R.J. (1995). The community. In J.Q. Wilson & J. Petersilia (Eds.), *Crime.* San Francisco: ICS Press.

Sampson, R.J., & Raudenbush, S.W. (2001 February). Disorder in urban neighborhoods—Does it lead to crime? [Research in Brief] Washington, D.C.: U.S. Department of Justice, National Institute of Justice.

Sampson, R.J., & Laub, J.H. (1997). A life course theory of cumulative disadvantage and stability of delinquency. In T.P. Thornberry (Ed.), *Developmental theories of crime and delinquency.* Advances in criminological theory, Vol. 7. New Brunswick, NJ: Transaction Publishers.

Schorr, L.B. (1999, July). Replicating complex community partnerships. In D. Kennedy, J.P. Thompson, L.B. Schorr, J.L. Edleson, & A.L. Bible (Eds.), *Viewing crime and justice from a collaborative perspective: Plenary papers of the 1998 Conference on Criminal Justice Research and Evaluation.* Washington, D.C.: National Institute of Justice.

Snyder, H.N. (1998). Serious, violent, and chronic juvenile offenders—An assessment of the extent of and trends in officially recognized serious criminal behavior in a delinquent population. In R. Loeber & D.P. Farrington (Eds.), *Serious and violent juvenile offenders: Risk factors and successful interventions.* Thousand Oaks, CA: Sage.

Snyder, H.N. (2001). Epidemiology of official offending. In R. Loeber & D.P. Farrington (Eds.), *Child delinquents: Development, intervention, and service needs.* Thousand Oaks, CA: Sage.

Stevens, A.B., Owen, G., & Lahti-Johnson, K. (July 1999). *Delinquents under 10: Targeted Early Intervention. Phase I Evaluation Report.* Minneapolis, MN: Wilder Research Center and Hennepin County Attorney's Office.

Stouthamer-Loeber, M., Loeber, R., VanKammen, W., & Zhang, Q. (1995). Uninterrupted delinquent careers: The timing of parent help-seeking and juvenile court contact. *Studies on Crime and Crime Prevention, 4*, 236-251.

Thornberry, T.P. (1996). Empirical support for interactional theory: A review of the literature. In J. D. Hawkins (Ed.), *Delinquency and crime: Current theories.* Cambridge, England: Cambridge University Press.

Tolan, P.H., & Thomas, P. (1995). The implications of age of onset for delinquency risk II: Longitudinal data. *Journal of Abnormal Child Psychology 23(2)*, 157-181.

Wasserman, G.A., & Seracini, A. (2001). Family risk factors and interventions. In R. Loeber & D.P. Farrington (Eds.), *Child delinquents: Development, intervention, and service needs.* Thousand Oaks, CA: Sage.

Wasserman, G.A., Miller, L.S., & Cothern, L. (2000 May). *Prevention of serious and violent juvenile offending.* [Juvenile Justice Bulletin] Washington, DC: U.S. Department of Justice, Office of Juvenile Justice and Delinquency Prevention.

Zimmerman, M.A., & Arunkuman, R. (1994). Resiliency research: Implications for schools and policy. *Society for Research in Child Development: Social Policy Report, 8*, 1-17.

Multi-Problem Violent Youth
R.R. Corrado et al. (Eds.)
IOS Press, 2002

207

The Under 12 Outreach Project for Antisocial Boys: A Research Based Clinical Program

Christopher D. Webster, Leena K. Augimeri and Christopher J. Koegl

For the past few years we have been striving to produce a simple, user friendly, manual for mental health practitioners dealing with children under the age of 12 who behave in seriously antisocial ways in their homes, their schools, and their communities. In doing this, we have been trying to capture recent scientific and professional advances, and to cast them in such a form that they might assist practitioners first to estimate risk of violence and then to take steps to attenuate it. This chapter describes previous research and clinical efforts in this direction and outlines the most recent version of what we call the *Early Assessment Risk List for Boys* (EARL-20B, Version 2; Augimeri, Koegl, Webster, & Levene, 2001). We conclude by showing that, having started on this track, we now face heavy responsibilities to continue to validate this device, which even after three years of effort we still view as provisional.

Case examples abound in our agency. Many of these have highly positive results. Over the years some children, with the help of our staff, their families and their communities, show a remarkable "turn-around." They complete our Under 12 Outreach Project (ORP), described below, perhaps even becoming counselors-in-training in their teens, and go on to make good vocational choices. Unfortunately, our agency is not successful in all cases. It could hardly be expected to be so. One case involving a young boy, here called Mark, which actually dates back to a time before the ORP came into existence in 1985, turned out particularly unfortunately. Now a man in his mid-20s, Mark was recently declared to be a Dangerous Offender (DO) under the *Criminal Code* of Canada. His adult offences involved serious sexual assaults and his DO status means that he will now be detained in a penitentiary indefinitely, perhaps for the rest of his natural life. Nearly twenty years ago he had been treated at ECFC, as well as other agencies. No matter what overall encouragement the grouped statistical data from outcome studies of the kind summarized below may indicate, a few individual cases of the sort just mentioned – ones that involve grave harm to victims on the one hand and lengthy imprisonment of perpetrators like Mark on the other – force continued attention to the serious suffering which can flow from failure to intercede effectively during the childhoods of children.

1. The ORP Program

A question which we can only answer rather impressionistically is this: If Mark were to be entering our Under 12 Outreach Project (ORP) today rather than the less-developed program offered many years earlier, would there have been a reduced chance of such future serious offending? Over the years we have gradually come to realize that children like Mark require comprehensive cognitive-behavioral interventions that extend into a variety of settings (i.e., home, school and community), and focus on impulse control and problem solving.

It would seem that the only sustained, multifaceted intervention designed for children under 12 engaging in delinquent and antisocial behaviors is the ORP of the Earlscourt Child and Family Centre in Toronto, Canada (Howell, 2001). Over the past 15 years the ORP has treated approximately 650 children and their families. Children are referred for engaging in a variety of behaviors such as shoplifting, vandalism, break and enter, assault, and firesetting. Between 1985 and 1999, the predominant presenting problem prompting referral to the ORP was theft (including shoplifting). Since its inception, the ORP has worked closely with the Toronto Police Service to ensure that children having police contact receive services with their families that are timely and effective. Since February 1999, with the implementation of the Toronto Centralized Police Protocol (Augimeri, Koegl, & Goldberg, 1999), children can obtain ORP-like services across the greater Toronto region through a single entry access system.

A social learning model is used within the ORP both to explain the development of aggressive and antisocial behavior and to suggest how it can best be curbed (Patterson, Reid, Jones, & Conger, 1975; Rose & Edelson, 1988). The central objective of the program is to reduce police contact among a population at risk of engaging in delinquent conduct (at least some of which would be considered criminal activity had the children turned 12 years of age). The program has two primary goals: to decrease the children's antisocial behavior and to increase their social competence. Interventions target the child, the parents, the school and the community, as required. The hub of the ORP is the Stop Now And Plan (SNAP™) program, a cognitive-behavioral strategy designed to teach children and their parents to stop and think before they act.

The ORP consists of eight major components: (1) SNAP™ Children's Group (formerly known as the Transformer Club), a 12-week after-school, structured group program focused on teaching self-control and cognitive-behavioral problem solving (Earlscourt Child and Family Centre, 2001a); (2) a concurrent 12-week SNAP™ Parenting Group focusing on effective child management (Earlscourt Child and Family Centre, 2001b); (3) family counseling based on SNAPP: Stop Now And Plan Parenting (Levene, 1998); (4) in-home academic tutoring; (5) school advocacy and teacher consultation; (6) victim restitution; (7) individual befriending of boys to help them hook up to structured community-based activities; and (8) a Friday Night Club for high-risk boys who have completed the SNAP™ Children's Group. Over the fifteen years of program operation, the ORP has become longer in duration (by adding the post-program club), increasingly "manualized" (Earlscourt Child and Family Centre, 2001a, 2001b; Levene, 1998; see, generally, Kazdin, 1997) and more comprehensive (through the offering of befriending, advocacy, victim restitution, and tutoring components).

2. The Research

Since the establishment of the ORP, we along with previous ORP researchers have attempted to address two major questions. First, is the ORP an effective intervention? And second, is it possible to develop a clinically relevant device that will allow tolerably accurate predictions of risk of future antisocial conduct? The following brief review presents some of the key findings from our research studies conducted at ECFC in the ORP since 1985. For more detailed accounts, readers are referred to the original studies cited below.

3. Studies on Intervention Effectiveness

Study 1. Our first attempt to establish program effectiveness involved gathering data on children at admission, discharge, and 6- and 12-month follow-up intervals (Hrynkiw-Augimeri, Pepler, & Goldberg, 1993) using the Child Behavior Checklist (CBCL) (Achenbach & Edelbrock, 1983). The study was based on an initial sample of 88 boys and 16 girls who were admitted to the ORP between 1985 and 1988. For the purposes of this chapter, we report on the 35 children for whom data were available on the Internalizing, Externalizing, and Total CBCL scores, and the 25 children for whom Social Competence CBCL scores were available for all four time intervals noted above.[1] The findings indicated a steady and statistically significant decrease in Externalizing and Internalizing problem sub-scale scores (i.e., indicating positive treatment gains). Total CBCL t scores followed this same pattern. Social competence scores increased over the period in accord with expectation (also indicating positive treatment gains). Although the treatment gain was statistically significant, it appeared to diminish at the 12-month CBCL follow-up interval. Results also indicated that only 1 in 5 of the children had further contact with the police (3 for running away, 2 for theft, 1 for vandalism, and 1 for assault) after being discharged from the ORP. Although encouraging, the overall positive results could not necessarily be attributed to the program due to the absence of comparable data from an untreated, contrast group.

Study 2. Building upon the promising results generated by Study 1, our second attempt to demonstrate the effectiveness of the ORP (Day & Hrynkiw-Augimeri, 1996) involved detailed study of 24 boys and 8 girls who met one or both of two ORP admission criteria: contact with the police within six months of referral, and high CBCL delinquency ratings by parents. The fact that nearly 90% met the first criterion alone gives some indication of the seriousness of the acts committed by the children in the sample. Children were referred for trespassing, public mischief, truancy, theft, severe defiance, fighting, vandalism, forgery, arson, and assault.

The most important design innovation for Study 2 entailed the inclusion of a random control group. Children were randomly assigned to two groups (matched on age and sex): an Immediate Treatment Group (ITG) in which children attended the 12-week Transformer Club, and a Delayed-Treatment Group (DTG) in which children participated in 12-weeks of non-clinical recreational activities. Groups were counterbalanced so that the DTG received Transformer Club sessions after 12 weeks and vice versa. CBCL measures were administered at pre and post admission, and at 6, 12, and 18-month follow up intervals. The only significant CBCL effects occurred in the Immediate Treatment Group. The DTG produced not a single positive effect. In contrast, the ITG yielded 8 out of 9 possible significant effects. Not only did positive changes occur in the CBCL scores following intervention, these effects were maintained at follow-up (i.e., replicating the basic Study 1 finding). When the ORP intervention was applied to the DTG group after the end of the 12-week period (during which time the ITG had been treated), it was not found to yield comparable statistically significant changes. Although this absence of such a positive result diminishes somewhat the strength of the overall major finding, it is entirely possible that the effect of waiting to receive the clinical Transformer Club groups seriously impaired the chance of effectively engaging the DTG children and their families in programs of intervention. In some cases, the crisis may have passed by the time children were accepted into the program and the need for intervention may have appeared remote at this stage (for other possible explanations see Day & Hrynkiw-Augimeri, 1996, pp. 79-80).

The ITG's improvement was not restricted to CBCL measures. Other pre-treatment to follow-up improvements were found in the children's self-reported delinquency, parental reports of their children's delinquency and their own levels of stress. As well, nearly 60% of the ITG children compared to a third of the DTG children scoring in the clinical range of the Externalizing sub-scale of CBCL pre-intervention were no longer scored within the clinical range at the second follow-up. During the course of the study, 14 of 29 children for whom data were available at follow up, had at least one contact with the police. Most of these dealings were for fairly minor incidents (e.g., throwing apples at parked cars, breaking windows). There was no significant difference between the two study groups in this respect. What did predict police contact in the pooled group was the child's report of delinquent activity. Those children who disclosed a wide range of anti-social acts were at especially high risk for police contact.

The overall results from Study 2 provided tolerably convincing evidence of the program's effectiveness. Based on the often called for, but rarely provided random assignment to groups, it encouraged us to think that the ORP is a cost-effective child-and-family focused intervention for children under 12 who commit offences. Although the main focus of Study 2 was on demonstrating program effectiveness, the report by Day and Hrynkiw-Augimeri (1996) does offer in its conclusion a list of 20 factors of possible importance in gauging risk of aggression and in providing services to children and families. This list, though different in many respects from that offered in the EARL-20B (see below), certainly foreshadows some of the items more fully described in Version 2 of the device. This underlies how concentration on program effectiveness issues leads almost ineluctably to considerations involving risk and protective factors.

4. Building a Risk Assessment Model

It would be extremely helpful to be able to create a device which would help clinical assessors forecast which children are at especially high risk of future anti-social conduct (i.e., like Mark described above). This has been an objective of many researchers and clinicians for a long time (e.g., Hellman & Blackman, 1966, who early proposed that firesetting, enuresis and cruelty to animals are critically important factors).

Study 3. Day and Hunt (1996) attempted to create a new model for predicting antisocial conduct based on five factors: Age of Onset (of antisocial behavior), Variety of Antisocial Behaviors, Variety of Settings (in which the child's behavior was a problem), Severity of Aggression, and Hyperactivity. These items were derived from earlier work by Loeber (1982, 1990, 1991). Scores on a 5-item Risk Assessment Instrument (RAI) were pitted against parent-rated Delinquent sub-scale CBCL scores, which served as the outcome criterion. There was an interval of 10 months between pre-treatment screening for the ORP and follow-up (i.e., 6 months after discharge from the ORP). Full data were available for 68 children who attended the ORP between 1990 and 1991. Of main interest in this study was the extent to which the items in the RAI linked to the outcome CBCL Delinquent subscale scores. The association between RAI items and outcome, as measured by CBCL Delinquency, was explored in two ways: via multiple regression and via Relative Improvement Over Chance (RIOC) statistics. Stepwise multiple regression revealed that Severity of Aggression (measured one month prior to admission to the ORP) accounted for 27% of the variance ($B = 0.52$, $r = 0.52$, $p < .001$). A second factor, Variety of Antisocial Behaviors, though reliable beyond the .05 level, accounted for an inconsequential 5% of the variance ($B = 0.22$, $r = 0.35$). The other three items failed to enter. Evaluation of predictive validity using RIOC analyses mirrored these findings. In general, the results from Study 3 lent support to the likely utility of developing a risk prediction instrument. Another finding was that ratings of Severity of Aggression, Variety of Antisocial Behaviors, and Variety of Settings were highly correlated, suggesting that it may be difficult to distinguish among these variables. In Version 2 of the EARL-20B discussed below, these factors are combined into the one item of "Antisocial Behavior" (Item C11, pp. 56-57).

Study 4. A limitation of the three studies previously discussed is the absence of an ecologically valid or "hard" outcome measure. What needed to be determined was the ability of the ORP to prevent youth court contact over a period of time. Scores on well-studied instruments like the CBCL, based as they are on parent reports, are subject to bias, and do not address an important area of interest, namely, establishing how many children have subsequent, documented, criminal justice system contact. Study 4 dealt with this topic. For the first time at ECFC we attempted to link possible predictor variables like family stressors, school problems, peer problems, history of abuse, and likeability to whether or not the child had subsequent contact with the law as an adolescent or adult. A total of 203 children, all treated within the ORP, were followed for an average of seven and a half years.

Forty-eight percent of these troubled children went on to have court contact, a finding which cannot be lightly dismissed.[2] This is particularly so when it is realized that many of these grown children would have received assistance from mental health and social service agencies in their teen and young adult years. There would have been a possibly large but unfortunately unknown mental health "dark figure". However this may be, the 48% "failure" figure, though likely inflated by some criminal charges resulting from minor or trivial circumstances, did not incline complacency.

Results from Study 4 showed via survival analyses that girls, as they grew older, were markedly less likely to have youth court contact by age 18 than their male counterparts (38 versus 66% respectively). As well, likeability, or actually "unlikeability", was associated with future court activity for boys (but not girls). Another factor, history of abuse, was predictive for girls (but not for boys). Although, as Day (1998) points out, these results would benefit from replication, they nonetheless influenced us as we later developed the EARL-20B and the EARL-21G (Levene et al., 2001).

While the above studies were in progress, the clinical program, was, of course, not standing still. Findings from Studies 1 through 4 provided some confidence about the usefulness of the ORP treatment model and seemed to offer general support for the idea of creating a structured risk assessment scheme. These findings helped inspire a risk assessment list consisting of 53 items (Augimeri & Levene, 1994, 1997). The items included the ones isolated in Study 4, along with many others. All were answered as Yes/No. This scheme proved highly useful for clinicians in "red flagging" high-risk children during the initial treatment planning phase, but its structure did not invite formal research (see Hrynkiw-Augimeri, 1998). For this reason we began in 1997 to create a 20-item device scored on a three-point scale.[3] This meant collapsing the many items of the above scheme into far fewer categories. The task took some time, patience, and persistence. It also required the joint efforts of clinicians, researchers, and administrators.

Fairly early in the development process we began to realize that it would be necessary to create two devices, not just one. We had already learned that girls may not exhibit the same kinds of problem behaviors as boys as they grow into adolescence and adulthood (Study 4) and that, conceivably, the kinds of risk factors present in childhood are either not the same or operate differently across sexes. For that reason we decided first to create a device for boys and then re-think our approach to risk assessment in girls. As well, starting in 1996, the ORP provided service only to boys. This move paralleled the establishment of the Earlscourt Girls Connection, a gender-specific program for girls with disruptive behavior problems (Levene, 1997).

The boys' version was published after a year or so of library research and consultations with experienced local clinicians. It was called the *Early Assessment Risk List for Boys*, Version 1, Consultation Edition (EARL-20B; Augimeri, Webster, Koegl, & Levene, 1998). Version 2 of the EARL-20B was recently published after extensive consultation with leading experts in the field. In Table 1 we show the items from the *current* version. Very recently, we have published the *Early Assessment Risk List for Girls*, Version 1—Consultation Edition (EARL-21G; Levene et al., 2001). Although it owes a good deal to the EARL-20B, painstaking efforts have been made to apply scientific literature and clinical experience unique to girls. The EARL-21G is not considered further in this chapter.

Table 1: Items in Version 2 of the Early Assessment Risk List for Boys (EARL-20B)

Family (F) Items		Child (C) Items		Responsivity (R) Items	
F1	Household Circumstances	C1	Developmental Problems	R1	Family Responsivity
F2	Caregiver Continuity	C2	Onset of Behavioural Difficulties	R2	Child Responsivity
F3	Supports	C3	Abuse/Neglect/Trauma		
F4	Stressors	C4	Hyperactivity/Impulsivity/-Attention Deficits (HIA)		
F5	Parenting Style	C5	Likeability		
F6	Antisocial Values and Conduct	C6	Peer Socialization		
		C7	Academic Performance		
		C8	Neighbourhood		
		C9	Authority Contact		
		C10	Antisocial Attitudes		
		C11	Antisocial Behaviour		
		C12	Coping Ability		

Study 5. Since there is very little point in pursuing the construction of a risk assessment device unless or until it can be demonstrated that different raters, when using it, can achieve similar scores, Study 5 was designed as a simple exploratory attempt to determine the basic inter-rater reliability of the EARL-20B. The device was used prospectively on 21 boys and their families admitted into the ORP during the 1998 spring session. The overall results indicated that the EARL-20B proved to have clinical and research relevance. The device was especially helpful in providing clinicians "with a thorough assessment procedure, a guide to gear the treatment interventions, and a barometer to evaluate whether a child was still considered high-risk at post intervention" (Hrynkiw-Augimeri, 1998, p. 31). Always recognizing that the perfectly acceptable inter-rater correlations based on total EARL-20B scores would doubtless diminish when studied item-by-item, Study 5 demonstrated that there was probably enough stability in the device to merit fuller development.

Study 6. With some confidence in the EARL-20B built from the results of Study 5, it now seemed worthwhile to attempt in Study 6 a large-scale follow-up of previously ORP-treated children using the EARL-20B as a risk assessment base. The design also called for a "hard" outcome measure by way of a search of correctional records obtained through a special court order. This order allowed us to establish whether or not there had been actual findings of criminal guilt in each and every one of the 447 ORP children treated between 1985 and 1999 (378 boys, 69 girls). This sample included nearly all of the cases examined in the studies reported above. Files were coded for EARL-20B scores by three different raters (see Appendix C in Augimeri et al., 2001). An analysis of total EARL-20B scores revealed a highly acceptable level of correspondence between the three raters with correlation coefficients ranging from .79 to .97. The follow-up data obtained from the correctional authorities showed that nearly 60% of the ORP children (85% boys, 15% girls) had never been found guilty of committing an offence. The mean age of the sample at the time of follow-up was 17.5 years. The question of main interest was, how do EARL-20B total scores relate to future convictions? Much work remains to be done on these data. But it is possible here to report that chi-square analyses showed that children scoring high on the EARL-20B were significantly more likely to have future criminal charges against them than those who had a low EARL-20B score. This effect was statistically significant for both the boys and girls in the sample.

Study 6 has been our most ambitious to date. Its limitations (e.g., failure to include a non-clinical comparison group, absence of prospective design component, unavailability of mental health outcome data) were offset by some strengths (e.g., large number of participants, multiple predictor variables, hard outcome data, demonstrated inter-rater reliability of EARL-20B scores). Curiously perhaps, after 15 years of struggle, only now do we have a full appreciation of what it takes to produce an assessment device like the EARL-20B, one that may come to have the strength and durability necessary for the efficient conduct of long-term prediction follow-up studies. Since the new version of the EARL-20B draws together not just our own research but that of others, and since it is based on extensive consultation with practicing clinicians and acknowledged experts, it is now necessary to describe it in more detail.

5. The EARL-20B

The scheme arises directly from the studies noted above, along with the outcomes of many articles and books published in recent years. It is based on the assumption that there now exists enough scientific information about aggression and violence in children under 12 years of age to put forward practical, researchable, assessment schemes (e.g., Brennan & Mednick, 1997; Dishion & Patterson, 1997; Frick & Ellis, 1999; Hinshaw & Zupan, 1997; Kazdin, 1987; Loeber & Farrington, 1997; Moore & Tonry, 1998; Rutter, 1997). Over the past quarter century, appreciable advances have been made in our understanding of violence prediction in adults suffering from mental and personality disorders (Bartel, Borum, & Forth, 2000; Boer, Hart, Kropp, & Webster, 1998; Ennis & Litwack, 1974; Monahan & Steadman, 1994; Rice, 1997; Webster, Douglas, Eaves, & Hart, 1997). Many of these conceptual and practical achievements have direct applicability to children. Within the developmental-clinical field itself, noticeable strides have also been made. There is no longer any notion that a very small number of negatively-loaded variables like firesetting, enuresis, and cruelty to animals can suffice to make accurate predictions of future antisocial behavior (Hellman & Blackman, 1966). There is increasing recognition that many factors, some positively- or protectively-loaded (Hawkins, Arthur & Olson, 1997; Rogers, 2000), tend to be multi-layered and that these interact together in complex ways (see generally, Hawkins et al., 1997; Loeber & Farrington, 1997; Pepler & Rubin, 1991; Stoff, Breiling, & Maser, 1997). A main aim of the EARL-20B, Version 2, is to render current scientific findings and concepts useable and to help overcome the often evident disjunction between textbook authority on the one hand, and the actualities of day-to-day practice on the other (Hetherington, 1998; Waggoner, 1983-1984).

The EARL-20B, Version 2, is expected to help inspire well-conceived but practically-helpful research. As an evolved version of earlier schemes noted above, the present items in the manual have been found to be both reliable (as remarked in the previous section) and capable of indexing change in risk over time. The issue of longitudinal monitoring of precise changes in the individual case has recently been shown to be of exceptional importance both from research and clinical points of view (Kraus, Sales, Becker, & Figueredo, 2000). The framework of the EARL-20B is designed such that the device will be useful in research and clinical practice (Loeber & Stouthamer-Loeber, 1998). As well, it is expected to stimulate study across different sites (e.g., as in the MacArthur risk study; see Steadman et al., 2000).

Assessment should not be seen as an end in itself. Our purpose through the EARL-20B is to stimulate action in the creating of effective treatment interventions under conditions that are often adverse (Frick, 1998) and to assist in the allocation of scarce resources (Ramey & Ramey, 1998). For this to be done, it is essential that basic terms be defined unambiguously. The obvious advantage of such "transparency" (Baker, 1993) is that it sets the stage for detailed research into treatment effectiveness (e.g., Achenbach, 1991, with respect to the CBCL; Day & Hunt, 1996, with respect to precursors of the EARL-20B, Version 2).

The EARL-20B is expected to help improve general standards of clinical practice (Webster & Cox, 1997). As a fairly straightforward *aide mèmoire* based both on scientific research and clinical practice (Hinshaw & Zupan, 1997), it helps practitioners to think about the criticality of some of the information that is needed (Kropp & Hart, 2000; Webster, Menzies, & Jackson, 1982), the way it is best and more fairly collected (Bruck, Ceci, & Hembrooke, 1998), and the methods by which its accuracy can be confirmed with parents and others (Webster, 1997). It also provides a framework within which it can be determined which clinicians have the best ability to make accurate forecasts (Menzies, Webster, McMain, Stahley, & Scaglione, 1994; Shah, 1981).

Researchers, clinicians, and policy makers are naïve if they think that there is a particular "right way" to go about assessing risk in young children and their families. The concepts we use, both in professional and scientific discourses, are constantly altering. Even "childhood" has been viewed very differently across decades (Aries, 1962). Our current, perhaps rather linear, theorizations may yield over time to forms only dimly conceived at present (Marks-Tarlow, 1993). Experience with devices like the EARL-20B, Version 2, may lead us eventually to acquire as much expertise in assessing situations and circumstances as in evaluating individuals and personalities (Ben & Allen, 1974). All of this means that the present manual will, in due course, have to undergo alteration as it accommodates not only to scientific advances but to general changes in society.

6. Next Steps

Having now launched the EARL-20B and the EARL-21G, we find that we have opened a number of new possibilities. These merit consideration in turn. First, the device ought to enable researchers to determine which kinds of children benefit most, if at all, from ORP-type programs. Could it be that children exhibiting a particular pattern of EARL-20B or EARL-21G scores are particularly inclined or disinclined to show benefits from systematically planned interventions? Or, to put it a little differently, what kinds of components in an ORP-type program are essential for what kinds of children? Since it seems a grave mistake to think that ORP-type programs necessarily have some kind of universal applicability, there are good reasons to begin to think seriously about matching children to programs. Second, the device ought to encourage continued thought about what constitutes ideal, yet practically feasible, measures of outcome. What measures are suitable for boys? Which ones are most applicable to girls? Now that we know a little about the fate of our children within the criminal justice system as they have aged, how can comparable mental health system information be obtained and incorporated into our datasets? Although the prime focus at ECFC has always been on averting criminal justice system involvement, does it make sense to ignore the "other side" (especially when it is well known that whether individuals are routed one way or the other contains a strong chance element; Quinsey, 1981). Third, the device, if it is to achieve any kind of proper use by professionals, has got to be presented in a way that colleagues will come to appreciate its strengths. This raises the whole issue of dissemination. How can the device be presented in such a way that it engages the attention of workers as they try to help children and families? Is there a science, of a kind, having to do with how best to impart information to colleagues whether they are highly experienced or otherwise? Fourth, the device should enable participation in research and clinical projects across sites. This is important because prospective studies, much favored for obvious reasons, take a long time to complete. Grouping data across sites can help overcome this difficulty. Accordingly, how can colleagues, especially if geographically separated, achieve consistency in data recording and how can cross-site information be integrated in order to obtain the requisite statistical power? Fifth, because it can be administered repeatedly over time, the device is suitable for application at the level of the individual participant. Massed statistical data are of some, but not complete, use to practicing clinicians. Researchers, unfortunately, usually do not occupy themselves with "outliers". Yet clinicians, in the light of everyday work, seem to be dealing with heavy case loads made up of cases that are largely heterogeneous. Although single-participant research methodology is hardly new (Morgan & Morgan, 2001), it would seem that practical steps must be taken to establish it as a clinically useful complementary task. Sixth, the device could lend itself to simplification for use at the "front end" by law enforcement officials and the like. In a reduced, more compact format, it could be helpful to police both as a screening and educational tool. Determining which items are *really* relevant to the kinds of *urgent* decisions often required of police officers could be accomplished with relative ease through research-based consultation with law enforcement authorities. Seventh, the devices could be extended to deal properly with boys and girls under the age of 6 and between 12 and 18. Although certain moves have been made in these directions of late (e.g., Bartel et al., 2000), much remains to be done. How do variables, ones presumed important in the 6 to 11 range, apply to older children, and what new variables need to be incorporated to take account of this very large and important group? Eighth, with the current 20/21 items spelled out, there is now good opportunity to determine which items are important in a predictive or clinical sense, or both (see Witt, 2000, p. 794, who urges "detailed item-analysis studies" for a related device).

Along with these and other exciting possibilities, come additional responsibilities. As already noted, manuals like the EARL-20B need periodic revision. This time-consuming work is best done face-to-face with colleagues who, very often, are at a distance. With the schemes published, there also comes a professional obligation to ensure their consistent and proper use. This means a big task in training, education, and dissemination. And, regrettably, no matter how much effort is expended in this direction, there can never be a guarantee that the instrument will not be misapplied.[4] The research and professional bases which underlie the manuals require constant consideration. New studies are published. Clinical practices change. Diagnostic systems alter. Even when manuals are published for use by the broader professional community, complex issues of "ownership" arise both with respect to the original documents and translations of them. These have to be dealt with carefully, patiently, and constructively. Lastly, there is a need to buttress assessment devices with up-to-date, carefully controlled, research findings on reliability and the various forms of validity. Since it costs considerable money to develop and test such instruments, a lot of time has to be spent competing for grants and contracts.

In very general terms, then, it seems that research and clinical practice at ECFC have gradually coalesced to yield, among many other advances, the EARL-20B and the EARL-21G manuals. These devices may come to have fairly wide applicability. This remains to be seen. Their present fragile state of development means that we and our colleagues working in the same general area will have no shortage of work over the next several years.

Notes

[1] Of the initial pool of 104 children, data were missing due to participant dropout ($N = 16$) and incomplete discharge and follow-up data see Hrinkiw-Augimeri et al. (1993) for a complete discussion of participant attrition.

[2] It needs to be noted that "court contact," though a very useful outcome measure, is less stringent than the criterion of having an actual criminal conviction (as we used in Study 6).

[3] This basic idea was adapted from other recent attempts to devise structured clinical guides, such as the Hare Psychopathy Checklist-Revised (Hare, 1991) and the HCR-20 (Webster et al., 1997). For a general review of the clinical and research utility of this 20-item, 0/1/2 approach, see Mossman (2000).

[4] It is surprisingly difficult to meet scientific and professional ethical standards. For an incisive discussion see Campbell (2000). This author's main points, though dealing with sex offender assessment devices, have applicability to the present topic.

References

Achenbach, T. M. (1991). *Manual for the Child Behavior Checklist and 1991 Profile*. Burlington, Vermont: Department of Psychiatry, University of Vermont.

Achenbach, T. M., & Edelbrock, C. (1983). *Manual for the Child Behavior Checklist and Revised Behavior Profile*. Burlington, Vermont: University Associates in Psychiatry.

Aries, P. (1962). *Centuries of childhood* (R. Baldick, Trans.). New York: Knopf.

Augimeri, L. K., Goldberg, K., & Koegl, C. J. (1999). *Canadian children under 12 committing offences: Police protocols*. Toronto: Earlscourt Child and Family Centre.

Augimeri, L. K., Koegl, C. J, Webster, C. D., & Levene, K. S. (2001). *Early Assessment Risk List for Boys*, Version 2. Toronto: Earlscourt Child and Family Centre.

Augimeri, L. K., Webster, C. D., Koegl, C. J, & Levene, K. S. (1998). Early Assessment Risk *List for Boys*, Version 1 Consultation Edition. Toronto: Earlscourt Child and Family Centre.

Augimeri, L., & Levene, K. (1994, Revised 1997). *Outreach Programme: Risk factors associated with possible conduct disorders and non-responders*. Toronto: Earlscourt Child and Family Centre.

Baker, E. (1993). Dangerousness, rights, and criminal justice. *Modern Law Review, 56*, 528-547.

Bartel, P., Borum, R., & Forth, A. (2000, March). *The development and use of the Structured Assessment of Violence Risk in Youth*. Paper presented at the American Psychology-Law Society Meeting, New Orleans, Louisiana.

Bem, D., & Allen, A. (1974). On predicting some of the people some of the time: The search for cross-situational consistencies in behavior. *Psychological Review, 81*, 506-520.

Boer, D. P., Hart, S. D., Kropp, P. R., & Webster, C. D. (1998). *Manual for the Sexual Violence Risk–20. Professional guidelines for assessing risk of sexual violence*. Burnaby, British Columbia: Mental Health, Law, and Police Institute, Simon Fraser University.

Brennan, P.A., & Mednick, S.A. (1997). Medical histories of antisocial individuals. In D. M. Stoff, J. Breiling, & J. D. Maser (Eds.), *Handbook of antisocial behavior* (pp. 269-279). New York: Wiley.

Bruck, M., Ceci, S. J., & Hembrooke, H. (1998). Reliability and credibility of young children's reports. *American Psychologist, 53*, 136-151.

Campbell, T. W. (2000). Sexual predator evaluations and phrenology: Considering issues of evidentiary reliability. *Behavioral Sciences and the Law, 18*, 111-130.

Day, D. M. (1998). Risk for court contact and predictors of an early age for a first court contact among a sample of high risk youths: A survival analysis approach. *Canadian Journal of Criminology, 40*, 421-443.

Day, D. M., & Hrynkiw-Augimeri, L. (1996). *Serving children at risk for juvenile delinquency: An evaluation of the Earlscourt Under 12 Outreach Project*. Unpublished report, Earlscourt Child and Family Care Centre, Toronto, Canada.

Day, D. M., & Hunt, A. C. (1996). A multivariate assessment of a risk model for juvenile delinquency with an under 12 offender sample. *Journal of Emotional and Behavioural Disorders, 4*, 66-72.

Dishion, T.J., & Patterson, G.R. (1997). The timing and severity of antisocial behavior: Three hypotheses within an ecological framework. In D. M. Stoff, J. Breiling, & J. D. Maser (Eds.), *Handbook of antisocial behavior* (pp. 205-217). New York: Wiley.

Earlscourt Child and Family Centre (2001a). *SNAP™ Children's Group Manual*. Toronto: Author.

Earlscourt Child and Family Centre (2001b). *SNAP™ Parent Group Manual*. Toronto: Author.

Ennis, B. J., & Litwack, T. R. (1974). Psychiatry and the presumption of expertise: Flipping coins in the courtroom. *California Law Review, 62*, 693-752.

Frick, P. J. (1998). *Conduct disorders and severe antisocial behavior*. New York: Plenum.

Frick, P. J., & Ellis, M. (1999). Callous-unemotional traits and subtypes of conduct disorder. *Clinical Child and Family Psychology Review, 2*, 149-168.

Hawkins, J. D., Arthur, M. J., & Olson, J. L. (1997). Community interventions to reduce risks and enhance protection against antisocial behavior. In D. M. Stoff, J. Breiling, & J. D. Maser (Eds.), *Handbook of antisocial behavior* (pp. 365-374). New York: Wiley.

Hetherington, E. M. (1998). Relevant issues in developmental science: Introduction to the special issue. *American Psychologist, 53*, 93-94.

Hellman, D., & Blackman, J. (1966). Enuresis, firesetting and cruelty to animals: A triad predictive of adult crime. *American Journal of Psychiatry, 122*, 1431-1436.

Hinshaw, S. P., & Zupan, B. A. (1997). Assessment of antisocial behavior in children and adolescents. In D. M. Stoff, J. Breiling, & J. D. Maser (Eds.), *Handbook of antisocial behavior* (pp. 36-50). New York: Wiley.

Howell, J. C. (2001). Juvenile justice programs and strategies. In R. Loeber & D. P. Farrington (Eds.), *Child delinquents* (pp.305-321). Thousand Oaks, CA: Sage.

Hrynkiw-Augimeri, L. K. (1998). *Assessing risk for violence in boys: A preliminary risk assessment study using the Early Assessment Risk List for Boys (EARL-20B)*. Unpublished master's thesis, Ontario Institute for Studies in Education, University of Toronto, Ontario, Canada.

Hrynkiw-Augimeri, L. K., Pepler, D., & Goldberg, K. (1993). An outreach project for children having police contact. *Canada's Mental Health, 41*, 7-12.

Kazdin, A. E. (1987). Treatment of antisocial behavior in children: Current status and future directions. *Psychological Bulletin, 102*, 187-203.

Kazdin, A. E. (1997). A model for developing effective treatments: Progression and interplay of theory, research and practice. *Journal of Clinical Child Psychology, 26*, 114-129.

Kraus, D. A., Sales, B. D., Becker, J. V., & Figueredo, A. J. (2000). Beyond prediction to explanation in risk assessment research. *International Journal of Law and Psychiatry, 23*, 91-112.

Kropp, P. R., & Hart, S .D. (2000). The Spousal Assault Risk Assessment (SARA) Guide: Reliability and validity in adult male offenders. *Law and Human Behavior, 24*, 101-118.

Levene, K. (1997). The Earlscourt Girls Connection: A Model Intervention. *Canada's Children, 4(2)*, 14-17.

Levene, K. (1998). *SNAPP Stop-Now-and-Plan Parenting: Parenting children with behaviour problems*. Toronto: Earlscourt Child and Family Centre.

Levene, K. S., Augimeri, L. K., Pepler, D., Walsh, M., & Webster, C. D., & Koegl, C. J. (2001). *Early Assessment Risk List for Girls – Version 1, Consultation Edition (EARL-21G)*. Toronto: Earlscourt Child and Family Centre.

Loeber, R. (1991). Antisocial behavior: More enduring than changeable? *Journal of the American Academy of Child and Adolescent Psychiatry, 30*, 393-397.

Loeber, R. (1990). Development and risk factors of juvenile antisocial behaviour and delinquency. *Clinical Psychology Review, 10*, 1-41.

Loeber, R. (1982). The stability of antisocial and delinquent child behavior: A review. *Child Development, 53*, 1431-1446.

Loeber, R., & Farrington, D. P. (1997). Strategies and yields of longitudinal studies on antisocial behavior. In D. M. Stoff, J. Breiling, & J. D. Maser (Eds.), *Handbook of antisocial behavior* (pp. 125-139). New York: Wiley.

Loeber, R., & Stouthamer-Loeber., M. (1998). Development of juvenile aggression and violence: Some common misconceptions and controversies. *American Psychologist, 53*, 242-259.

Marks-Tarlow, T. (1993). A new look at impulsivity: Hidden order beneath apparent chaos? In W. G. McCown, J. L. Johnson, & M. B. Shure (Eds.), *The impulsive client: Theory, research, and treatment* (pp. 119-138). Washington, DC: American Psychological Association.

Menzies, R. J., Webster, C. D., McMain, S., Staley, S., & Scaglione, R. (1994). The dimensions of dangerousness revisited. *Law and Human Behavior, 18*, 1-28.

Monahan, J., & Steadman, H.J. (Eds.). (1994). *Violence and mental disorder: Developments in risk assessment*. Chicago: University of Chicago Press.

Moore, M. H., & Tonry, M. (1998). Youth violence in America. In M. Tonry & M. H. Moore (Eds.), *Youth violence* (pp. 1-26), Chicago: University of Chicago Press.

Morgan, D. L., & Morgan, R. K. (2001). Single-participant research design. *American Psychologist, 56*, 119-127.

Mossman, D. (2000). Evaluating violence risk "by the book" [Review of the books, *HCR-20: Assessing Risk for Violence*, Version 2, and *Manual for the Sexual Violence Risk–20*]. *Behavioral Sciences and the Law, 18*, 781-789.

Patterson, G. R., Reid, J. B., Jones, R. R., & Conger, R. E. (1975). *A social learning approach to family intervention; I, Families with aggressive children*. Eugene, Oregon: Castalia Publishing.

Pepler, D. J. & Rubin, K. H. (Eds.). (1991). *The development and treatment of childhood aggression*. Hillsdale, New Jersey: Erlbaum.

Quinsey, V .L. (1981). The long term management of the mentally disordered offender. In S. J. Hucker, C. D. Webster, & M. H. Ben-Aron (Eds.), *Mental disorder and criminal responsibility* (pp. 137-155). Toronto: Butterworths.

Ramey, C. T., & Ramey, L. S. (1998). Early intervention and early experience. *American Psychologist, 53*,

109-120.

Rice, M. E. (1997). Violent offender research and implications for the criminal justice system. *American Psychologist, 42*, 414-423.

Rogers, R. (2000). The uncritical acceptance of risk assessment in forensic practice. *Law and Human Behavior, 24*, 595-605.

Rose, S. D., & Edelson, J. L. (1988). *Working with children and adolescents in groups*. San Francisco: Jossey-Bass.

Rutter, M. (1997). Nature-nurture integration: The example of antisocial behavior. *American Psychologist, 52*, 390-398

Shah, S. (1981). Dangerousness: Conceptual, prediction, and public policy issues. In J. R. Hays, T. K. Roberts, & K. S. Solway (Eds.), *Violence and the violent individual* (pp. 151-178). New York: S. P. Medical and Scientific Books.

Steadman, H., Silver, E., Monahan, J., Applebaum, P., Robbins, P., Mulvey, E., Grisso, T., Roth, L., & Banks, S. (2000). A classification tree approach to the development of actuarial violence risk assessment tools. *Law and Human Behavior, 24*, 83-100.

Stoff, D. M., Breiling, J., & Maser, J. D. (Eds.). (1997). *Handbook of antisocial behavior*. New York: Wiley.

Waggoner, S. G. (1983-1984). First impressions. *Child Care Quarterly, 12*, 247-255.

Webster, C. D. (1997). A guide for conducting risk assessments. In C. D. Webster & M. A. Jackson (Eds.), *Impulsivity: Theory, assessment, and treatment* (pp. 343-357). New York: Guilford Press.

Webster, C. D., & Cox, D. (1997). Integration of nomothetic and idiographic positions in risk assessment: Implications for practice and the education of psychologists and other mental health professionals. *American Psychologist, 11*, 1245-1246.

Webster, C. D., Douglas, K. S., Eaves, D., & Hart, S. D. (1997). *HCR-20: Assessing Risk for Violence*, Version 2. Burnaby, British Columbia: Mental Health, Law, and Policy Institute, Simon Fraser University.

Webster, C. D., Menzies, R., & Jackson, M. A. (1982). *Clinical assessment before trial: Legal issue and mental disorder*. Toronto: Butterworths.

Witt, P. H. (2000). Book review: A practitioners view of risk assessment: The HCR-20 and SVR-20. *Behavioral Sciences and the Law, 18*, 791-798.

Monitoring Vital Signs:
Integrating a Standardized Assessment into
Washington State's Juvenile Justice System

Robert Barnoski

1. Legislative History

Responsibility for Washington State's juvenile justice system is shared between state and county governments. County governments operate the juvenile courts, detention facilities, and diversion programs. The most serious offenders are sentenced to incarceration in state correctional institutions managed by the Juvenile Rehabilitation Administration (JRA). The state also distributes supplemental funding for county-based services to the juvenile courts.

Washington is the only state with a determinate sentencing system for juvenile offenders. The point-based sentencing system was enacted in 1977 and measures the youth's prior record and seriousness of the current offense. Because the sentencing guidelines principally focus on terms of confinement, the state's sentencing system has less emphasis on rehabilitation-focused sentences than commonly found in state juvenile systems.

In 1997, the state Legislature embarked on a bold experiment to encourage use of research-proven programs for juvenile offenders. A new funding program was created that required juvenile courts to significantly change several of their operating features in order to qualify. These included:

- Develop and use a statewide risk assessment instrument to determine which programs are most likely to be effective with particular juvenile offenders;
- Emphasize services that reduce risk factors associated with juvenile offending and have demonstrated effectiveness; and
- Participate in a statewide evaluation.

The juvenile courts' statewide organization, the Washington State Association of Juvenile Court Administrators, requested that the Washington State Institute for Public Policy (Institute) develop the assessment specified in the legislation. The Institute is a non-partisan research organization created and funded by the Washington State Legislature.

The juvenile court administrators envisioned an assessment that could accomplish several goals:

- Determine the youth's level of risk for re-offending as to target resources at higher-risk youth;
- Identify the risk and protective factors that are linked to criminal behavior, so the rehabilitative effort can be tailored to address the youth's assessment profile.
- Develop a case management approach that focuses on progress in reducing risk factors and increasing protective factors;

- Streamline assessment and recording procedures through the use of sophisticated assessment software; and
- Allow managers to review the performance of court staff and service providers.

The court administrators and the Institute worked from 1997 to 1998 to develop the risk assessment instrument and establish procedures to implement the legislation. The Washington State Juvenile Assessment (WAJA) was a collaborative process to meet goals of both the legislation and court administrators.

2. Research Basis For Assessment

The development of the WAJA relied on a variety of sources: Recidivism prediction literature and instruments; Theoretical models for juvenile delinquency; Risk and protective factor research; Resiliency research; Research on effective juvenile delinquency programs; Review by an international team of experts; and a series of reviews by Washington State juvenile court professionals including testing of a draft assessment with 150 youth.

The science of risk prediction for juvenile offenders has been evolving for over 30 years. Hoge and Andrews' (1996) book, *Assessing the Youthful Offender, Issues and Techniques,* provides a good overview of this topic. The Office of Juvenile Justice and Delinquency Prevention's *Guide for Implementing the Comprehensive Strategy for Serious, Violent and Chronic Juvenile Offenders* is an excellent source of practical information, and Farrington and Tarling (1985) and Jones (1994) summarize methodology issues.

This research clearly indicates the value of two types of information in an assessment: risk and protective factors. *Risk factors* are circumstances or events in the youth's life that increase the likelihood that the youth will start or continue criminal activities. Historically, risk factors have been the focus of most assessments.

Protective factors are events or circumstances in the youth's life that reduce the likelihood of the youth committing a crime. However, the absence of a protective factor does not always indicate an increased risk for re-offending. Protective factors are those positive things in a juvenile's life that help the youth overcome adversity. An example is having a good relationship with a positive adult role model, assuming the youth's family is unable to fulfill that role. Protective factors are generally based on the resiliency research in delinquency. In addition, efforts were made to ensure that responses to the risk items include the positives as well as negatives. For example, attachment to school is measured not only by a lack of involvement in activities but also by the presence of involvement.

In addition to the risk and protective factor classification, assessment information can be either static or dynamic. *Dynamic factors* are circumstances or conditions in a youth's life that can potentially be changed, such as the youth's friends or school performance. *Static factors* are events in a youth's life that are historic and cannot be changed, such as the youth being physically abused. The factors typically employed in assessments have measured static risk factors. Inclusion of dynamic factors in a comprehensive assessment is critical, since these are the factors used to guide the rehabilitative effort.

The review of theoretical delinquency models revealed a consistent set of risk factors identified by theorists as Andrews and Bonta (1994), Andrews, Bonta, and Hoge (1990), Elliott, Huizinga, and Ageton (1985), Henggeler (1989), LeBlanc, Quimet, and Tremblay (1988), and Patterson, DeBaryshe, and Ramsey (1989). The risk and protective factor research from Hawkins, Catalano, and Miller (1992), and Loeber and Farrington (1998) was also reviewed.

The assessment was also influenced by research on treatment programs. Presumably, treatment approaches are successful when they target and change factors that significantly influence continued criminal behavior. From literature reviews and meta-analyses, the comparative success of cognitive behavioral programs is clear. These programs include family or ecologically focused efforts such as Functional Family Therapy and Multi-systemic Therapy, as well as life or social skill enhancement programs.

Numerous prediction instruments were reviewed during the effort, including an examination of tools such as Lerner's (1986) *Strategies for Juvenile Supervision*, Baird's (1984) *Wisconsin Risk Scale*, and the *Youth Level of Service and Case Management Inventory* by Hoge and Andrews (1994).

During the summer of 1997 a group of international experts reviewed a draft version of the assessment and provided written comments.[1] In addition, more than 40 juvenile court professionals from Washington State worked with the Institute to develop the assessment. In particular, probation counselors helped in refining the terminology and examples. After a series of focus group sessions with juvenile court professionals and a two-day training session in the spring of 1998, a draft instrument was piloted in a dozen Washington State juvenile courts.

3. The Vision: Assessment as a Key Component of Probation

Throughout this process, the team was motivated by a vision of an instrument that played a vital role in the juvenile courts' decisions. This section lays out this vision in detail. In some areas, the vision succeeded, and in others, it fell short; later sections will analyze the relative degrees of success.

For an assessment to become integral to an organization, it must have value to the employees; in this case, the probation officers. That is, the assessment must be more than a form that the probation counselor completes, files, and then ignores in day-to-day decisions. The development team understood that to achieve the goal of reduced re-offending, probation activities must target and change those risk and protective factors related to re-offending. This goal cannot be achieved by a single event; rather it must be part and parcel of the on-going rehabilitative effort. The assessment process includes an Initial Assessment and goal setting, continual Re-Assessment and goal monitoring, case management reviews, and a final assessment.

3.1 Intake or Pre-Screen Assessment

The assessment process begins when a youth is brought to the court for a new offense. During intake, routine criminal and social history data are typically collected by court staff. These data must be recorded and then organized by the assessment software into the major domains on the assessment. That is, the assessment's structure is the guide to collecting and organizing the youth's social file in the database. This enables anyone looking at the youth's computerized file to find, for example, school information by clicking on the school domain tab and viewing all relevant data.

The intake data must be sufficient to provide a good estimate of the youth's risk for re-offense. This Pre-Screen Assessment is a shortened version of the full assessment. The Pre-Screen quickly indicates whether the youth is of low, moderate, or high risk to re-offend. The information collected during intake is then carried forward by the software for use in the full assessment.

3.2 Initial Full Assessment

Juvenile courts were concerned that completing a comprehensive assessment on each youth referred to juvenile court would overwhelm their personnel resources. Courts were to complete the comprehensive assessment only for those youth rated as posing moderate and high risk on the Pre-Screen. Low-risk youth by definition have few problems; so not completing a full assessment presumes the absence of significant risk factors and the presence of some protective factors. In addition, there is no point to devoting resources to low-risk youth because "you cannot fix something that is not broken." As a result of implementing the assessment, many Washington courts now assign low-risk youth to minimum supervision caseloads. This group generally includes youth who are not anti-social, but rather, made an error in conduct and have a very low probability of re-offending.

For moderate- and high-risk youth, a structured motivational interview is then conducted with the youth and youth's family. During this interview, additional information related to risk and protective factors is gathered. The juvenile probation counselor must use his or her professional judgment and training to interpret the Pre-Screen and interview information to complete the full assessment. This analysis combines a thorough understanding of the assessment concepts with the ability to elicit and interpret information. This interview is also the first step in the rehabilitative process in which the probation counselor lets the youth and family know the counselor is interested in getting to know them and their problems.

Motivational interviewing techniques at this stage are intended to clarify the problem for both the youth and family and facilitate a commitment for change. Based on the Initial Assessment, the youth and family work with the juvenile probation counselor to set rehabilitation goals for the youth, and place the youth into an intervention designed for the youth's risk profile.

When new or revised information is obtained after the Initial Assessment, a correction may be necessary. For example, the criminal history domain could need modification because the youth was recently adjudicated for an offense committed prior to the Initial Assessment. The corrected Initial Assessment then becomes an accurate baseline for measuring change.

3.3 Selecting an Intervention That Best Meets the Youth's Risk Profile

Clearly, an assessment alone will not reduce a youth's potential for re-offending. The courts need validated interventions that are designed to address youth with certain risk profiles. A principle use of the assessment is to use the risk level and profile to assign youth to a relevant program. This strategy assumes the availability of a menu of effective programs designed to address various youth profiles. In conjunction with the assessment, Washington's legislators also provided funding for effective services. The Institute, working with juvenile court representatives, reviewed the research literature to identify interventions with demonstrated track records. Only programs with sufficient scientific validity were included in the final set of programs. Five programs were identified from this effort: Multi-Systemic Therapy (MST), Functional Family Therapy (FFT), Aggression Replacement Training (ART), Adolescent Diversion with Advocacy and Mentoring, and coordination of services.

The developers identified relevant assessment profiles that matched these programs. For example, to be eligible to receive Functional Family Therapy, a youth must be at least moderate risk and have high family risk factors (6 out of 24 points on the family functioning scale). Aggression Replacement Training is available only for moderate- to high-risk youth with aggression problems. Youth are assigned to programs designed to match their particular risk factors.

3.4 Re-Assessment

The Re-Assessment is the probation counselor's principle tool to track a youth's progress on factors related to re-offending. The Re-Assessment detects change in a youth's risk or protective factors during supervision. A Re-Assessment does not repeat the previous steps. Rather, the juvenile probation counselor reviews the risk and protective factor information, and following a conversation with the youth, records any changes. If the factors are changing in a positive direction, the counselor has reason to continue the plan. If there is no progress, the probation counselor may need to change the approach.

For this process to be efficient, the assessment software must concisely display the assessment information within each domain and allow easy updating by clicking on the out-of-date information. Changes recorded for each factor need to be maintained in the system so the counselor has access to the full history. Without sophisticated software, this process of monitoring and recording changes is too burdensome.

The assessment software also needs to help the counselor monitor tasks associated with the youth's goals, record progress, and set new goals and tasks. These goals and tasks can be court order obligations, directives of the juvenile probation counselor, or mutually agreed upon plans.

3.5 Assessment Reviews

Periodically, the probation counselor must systematically examine the youth's progress by reviewing all assessment domains. This step is best done in a team setting with other probation counselors or with a supervisor to share successful outcomes and determine alternative strategies.

3.6 Final Assessment

When the youth completes supervision, either successfully or unsuccessfully, all final changes to the assessment information are recorded, and the assessment is closed and archived.

4. Washington's Assessment: Structure and Organization

Washington's legislation required juvenile courts to develop and use a comprehensive assessment to map a youth's risk and protective factor profile to an effective program designed to address these factors. The Washington State Juvenile Assessment was developed to cover major domains related to juvenile delinquency and continued criminal activity. These domains are: 1) Criminal History, 2) School, 3) Use of Free Time, 4) Employment, 5) Relationships, 6) Family, 7) Alcohol and Drugs, 8) Mental Health, 9) Attitudes, and 10) Social Skills.

A summary of the assessment is included at the end of this chapter. Although community risk and protective factors are correlated with juvenile delinquency, Washington's assessment does not include this domain. The juvenile court administrators decided not to include this domain, believing it was not fair to increase a youth's risk score based on his or her neighborhood. Once the assessment domains and concepts within each domain were determined, the detailed wording of items and responses needed to be addressed. For an assessment to be reliable, each concept must be well defined and understood by those completing the assessment. Care must be taken in the wording and definition for each item and response category. In practice, even seemingly simple items, like the number of convictions, can become complicated to measure. Does this term reference offenses, adjudications, or sentences?

To accomplish the necessary clarity, training manuals and sessions are needed to ensure those completing the assessments understand the concepts being measured. A periodic review of the assessment system is required to ensure the practice is not slowly moving away from the original definitions and principles.

5. Validating the Instrument

During the fall of 1998, the juvenile courts started implementing the WAJA. The courts needed an empirically validated assessment to estimate the proportion of youth likely to fall into various risk levels. Fortunately, two sources of information could be used for this purpose: Washington State's Office of the Administrator for the Courts statewide database of juvenile criminal history information, and data for the Wisconsin Risk Scale from a 1996 evaluation of a juvenile court early intervention program. Since the WAJA Pre-Screen is based on the Wisconsin Risk Scale, these data were used to establish empirically validated recidivism risk levels for the WAJA.

The following summarizes the Pre-Screen items, which are a subset of items from the full assessment.

Criminal History

1. Age at first offense
2. Misdemeanor referrals
3. Felony referrals
4. Weapon referrals
5. Against person misdemeanor referrals
6. Against person felony referrals
7. Confinement orders to detention
8. Confinement orders to state institution
9. Escapes
10. Warrants for failure to appear

Social History

1. Male gender
2. School attendance, grades, and misconduct
3. Friends, pro-social, anti-social, and gang
4. Court-ordered/DSHS voluntary out-of-home and shelter care placements exceeding 30 days
5. Runaways or times kicked out of home
6. Family members have jail/prison record
7. Current parental rule enforcement and control
8. Alcohol/drugs disrupt functioning
9. Victim of physical/sexual abuse
10. Victim of neglect
11. Mental problems

The validity of the criminal history domain on the WAJA was established using the court's criminal history database for youth adjudicated during 1995. Washington law dictates that youth who commit minor offenses are first diverted to community accountability boards rather than being sent through a formal court process. Youth who commit more serious crimes or have a history of court involvement are prosecuted in juvenile court and placed on probation. The diversion population is twice as large as the probation population and is viewed as a lower-risk group of youth.

Figure 1 illustrates the relationship between the criminal history score and 18-month recidivism for youth granted diversion and youth placed on probation.

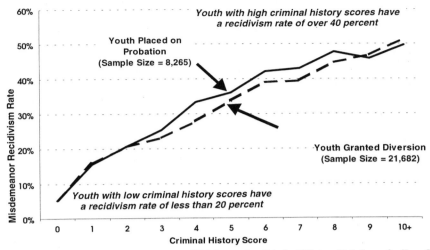

Figure 1: 18-Month Recidivism Rate Increases With Increasing Criminal History Risk Score for Two Groups of Youth Adjudicated in 1995

For a given criminal history score, the 18-month recidivism rate for the diversion and probation groups are nearly identical. That is, the criminal history score is a valid predictor for both groups of youth. A multivariate statistical analysis (logistic regression) was conducted to determine which criminal history variables contributed to prediction. All the variables except escapes make a statistically significant contribution (escapes were a rare event). The Area Under the Receiver Operator Characteristic (AUROC) is .67. Through this analysis, the WAJA criminal history domain was found to be a valid predictor of 18-month recidivism.

Figure 2 illustrates the distribution of criminal history scores for the two groups. Seventy-one percent of the youth placed on diversion have criminal history scores below 3 compared to 19% for youth placed on probation. Forty-three percent of youth placed on probation had scores above 7 compared to 2% of the youth given diversion.

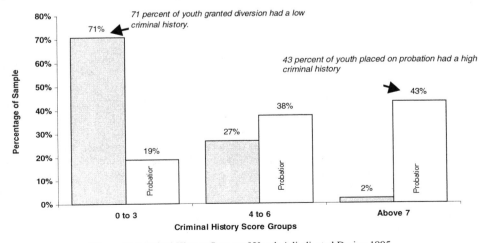

Figure 2: Criminal History Scores of Youth Adjudicated During 1995

The criminal history domain of the WAJA is a valid predictor of recidivism for both youth granted diversion and youth placed on probation. The two groups have different distributions of criminal history risk scores. Because twice as many youth are given diversion as placed on probation, there are actually more moderate-risk youth in the diversion population than in the probation population.

Figure 3 illustrates the correspondence between the predictive capability of the criminal history domain of the WAJA for youth adjudicated in 1995 and youth given the Wisconsin Risk Scale for the 1996 Early Intervention Program evaluation (EIP) sample. The recidivism rate for the EIP sample is lower than the 1995 sample since the EIP rate is based on a six-month follow-up period. The six-month recidivism rate of the youth given the EIP risk assessment, however, is nearly identical to the six-month rate for the 1995 adjudication group. Few youth in the EIP group have criminal history risk scores above 7 because the EIP sample includes only youth placed on probation for the first time.

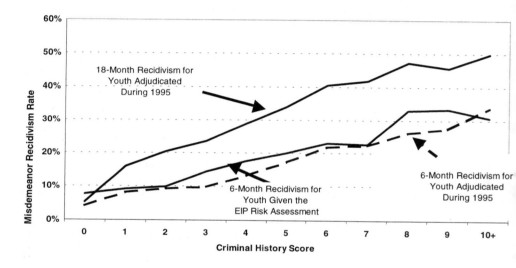

Figure 3: WAJA Criminal History Scores and Recidivism for Youth Adjudicated in 1995 Youth and given the EIP Risk Assessment

The six-month recidivism rates of the EIP sample and the 1995 adjudication group are nearly identical. The difference between the six-month and 18-month recidivism rates increase as criminal history increases. The higher the risk level of a group, the more the six-month recidivism rate underestimates the 18-month rate. The WAJA criminal history domain is a valid predictor of recidivism for youth in the EIP sample as well as the 1995 sample on which it was developed.

The next validation issue is whether the social history items on the WAJA add to the predictive capability of the criminal history items. These items include personal, school, family, and peer risk factors. The social history items of the WAJA Pre-Screen were validated using data for youth administered the Wisconsin Risk Scale in 1996 for the EIP evaluation study.

Figure 4 illustrates how social history scores and criminal history scores are associated with recidivism rates, for youth given the Wisconsin Risk Scale. The dashed line in Figure 4 represents the recidivism rates of all youth in the EIP sample regardless of their social history risk score. Youth with a moderate social history risk score of 6 to 9 have a recidivism rate identical to that of the entire group of youth. Youth with low social history scores of 0 to 5 have lower recidivism rates, while youth with high social history scores of 10 to 17 have high recidivism rates. For this analysis, the AUROC is .68.

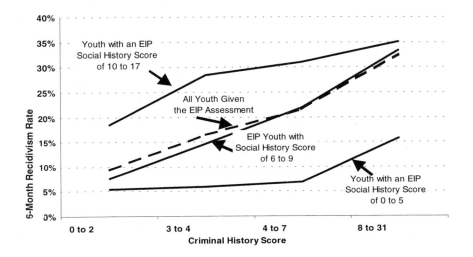

Figure 4: Relationship Between Recidivism Rates and Criminal History is Affected by Social History

If the influence of the social history risk score with these data was additive, all the lines would be parallel and differ by an average of 10 percentage points. However, the recidivism rate for youth with low social history risk scores remains low as the criminal history score increases. The recidivism rates for high social history risk scores remain high even at low criminal history scores.

Table 1 presents the number and percentage of youth in the EIP sample by each combination of criminal history and social history risk scores displayed in Figure 4. Examining the distribution of criminal history scores reveals that 50% of the youth have scores of 5 to 7. That is, for the majority of youth, the criminal history score is related to moderate risk. Similarly, the majority of youth have social history risk scores of 6 to 9. The goal of prediction tools is to accurately determine the characteristics that distinguish those with a high probability of re-offending from those with a low probability, having few persons in the gray area of moderate probabilities. By using social history to partition youth with moderate criminal scores, more youth are placed into the high- and low-risk categories. That is, social history allows a finer categorization than using just criminal history.

Table 1: Number and Percentage of EIP Youth for Each Combination of Criminal History and Pre-Screen Social History Risk Scores

Criminal History Score	Social History Risk Score							
	0 to 5		6 to 9		10 to 17		Total	
0 to 2	55	(4%)	65	(5%)	38	(3%)	158	(11%)
3 to 4	101	(7%)	136	(10%)	109	(8%)	346	(25%)
5 to 7	159	(11%)	306	(22%)	235	(17%)	700	(50%)
8 to 31	19	(1%)	84	(6%)	97	(7%)	200	(14%)
Total	334	(24%)	591	(42%)	479	(34%)	1,404	(100%)

Table 2 illustrates how the WAJA criminal history and social history risk scores are combined to define levels of risk. Groups of youth with 6-month recidivism rates of 10% or less are defined as low risk. Recidivism rates of 11 to 25% are defined as moderate risk, and 6-month recidivism rates above 25% are high risk.

Table 2: Risk Level Definitions Using Criminal History andEIP Social History Risk Scores

Criminal History Score	EIP Social History Risk Score		
	0 to 5	6 to 9	10 to 18
0 to 2	Low (5%)	Low (8%)	Moderate (18%)
3 to 4	Low (6%)	Moderate (15%)	High (28%)
5 to 7	Low (7%)	Moderate (22%)	High (31%)
8 to 31	Moderate (16%)	High (33%)	High (35%)

The Institute has empirically determined that the criminal history domain and items from the WAJA on the EIP Risk Assessment can determine risk levels. These items may be used as a Pre-Screen to identify low-risk youth early in the juvenile justice process.

6. From Development to Implementation

Since the WAJA was developed in 1998, the juvenile court administrators have established an infrastructure for training staff in its use. A training consultant was hired to develop a training manual and curriculum. An experienced probation manager was assigned to oversee the training effort on a full-time basis. Next, probation staff from courts across the state volunteered to become assessment trainers. These individuals received training from the consultant and the Institute. To become certified trainers, staff must be videotaped and critiqued while conducting the assessment. In 1999, these certified trainers began training court staff across the state. Each court must designate at least one person to become a certified assessment specialist for their court; this specialist must also be videotaped and critiqued.

In addition to training court staff in conducting the assessment, staff were also trained to use the assessment to guide the rehabilitative effort. This training is critical to the probation counselors since it provides the justification for gathering the assessment information.

As of August 2000, Washington had eight certified assessment trainers, and a certified assessment specialist in 29 of the 34 courts; more than 700 court staff had received assessment training. Ten thousand youth were assessed in 1999 and 12,000 in 2000. The assessment data have been sent to the Institute for use in the validation and study of the WAJA. This is an extraordinarily rich data set for analyzing delinquent behaviors.

The enabling legislation required the assessment be used to assign youth to programs based on their level of risk and risk profile. As of August 2000, approximately 200 high-risk youth with family-functioning problems have been assigned to Multi-Systemic Therapy, 700 moderate- to high-risk youth with family functioning problems have been assigned to Functional Family Therapy, and approximately 1,000 moderate- to high-risk youth with aggression problems have been assigned to Aggression Replacement Training.

A third version of the assessment is being developed based on staff feedback and analyses of the WAJA data collected. This version may include items more detailed in the areas of aggression, sexual conduct, mental health, suicide, and drug and alcohol use. The purpose of this information is to predict specific types of offending such as violence, sex, or substance abuse.

Software was implemented to record the Initial Assessment. However, the one area that has not been implemented as initially conceived is the sophisticated computer software to support the Re-Assessment process and the guidance of the rehabilitative effort. This continues to plague the process.

7. The Future

An empirical examination of the instrument's predictive validity will be conducted in 2002. The results of this validation study will be used to revise the assessment. The Institute hopes that modified versions of the WAJA and associated programs will be used for a variety of at-risk youth and family populations within Washington State.

Comprehensive assessments can play a significant role in guiding rehabilitative efforts by delivering the most appropriate programs to troubled youth and families. The Institute encourages the formation of a research group to work as a team in the development, study, and dissemination of assessments for anti-social behavior.

8. Summary of WAJA by Domains

SECTION 1: Criminal History
1. Age at first offense
2. Misdemeanor referrals
3. Felony referrals
4. Weapon referrals
5. Against person misdemeanor referrals
6. Against person felony referrals
7. Confinement orders to detention
8. Confinement orders to state institution
9. Escapes
10. Failure to appear warrants

SECTION 2: School
1. Current school enrollment status
2. Type of school in which youth is enrolled
3. Special education student or has a formal diagnosis of a special education need
4. Believes there is value in getting an education
5. Believes school provides an encouraging environment for him or her
6. Number of expulsions and suspensions since the first grade
7. Age at first expulsion or suspension
8. Teachers/staff/coaches the youth likes or feels comfortable talking with
9. Involvement in school activities during most recent term
10. Conduct in most recent term
11. Attendance in most recent term
12. Performance in most recent school term
13. Interviewer's assessment of the youth staying in and graduating from high school or an equivalent vocational education

SECTION 6B: Current Living Arrangements
1. Currently living with family in which primarily raised; or length of time living with current family
2. Current living arrangements
3. Family annual income
4. Health insurance and Title 19 eligibility
5. Support network for family; extended family, and friends that can provide additional support
   Complete only if different from family in which raised:
7. Problems of family members in household
8. Love and support for youth
9. Family member(s) has good relationship with
10. Family provides opportunities for youth to participate in decisions affecting the youth
11. Level of conflict between parents, between youth and parents, among siblings
12. Supervision
13. Rule enforcement and control
14. Consistent appropriate discipline
15. Characterization of discipline
16. Disapproval of youth's anti-social behavior

SECTION 7: Alcohol and Drugs
1. Alcohol abuse
2. Drug abuse
3. Alcohol contributes to criminal behavior
4. Drugs contribute to criminal behavior

SECTION 3: Use of Free Time
1. Structured recreational activities
2. Unstructured recreational activities

SECTION 4: Employment
1. History of successful employment
2. Total number of times youth has been employed
3. Longest period of employment
4. Positive personal relationship(s) with employer(s) or adult coworker(s)
5. Youth is currently employed

SECTION 5: Relationships
1. Existing positive adult non-family relationships
2. Pro-social community ties
3. Friends the youth spends his or her time with
4. Role of youth among peers
5. Admiration/emulation of tougher anti-social peers
6. Length of association with anti-social friends/gang
7. Amount of free time spent with antisocial peers
8. Strength of loyalty to anti-social peers
9. Strength of anti-social peer influence

SECTION 6A: Environment in Which the Youth Was Primarily Raised
1. Age when last living with biological parents
2. Problems of family members living in household
3. Court ordered or voluntary out-of-home and shelter care placements exceeding 30 days
4. Runaways or times kicked out of home
5. Petitions filed
6. Love and support for youth
7. Family member(s) has good relationship with
8. Family provides opportunities for youth to participate in activities and decisions
9. Level of conflict between parents, between youth and parents, among siblings
10. Supervision
11. Rule enforcement and control
12. Consistent appropriate discipline
13. Characterization of discipline
14. Disapproval of youth's anti-social behavior

SECTION 8: Mental Health
1. Victim of physical or sexual abuse
2. Victim of neglect
3. Mental health problems
4. Violence/anger
5. Sexual aggression
6. Sexual vulnerability/exploitation

SECTION 9: Attitudes/Behaviors
1. Attitude before, during, and after crime(s)
2. Purpose for committing crime(s)
3. Accepts responsibility for anti-social behavior
4. Empathy, remorse, sympathy, or feelings for the victim(s) of criminal behavior
5. Fatalistic attitude
6. Loss of control over antisocial behavior
7. Hostile interpretation of actions and intentions of others in a common non-confrontational setting
8. Pro-social values/conventions
9. Respect for authority figures
10. Tolerance for frustration
11. Belief in use of aggression to resolve a disagreement or conflict
12. Readiness for change
13. Successfully meet conditions of supervision

SECTION 10: Skills
1. Consequential thinking skills
2. Critical thinking skills
3. Problem-solving skills
4. Self-monitoring skills for triggers that can lead to trouble
5. Self-control skills to avoid getting into trouble
6. Interpersonal skills

Notes

[1] These individuals included: Robert DeComo, Donna Hamparian, and Patricia Hardyman (National Center on Crime and Delinquency); Delbert Elliot and Jennifer Grotpeter (University of Colorado); Scott Henggeler (Medical University of South Carolina); Mark Lipsey (Vanderbilt University); Patrick Tolan (University of Illinois at Chicago); Robert Hoge (Carleton University, Ontario); Vern Quinsey (Queen's University, Ontario); and David Farrington (Cambridge University, England).

References

Andrews, D. A., & Bonta, J. (1994). *The psychology of criminal conduct.* Cincinnati: Anderson.

Andrews, D. A., Bonta, J., & Hoge, R. D. (1990). Classification for effective rehabilitation: Rediscovering psychology. *Criminal Justice and Behavior, 17,* 19-52.

Baird, S. C. (1984). *Classification of juveniles in corrections: A model systems approach.* Washington D.C.: Arthur D. Little.

Elliot, D. S., Huizinga, D., & Ageton, S. S. (1985). *Explaining delinquency and drug use.* Beverly Hills: Sage.

Farrington, D., & Tarling, D. (1985). *Prediction in criminology.* Albany: State University of New York Press.

Hawkins, J. D., Catalano, R. F., & Miller, J. Y. (1992). Risk and protective factors in adolescence and early adulthood: Implications for substance abuse prevention. *Psychological Bulletin,112,* 64-105.

Henggeler, S. W. (1989). *Delinquency in adolescence.* Newbury Park, CA: Sage.

Henggeler, S. W. (1991). Multidimensional models of delinquent behavior and their implications for treatment. In R. Cohen & A. W. Siegel (Eds.), *Context and development* (pp. 211-231). Hillsdale, NJ: Erlbaum.

Hoge, R. D., & Andrews, D. A. (1994). *The Youth Level of Service/Case Management Inventory and Manual.* Ottawa, Ontario: Department of Psychology, Carleton University.

Hoge, R., & Andrews, D. (1996). *Assessing the youthful offender: Issues and techniques.* New York: Plenum Press.

Jones, P. R. (1994). *Risk prediction in criminal justice.* National Institute of Corrections Conference, Public Protection Through Offender Risk Management.

LeBlanc, M., Quimet, M., & Tremblay, R. E. (1988). An integrative control theory of delinquent behavior: A validation of 1976-1985. *Psychiatry, 51,* 164-176.

Lerner, K., Arling, G., & Baird, S. C. (1986). Client management classification strategies for case supervision. *Crime and Delinquency, 32,* No. 3.

Loeber, R., & Farrington, D. P. (Eds.). (1998). *Serious and violent juvenile offenders: Risk factors and successful interventions.* Thousand Oaks, CA: Sage.

Office of Juvenile Justice and Delinquency Prevention. (1995). *Guide for implementing the comprehensive strategy for serious, violent and chronic juvenile offenders.* Washington, D.C.: Office of Juvenile Justice and Delinquency Prevention.

Patterson, G. R., DeBaryshe, D. D., & Ramsey, E. (1989). A developmental perspective on antisocial behavior. *American Psychologist, 44,* 329-335.

Multi-Problem Violent Youth
R.R. Corrado et al. (Eds.)
IOS Press, 2002

232

The Development and Implementation of the BARO: A New Device to Detect Psychopathology in Minors with First Police Contacts

Theo A.H. Doreleijers and Maarten Spaander

Minors in the Netherlands who are arrested by the police for criminal behavior run the risk of being brought before a criminal court if their offence is serious enough. Youth from 12 to 18 years of age in the Netherlands fall under the juvenile criminal justice system. This system, more than the general penal system, takes into consideration the developmental perspectives of the offender, and seeks to treat the juvenile in an educationally responsible fashion. This means that in addition to punishment in the narrow sense of the term, counseling and treatment (if necessary, in a penal context) are possible when the offender's character or social situation warrants it.

When an underage suspect is taken into custody, the police are legally obliged to report this to the Child Welfare Council. In addition, the police are expected to inform the Council of all other cases in which an offence has been registered. Such a report results in a brief contact with the minor (and with his or her parents if possible) aimed at giving information and advice to the judicial authorities. Another important function of this preliminary contact is to select those juveniles whose criminal behavior may be a signal of a more fundamental problem. The Council can then decide to carry out a more detailed examination of the character of the juvenile and his or her family and social environment to advise those involved as to the nature and intensity of the (voluntary) treatment indicated. Most exceptionally, this treatment can be imposed in the form of a child welfare measure.

In the 1990s the Child Welfare Council in the Netherlands became increasingly involved with criminal cases against juveniles, leading to a vast expansion of its capacity. In the meantime, most of the Council's offices had developed their own methods because the standard form that had been used as a checklist for interviewing juveniles no longer sufficed. Research (Doreleijers, Moser, Thijs, van Engeland, & Beyaert, 2000) also showed that the items on this form did not yield any distinction between cases of more or less age-appropriate misbehavior and those cases that the Council did need to be concerned about because of eventual developmental risks.

In the late 1990s the Council was determined to improve its quality by initiating various projects. One of these was the establishment of standards and protocols for the preliminary investigation in criminal cases. In 1997, the first author was commissioned to develop a new instrument that had to meet the following requirements: selection of juveniles with an unfavorable developmental perspective; prompt advice to the judicial authorities; optimal relationship between diagnostic quality and investment of time; greater objectivity and quality: standardization and verification facility; taking into account the nature and severity of the offence and the diversity of the suspects. The second author was then commissioned to train staff members to use this new instrument.

Structure of the Child Welfare Council in the Netherlands

The Child Welfare Council is a public body responsible for investigating civil as well as criminal cases. The council is thus officially not an agency for professional assistance. The State Secretary of Justice is politically responsible for the Council. Each year the Council submits an annual plan along with a budget as a guide to the policy agreements made with the Ministry of Justice.

The Council has 22 offices spread over five court jurisdictions. Together with the five directors of the court jurisdictions, the General Director forms the National Management Team (NMT) where the decision process takes place. Each office's management consists of one or more unit managers whose tasks include overseeing the supervisors and psychologists. In turn, the supervisors oversee the investigators who administer the BARO in practice. The psychologist, the supervisor and the council investigator comprise the Multidisciplinary Consultation Team.

Aim of the Preliminary Investigation and Duties of the Council Investigator

The preliminary investigation in criminal cases involving minors has two aims: advising the judicial authorities and selecting juveniles with unfavorable developmental perspectives. With regard to the first aim, the judicial authorities consulted for the present project emphasized that they expect a traditional report with an account of the juvenile's character and his or her family and social situation.

When the Council investigator initiates a preliminary report of the juvenile's character and social situation, he or she is expected to place the offence in its proper context, making an inventory of both risk and protective factors: school behavior and achievement; the juvenile's developmental level and physical health; suspicion of developmental or other mental disorders; leisure activities and friends; parental child-rearing abilities; the influences of culture and religion on the family; societal factors. Moreover, the Council investigator has to assess the risk of recidivism as well as the possible consequences of the offence for the juvenile's development. In effect, he or she has to formulate a judicial recommendation as well as advice for more extensive examination. Until now, Council investigators were expected to accomplish all of this in approximately six hours without the aid of any reliable instrument to achieve standard results. When a juvenile was taken into custody (and therefore had to be arraigned before the examining magistrate within three days), the report had to be produced very rapidly as well.

1. The BARO

The BARO comprises sociodemographic and offence-related information, personal, school and counseling history, family composition, police and judicial case history, and child welfare measures, if any.[1]

1.1 Protocol for the Preliminary Investigation

Council investigators formulate a protocol meant to standardize the use of the instrument as much as possible. For example, it is required that the rights of those involved are discussed during the introduction: the right not to cooperate, to access information and correct factual errors, and the right to be informed about further procedures as well as to whom the report will be sent. In general, there is an appreciable increase in the quality of the answers when the interviewees are adequately informed before the interview about what to expect (it is useful to summarize the topics of the interview) as well as what the safeguards of the investigation.

1.2 Interview and Ranking of Each Area

The interview covers nine areas. About 20 questions are asked in each area, selected on the basis of literature data and clinical experience. During the project, certain questions were revised according to the researchers' assessment of their reliability. The questions are asked of the juvenile involved, his or her parent(s) or guardian(s) and preferably a third party such as a teacher, family guardian, juvenile probation officer or a police officer. The more informants, the more reliable the information. Well-trained professionals should carry out the interviews. The results provide reliable screening and selection of juveniles with unfavorable developmental perspectives.

Eight areas cover the development and functioning of the juvenile and one area his or her living conditions: family, school, neighborhood. We have chosen to place more emphasis on the development and functioning of the juvenile than on the conditions of his or her upbringing. In juvenile criminal law it is the juvenile who is being addressed, even though juvenile criminal law takes more account of pedagogical conditions in particular than does general criminal law.

The investigator can choose from four scores (no concern, some concern, much concern, very much concern) to rank each question, according to how serious the problem is in the relevant area. The following must always be expressly considered: risk vs. protective factors, progressive vs. regressive developmental tendencies and strong vs. weak personality traits. The ranking scheme can also indicate how the investigator has reached the conclusions based on the data of the number of informants. This approach offers at a glance the opportunity of becoming aware of the seriousness of the situation in each area and the reliability of the ranking.

1.3 Problem Index

The problem index has been developed to make the rather subjective judgment of the Council investigator more objective. To this purpose, secondary analyses have been carried out on research data compiled several years earlier from juveniles who were suspected of committing an offence and brought before the juvenile court magistrate. From the interviews employed at that time, questions were singled out which yielded responses that in combination with each other best predicted the presence of psychopathology. This was based on the premise that juveniles with the most severe psychopathology were the most threatening and therefore a request for an extensive diagnostic examination should be considered in a criminal procedure. Moreover, it is evident from the literature that psychopathology considerably increases the risk of recidivism.

The secondary analyses yielded 10 risk factors (called indicators) which, in combination with each other, gave a certain rating of severity—the problem index. The following indicators from the interviews with the juveniles were found to be important: a history of institutionalization; behavior as well as learning disorders at school; having been a victim of or witness to violence outside of the family; physical complaints without a medical cause; problems resulting from the use of alcohol and/or drugs (thus not the use itself).

The following indicators from the interviews with parents were found to be

the BARO form, a lack of clarity regarding the index factors, guidelines for determining when the BARO should not be administered and when it should be readministered when the youth recidivate.

On an intrinsic level, the BARO is experienced as too unilaterally focused on psychopathology and too much an instrument for the scientist/researcher instead of the helping council investigator. Other concerns noted were a lack of clarity as to the 'objective' compilation of information and the 'subjective' interpretation in the ranking, the case histories for the actors should be better keyed to examples from the practice, a lack of guidelines for how to proceed on obtaining the first disclosure of sexual abuse, a lack of sample reports based on a previously administered BARO, and a lack of clarity as to the possibilities of extensive examinations, internal or external.

2. Conclusion

The evaluations of the BARO training sessions in all of the offices were strikingly positive. At the beginning 9% of the trainees were opposed to the BARO. At the end of the sessions the opposition had dropped to 1.5%. The proportion of trainees that designated the BARO as a significant improvement for the client(s) rose from 7% to 31%. At the beginning 18% considered the BARO a qualitative improvement, which rose to 41% at the end of the training. Meanwhile there is a close knit and enthusiastic team of trainers, which guarantees the uniformity and quality of the training. The "train the trainer" principle has taken shape and can be further optimized.

As to improvement of skills, 22% of the trainees indicated that they were more aware of their blind spots, 12% of their creative potential, 37% of pitfalls during interviews and 23% of their distance and closeness during the interview. That part of the training dealing with psychopathology is reserved for the child and adolescent psychiatrist (the main trainer).

The first data obtained from monitoring the administration of the BARO in the offices which have had BARO training indicate that the problems observed are mainly organizational and not intrinsic. Where the BARO has brought forward or increased shortcomings or insecurities in writing skills, the Council has decided to offer training sessions.

2.1 Future Expectations

The NMT resolution of January 2000 to implement the BARO in stages included the decision to reduce the council investigator's caseload from over 203 cases annually to 170 basic investigations per full-time council investigator. If the productivity agreements with the Ministry stay the same, the result of the study means that there will have to be a one-third increase in the number of council investigator positions. As long as the number of new staff members necessary have not been found, a temporary solution will have to be sought with regard to the BARO. Lowering the caseload will take away much of the resistance of the staff carrying out the work.

The first example of a demonstration film featuring the administration of the BARO has produced practicable material for the BARO training sessions. In future, the film will be revised qualitatively into a definitive instruction example for the integration of the BARO training into the general basic training of council investigators. It is expected that in the future the BARO and the BARO training sessions will make an important auxiliary contribution to the process of change of the roles of the psychologists, supervisors and council investigators with respect to the desired quality improvement in the MDC.

The BARO is a useful instrument to educate the judicial authorities and select those juveniles who may need extensive diagnostic assessment. Although it is somewhat more time-consuming than the presently used methods, it yields much more (higher quality) information relatively rapidly. Objectivity and verifiability are increased by standardization and establishing a protocol. The BARO takes into consideration diversity of offences and suspects, but intensive training is needed for its use. A separate section needs to be developed for sex offenders, and juveniles and their parents with a non-Dutch background require extra attention due to their poor knowledge of the Dutch language.

The implementation of the BARO necessitates an increase in the number of investigators. BARO training often appears to be more effective with investigations who have only recently begun their activities. After the BARO training, the investigator has to rely upon a psychologist for the ranking which will lead to a recommendation. These psychologists should be specially trained in ranking for further assessment.

Notes

[1] For an extensive description of the factor analysis that is the basis of the instrument and of the validity study, please contact the first author.

Reference

Doreleijers, Th.A.H., Moser, F., Thijs, P., van Engeland, H., & Beyaert, F.H.M. (2000). Forensic assessment of juvenile delinquents: Prevalence of psychopathology and decision-making at court in The Netherlands. *Journal of Adolescence, 23,* 263-275.

Multi-Problem Violent Youth
R.R. Corrado et al. (Eds.)
IOS Press, 2002

241

Towards Valid Cross-cultural Measures of Risk

David J. Cooke and Christine Michie[*]

The irresistible rise of risk assessment for violence is an international phenomenon that has highlighted the importance of cross-cultural perspectives. It is a fundamental, and dangerous, error to assume that a risk assessment procedure developed in Canada or the United States will function in the same way in Spain, Sweden, or Scotland. The focus of this book, the development of a risk assessment instrument that functions in comparable ways in different cultures, is an ambitious undertaking. A cross-cultural perspective is essential.

A cross-cultural perspective contributes both at the practical level and at the level of substantive theory. In this brief chapter we will consider two broad themes: first, how do cultural differences affect the manner in which risk instruments should be designed, and second, what analytic strategies can be brought to bear to ensure that relevant constructs are measured in the same way across cultures.

Culture and Variation in Violence

On a substantive level, culture, however we define this term, can be part of the problem. Cultural factors can promote or inhibit violence: Risk for violence varies with culture. There is ample evidence to indicate that aggressive behavior occurs in nearly every society, but there is also ample evidence of substantial variation in norms and values concerning violence and aggression across cultures (e.g., Ekblad, 1988, 1990; Moghaddam, Taylor, & Wright, 1992). Moghaddam et al. (1992), for example, argued that cultural groups such as the Inuit of the Arctic, the Blackfoot Nation of North America and the Pygmies of Africa rarely engage in physical violence; their preference is to resolve disputes through withdrawal or negotiation. This behavior contrasts starkly with accounts of the behavior of the Yanomano of the Upper Amazon. Chagnon (1974) reported that, not only is there constant warfare between villages, but there is also constant fighting and argument within villages. A male's status within his village is determined by his level of aggressiveness.

Cross-cultural variation in the level of violence is not a peculiarity of so-called primitive groups. Groups such as the Mennonites, the Amish and the Hutterites exist within contemporary North American society, where rates of aggression are generally high, but these groups display low levels of aggression, presumably as a consequence of their strong norms and values against aggression (Bandura & Walters, 1963).

Both the processes of enculturation and the processes of socialization influence levels of observed aggression. In some cultures, or subcultures, violence is the norm. Glasser (1987) provided a graphic account of his early life and the enculturation of violence in the Gorbals area of Glasgow, an area once notorious for its violence.

> We grew up with violence. It simmered and bubbled and boiled over in street
> and close, outside the pubs, at the dance halls...violence settled private

accounts, transgressions of codes, the spilling over of grievance or spleen. It was so closely intertwined with everyday life, its inescapable rough edge, logical, cathartic, that its occurrence, like rain and cold and frequent shortage of food, was recognized with equal fatalism. (p. 61)

Cultural explanations have been used to explain the North/South divide in the rate of homicide within the United States: The rate being substantially higher in the South. Cohen and Nisbett (1996) argued that a *culture of honor* exists in the South that requires men to maintain their reputation of toughness by responding violently to people who insult them. "Honor in this sense is based not on good character but on a man's strength and power to enforce his will on others" (p. 5). In order to explain this North/South divide, Cohen and Nisbett (1996) argued that it was founded on the different cultural norms carried by immigrants to the different regions of the United States. They commented that

The northern United States was settled by farmers—Puritans, Quakers, Dutch, and Germans. These people were co-operative, like farmers everywhere, and modern in their orientation towards society. They emphasized education and quickly built a civilization that included artisans, tradespeople, businesspeople, and professionals of all sorts. In contrast, the South was settled primarily by people from the fringes of Britain—the so-called Scotch-Irish. (p. 7)

Being a Scot of Irish descent the first author finds it hard to agree with their thesis! Nonetheless, I think that they illustrate that the risk assessment enterprise must be grounded in cultural norms and take into account cross-cultural variations.

In tackling the cross-cultural problem we need to think about both the design of our instruments and about the analytic techniques that we employ. We will examine some of the problems that can arise with the design of a simple interview before considering issues of analysis.

1. Some Issues of Design

There are many issues to be considered in the design of cross-culturally valid instruments. Van de Vijver and Leung (1997) provide an excellent introduction to the field. We will merely focus on three issues: the problem of differential cultural relevance, the problem of self-disclosure, and the problem of translation.

1.1 The Problem of Differential Cultural Relevance

The cross-cultural generalizability and validity of risk assessment methods may founder if the relevance of responses to questionnaires or interviewers varies across cultures. There may be variations in the relevance, significance or psychological meaning of behaviors across cultures. Cultural factors may influence the significance of responses even to a deceptively simple set of questions such as the MacArthur Community Violence Screen, a key measure in the MacArthur violence risk study (Steadman et al., 1998). The following procedure is used by an interviewer to discover the extent to which an individual has engaged in violence. The participant is asked if someone has been violent to them in a particular manner, then is asked if he or she has committed the same act. For example, the first question is "Has anyone thrown something at you?" The participant is then asked "Have you thrown something at anyone?" The process continues in that manner until the ninth question, which is "Have you done anything else that might be considered violent?"

We used the same procedure unmodified in a study of 255 prisoners in a Scottish prison (Michie & Cooke, 2001). On reflection, however, it may have been sensible to revise the measure in the light of cultural variables. One of the items on the scale is "Have you used a knife or fired a gun at anyone?"; within the Scottish context this item should have been split in two in order to produce an item referring to knifes and an item referring to guns. Knives are obviously available in Scotland but guns have always been rare, and since the murders at Dunblane they are extremely rare. By way of contrast, in many parts of the United States guns are relatively freely available, it is conceivable that someone in a rage will pick up a gun and use it. In Scotland, where it is harder to get access to a gun, using a gun suggests that you have planned your violent act. Gun use is indicative of a different type of violence and of different psychological processes. Violence perpetrated with a gun is more likely to be instrumental and mark more serious intent (Zimring & Hawkins, 1997). Thus the psychological meaning of a positive response to the "gun" item may well vary across cultures. Awareness of such cultural issues is important when developing procedures that have cross-cultural generalizability.

1.2 The Problem of Self-Disclosure

Risk assessment processes depend to a greater or lesser extent on extracting information from the individual being assessed; this may include information about current symptoms, sexual and violent fantasies or intentions, it may include information about past behavior. Cultural norms and expectations may influence the nature and quantity of the information provided: Levels of self-disclosure vary across cultures. For example, Chen (1995) demonstrated that American students disclosed more than Chinese students; Lewin (1948) found that Americans disclosed less than Germans; Barnlund (1975; 1989) demonstrated that Americans reveal significantly more about their financial affairs, their sexual adequacy and personal traits than do Japanese subjects; similar findings were obtained by Nakanishi (1987).

Cultures value talkativeness and self-disclosure to differing degrees. Chen (1995) suggested that Chinese society devalues the talkative, viewing such people as shallow and even dangerous. Chen quotes Lao Tze in support of his contention *"He who knows does not speak, he who speaks does not know."* The cultural norm in much of Northern Europe is against self-disclosure, and this may influence the quality and quantity of information available for making risk assessments as compared with North America. Systematic studies of self-disclosure across target countries, and evaluation of the relations between self-disclosure on risk relevant variables, may cast light on apparent cross-cultural differences.

1.3 The Problem of Translation

Many judgments about risk factors require sophisticated judgments about subtle psychological characteristics; ratings of PCL-R items including *Glibness/superficial charm*, *Lack of affect* and *Pathological lying* are examples. The subtleties of language translation could result in a lack of equivalence of the construct being assessed across cultures. Thus an instrument developed in one language may include particular denotations, connotations or meanings that are specific to that language and which are impossible to translate, with conciseness and clarity, into another language (Van de Vijver & Leung, 1997).

One standard technique for improving the translation of instruments is *back translation*. The instrument is written in one language and then translated into the target language. Another translator then translates the target version back to the source language. Differences in the original and the back-translated version are compared and adjusted. This is a good method for checking translations; unfortunately, it emphasizes the semantic aspects of the language at the expense of comprehensibility, naturalness and connotations (Van de Vijver & Leung, 1997).

Fortunately, guidelines are available to assist those who undertake the arduous task of developing equivalent versions in different languages (e.g., Brislin, 1986). Indeed, these guidelines would be helpful in the development of any instrument even if it is going to be applied in only one language. Brislin (1986) provides some examples: Use short, simple sentences, use the active voice as it is easier to understand, repeat nouns instead of using pronouns, avoid metaphors and colloquialisms and avoid "could" and "would." It is important to use specific rather than general terms (e.g., the term "Members of family" has different meanings across cultures) as more precise terms are less likely to be culturally biased.

Cultural decentering is the preferred method for developing cross-culturally valid tests (Van de Vijver & Leung, 1997). Cultural decentering entails the removal of *words* or *concepts* that are difficult to translate or that are too specific to a culture. Thus the instrument in the original language may be changed in order to provide comparability with the instrument in the target language. A clear example comes from the development of the European Value Survey. The results from Spain in relation to the word "Loyalty" were different from all other countries. Unlike the other languages the Spanish word used for "Loyalty" has connotations of sexual faithfulness (Van de Vijver & Leung, 1997).

The disadvantage of the decentering approach is that it is labor intensive. It requires a multicultural, multilingual team with expertise in the construct under study; just like the team who participated in the Advanced Research Workshop in Crakow. A further difficulty is that any data collected with previous versions of the instrument cannot be used for cross-cultural comparisons because the decentered instrument differs from the original instrument. Essentially, with the decentering approach a new instrument is developed simultaneously in many languages. There are many more concerns when developing a new risk instruments, space precludes their discussion here. We will now consider approaches to data analysis.

2. Some Issues of Analysis

Cross-cultural studies by their very nature are essentially quasi-experimental studies, clearly it is not possible to determine the impact of culture by random assignment of participants. Many data analytic techniques can be applied to explore differences observed across cultures, but due to space restriction we will just focus on two of the most powerful techniques, namely, confirmatory factor analysis (CFA) and Item Response Theory (IRT) approaches.

2.1 Assessing Measurement Bias Across Cultures

Cross-cultural work aims to establish equivalence of measurement, both structural and measurement equivalence, in different cultures. In cross-cultural research interest is not focused on the manifest variables, the test scores or the interviewer ratings per se, but on the underlying construct; for example, on intelligence, impulsivity, psychopathy or aggressivity. We want to know that the same latent construct is identified in both settings: In other words do the test scores or ratings have the same structural relationships with the underlying construct in both cultures. The second thing that must be established is that the construct is measured on the same scale, in terms of metric equivalence or measurement unit equivalence. We will consider these issues in some detail.

Traditionally, cross-cultural studies have used classical test theory (CTT) indices such as Cronbach's alpha and corrected item-to-total correlations to compare samples from different cultures. Exploratory factor analysis has been applied to assess the cross-cultural equivalence of the latent structures underpinning scales (e.g., Barrett & Eysenck, 1984). These factor analytic approaches are correlational and thus cannot ensure that there is measurement equivalence across cultures. Invariance of factor structures is a necessary but not sufficient condition for ensuring cross-group equivalence (Bijnen & Poortinga, 1988; Tanzer, 1995). An example from the physical sciences may illustrate the point: The Fahrenheit and Centigrade scales both measure the same underlying construct, temperature, but they are not metrically equivalent because they do not have the same zero points and their scale intervals are different. To ensure equivalence of measurement across groups, it is not only necessary to demonstrate that the indicators of interest have the same relationships with the underlying construct but also that the underlying construct is being measured on the same metric (Van de Vijver & Leung, 1997). Clearly, in the absence of metric equivalence it is meaningless to compare, for example, means and standard deviations or prevalence estimates based on a cut-off.

Item response theory (IRT) provides powerful methods for modeling the performance of items and tests (Embretson, 1996; Embretson & Reise, 2000; Santor & Ramsay, 1999). These methods are more appropriate than factor analytic or CTT approaches for examining cross-group equivalence. One particular advantage of IRT approaches is that they can provide measurement on an identical scale across groups. IRT methods provide a formal psychometric model that allows different groups to be matched on the underlying latent trait. By focusing on latent variables rather than manifest variables it is possible to distinguish between measurement bias and true group differences. Indeed, it has been argued that it is not possible to assess bias using manifest variables and that is essential to model manifest variables using latent variables (Meredith & Millsap, 1992).

In brief, IRT models specify the relationship between item or test scores and the underlying latent trait (θ) that is postulated to underpin item or test scores. Graphical methods can be used to map the probability of an item or test score against θ. Two features of item characteristic curves (ICCs) are relevant, their slope and the position of their maximum point of inflection. Items with steeper slopes are more discriminating. The maximum point of inflection (as measured by the threshold parameter) reflects the extremity or difficulty of the item. For example, in studies of the PCL-R it has been demonstrated that the PCL-R item *Irresponsibility* is usually rated positive at average levels of θ, whereas the PCL-R item *Glibness/Superficial charm* is generally rated positive only when higher levels of the underlying trait are present (Cooke & Michie, 1997, 1999). A participant would have to be high on the underlying trait before *Glibness/Superficial charm* would become apparent.

IRT methods are the methods of choice for detecting differential item functioning (DIF) or differential test functioning (DTF) across groups. DIF occurs when an item is more discriminating, or is more difficult or more extreme, in one group as compared with another. Careful consideration of ICCs can assist in identifying cross-cultural, ethnic, gender or other biases in an item, or indeed, biases in a test as a whole. Several features make IRT methods more suitable than CTT methods for examining test biases. First, CTT indices such as Cronbach's alpha and corrected item-to-total correlations are highly sensitive to variations in the range of test scores across samples (Van de Vijver & Poortinga 1994). ICCs are, in contrast, independent of the samples from which they are derived (Mellenbergh, 1996). Second, representative samples are not necessary in order to obtain unbiased estimates of item and test characteristics; non-representative samples may be utilized (Embretson, 1996). This is clearly a great advantage in the field of risk assessment where representative samples would be virtually impossible to obtain.

Third, direct comparison of the performance of items can be made across groups. For example, it is possible to distinguish between item differences in extremity (i.e., differences in the level of the underlying trait at which the inflection point occurs) and differences that reflect differences in the relevance of items across groups (i.e., differences in slopes). Differences in extremity demonstrate *uniform DIF*, indicating that the level of the underlying trait for which the item is useful differs across groups, but the item is still useful for both. Instances of uniform DIF indicate measurement bias but not necessarily true group differences in the composition of the underlying trait. By contrast, differences in slopes demonstrate *non-uniform DIF*, indicating that the discriminating power of the item differs from one group to another (Holland & Wainer, 1993), and that groups differ in the importance of specific components of a construct. Differences in item performance can provide important information about cross-group differences. Whereas unbiased items define common aspects of the construct, biased items denote cross-group idiosyncrasies (Bontempo, 1993; Holland & Wainer, 1993—; Reise, Widaman, & Pugh, 1993; Van de Vijver & Poortinga, 1994).

A fourth advantage of IRT over CTT is that the scale of θ is defined by the items and, when comparing groups, it is possible to ensure that a common metric is used for comparisons. This property of measurement invariance across groups can ensure, for example, that diagnostic cut-offs are equivalent (e.g., Cooke & Michie, 1999; Reise et al., 1993).

Waller, Thompson, and Wenk (2000) emphasized that research on bias should focus on latent rather than observed variables for three reasons. First, differences may be detected in manifest variables when no differences occur on the latent variable. Second, the opposite may also be true: Differences on the latent variable may be masked by a lack of differences on observed variables. Third, although biases may be present at the item level, aggregation across items may result in non-biased estimates of the underlying trait at the test score level. Cross-group DIF for multiple items may result in *amplification* or *cancellation* of DTF (Raju, Van der Linden, & Fleer, 1995). For example, positive ratings may be obtained in one culture (cf. other culture) at lower levels of θ, for some items, and at higher levels of θ for other items. Nonetheless, using an explicit IRT model it is possible to obtain unbiased estimates of the underlying latent trait from scales that contain some biased items.

2.2 Cross-cultural Variation — The Case of Psychopathy

We will try to illustrate the analytic principles outlined above by examining the cross-cultural validity of the PCL-R as a measure of psychopathy. We will do this for two reasons. First, psychopathy is one of the best predictors of future violence that we have available (e.g., Hemphill, Hare, & Wong, 1998). Indeed, Salekin, Rogers, and Sewell (1996) noted "The ability of the PCL-R to predict violence is unprecedented in the literature on dangerousness" (p. 211). There is now cross-cultural evidence to support the validity of the PCL-R as a predictor (e.g., Tengstrom, Grann, Langstrom, & Kullgren, 2000). Second, because of the widespread use of the PCL-R there are now many PCL-R data sets in different countries. We will now describe analyses designed to assess the structural and measurement equivalence of PCL-R ratings and total scores across cultures.

2.3 Structural Equivalence

In earlier work we demonstrated (Cooke & Michie, 2001) that PCL-R ratings are structurally complex and can be best represented by a hierarchical model in which 13 PCL-R items form a higher order factor underpinned by three distinct but correlated factors. We named these factors Arrogant and Deceitful Interpersonal Style, Deficient Affective Experience, and Impulsive and Irresponsible Behavioral Style. Examination of factor one indicated that it measures interpersonal style being specified by the PCL-R items *Glibness and superficial charm*, *Grandiose sense of self-worth*, *Pathological lying* and *Conning/manipulative*. Factor two represented an affective factor, being specified by the items *Shallow affect* and *Callous/lack of empathy*, *Lack of remorse or guilt* and *Failure to accept responsibility*. Factor three represented a behavioral factor specified by five items: *Need for stimulation/proneness to boredom*, *Impulsivity*, *Irresponsibility*, *Parasitic lifestyle* and *Lack of realistic, long term goals*. The higher order factor was shown to be clearly unidimensional in these samples. We are privileged in that many colleagues have given us access to their data sets to test the model in different settings. In assessing the goodness of fit of a factor model there are many indices that may be applied. For the simplicity of presentation we have tabulated the values of the Comparative Fit Index (CFI), the index regarded by Byrne (1994) as the index of choice. Values over .90 indicate an adequate fit. Examination of table 1 below indicates that the structure of the PCL-R as derived from North American data generalizes to many European countries.

Table 1: Comparative Fit Indices (CFI) for Three Factor Model for North American and European Samples

Researchers	Location	CFI
Development Samples	North America	.96
Cooke	Scotland	.94
Clark	England	.94
Shine et al.	England	.91
Grann et al.	Sweden	.90
Bender & Loesel	Germany	.93
Pham	Belgium	.92
Rasmussen et al	Norway	.88
Molto et al.	Spain	.84
Andersen et al.	Denmark	.92
All European Men	Europe	.95

It is perhaps noticeable that the fit tends to be rather better in the UK samples, perhaps because there is less impact of language differences on ratings.

2.4 Measurement Equivalence

CFA methods can be used to examine whether the underlying structures of risk assessment instruments are the same across cultures, IRT methods are necessary when issues concerning scaling are examined. We will illustrate how IRT methods can be used to answer two important questions: Do items perform the same way in one culture as in another, and do overall test scores mean the same thing in different cultures.

IRT methods were used to compare the performance of PCL-R items in Scotland and North America (Cooke & Michie, 1999). All the item parameters were constrained to be equal in Scotland and North America, but Generalized likelihood ratio testing (GLRT) revealed that the data could not be modeled adequately under this assumption of equal item parameters. However, it was found that when the slopes were constrained to be equal then the model fitted well. This indicated that the slope parameters were essentially equal and that the items discriminate as well in Scotland as they do in North America. However, the variation in the difficulty parameters revealed that the level of the underlying trait at which the characteristics of the disorder (e.g., *Glibness/superficial charm* and *Lack of remorse*) become apparent, differed in the two settings.

As noted above, although biases may be present at the item level, aggregation across items may result in non-biased estimates of the underlying trait at the test score level. Cross-group DIF for multiple items may result in *amplification* or *cancellation* of DTF (Raju et al., 1995). For example, positive ratings may be obtained by the minority group (cf. majority group) at lower levels of θ, for some items, and at higher levels of θ for other items. Nonetheless, using an explicit IRT model it is possible to obtain unbiased estimates of the underlying latent trait from scales that contain some biased items.

Because DIF does not always have an impact at the level of the test as a whole, we also examined DTF both graphically and numerically. Test characteristic curves (TCC) are the test score equivalents of ICCs in which test scores are plotted as a non-linear function of θ (Lord, 1980). The slope of a TCC describes the extent to which a change in the test score varies with the level of θ. Visual inspection of TCCs can be used to evaluate whether there is differential test functioning for two groups. The impact of metric inequivalence on test scores can be assessed graphically. To illustrate this point we have plotted data from over 600 Scottish cases and over 700 English cases. When Scottish and English PCL-R data are compared with North American data the difference the average difference is of the order of 3 points (Cooke, Michie, Clark, & Hart, 2001). When Scottish and English PCL-R data are compared with North American data (see Figure 1) the average difference is of the order of 3 points (Cooke, Michie, Clark, & Hart, 2001).

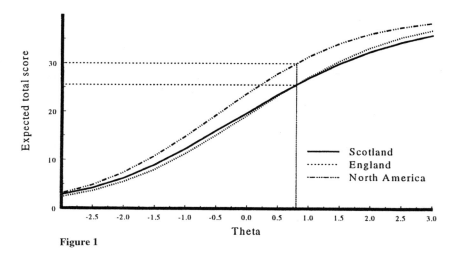

Figure 1

In conclusion, any endeavor designed to develop a risk assessment instrument that has both utility and validity in a range of diverse countries can only benefit form taking a cross-cultural perspective to the collection and analysis of data.

Notes

* Christine Michie received support from the Economic and Social Research Council grant Ref. No. L133222704 while carrying out these analyses. We would like to thank all our colleagues at home and abroad who have graciously given us access to their data for re-analysis.

References

Bandura, A., & Walters, R. (1963). *Social learning and personality development*. New York: Holt, Rinehart and Winston.

Barnlund, D.C. (1975). *Public and private self in Japan and the United States*. Tokyo: Simuli.

Barnlund, D.C. (1989). *Communicative styles of Japanese and Americans: Images and realities*. Belmont, CA: Wadsworth.

Barrett, P., & Eysenck, S.B.G. (1984). The assessment of personality factors across 25 countries. *Personality and Individual Differences, 5*, 615-632.

Bijnen, E.J., & Poortinga, Y.H. (1988). The questionable value of cross-cultural comparisons with the Eysenck Personality Questionnaire. *Journal of Cross-cultural Psychology, 19*, 193-202.

Bontempo, R. (1993). Translation fidelity of psychological scales: A item response theory analysis of an Individualism-Collectivism Scale. *Journal of Cross-cultural Psychology, 24*, 149-166.

Brislin, R.W. (1986). The wording and translation of research instruments. In W. J. Lonner & J. W. Berry (Eds.), *Field methods in cross-cultural research* (pp. 137-164). Newbury Park, CA: Sage.

Byrne, B.M. (1994). *Structural equation modeling with EQS and EQS/Windows*. London: Sage.

Chagnon, N.A. (1974). *Studying the Yanomano*. New York: Holt, Rinehart and Winston.

Chen, G. (1995). Differences in self-disclosure patterns among American versus Chinese: A comparative study. *Journal of Cross-cultural Psychology, 26*, 84-91.

Cooke, D.J., & Michie, C. (1997). An Item Response Theory evaluation of Hare's Psychopathy Checklist. *Psychological Assessment, 9*, 2-13.

Cooke, D.J., & Michie, C. (1999). Psychopathy across cultures: North America and Scotland compared. *Journal of Abnormal Psychology, 108*, 55-68.

Cooke, D.J., & Michie, C. (2001). Refining the construct of psychopathy: Towards a hierarchical model. *Psychological Assessment, 13, 171-188.*

Cooke, D.J., Michie, C., Clark, D.A., & Hart, S.D. *Psychopathy across cultures: North American and the United Kingdom compared*. (unpublished manuscript).

Ekblad, S. (1988). Influence of child-rearing on aggressive behavior in a transcultural perspective. *Acta Psychiatrica Scandanavica, 78*, 133-139.

Ekblad, S. (1990). The children's behaviour questionnaire for completion by parents and teachers in a Chinese sample. *Journal of Child Psychology and Psychiatry, 31*, 775-791.

Embretson, S.E. (1996). The new rules of measurement. *Psychological Assessment, 8*, 341-349.

Embretson, S.E., & Reise, S.P. (2000). *Item response theory for psychologists*. Mahwah, NJ: Lawrence Erlbaum, Associates.

Glasser, R. (1987). *Growing up in the Gorbals*. London: Pan.

Hemphill, J.F., Hare, R.D., & Wong, S. (1998). Psychopathy and recidivism: A review. *Legal and Criminological Psychology, 3*, 139-170.

Holland, P.W., & Wainer, H. (1993). *Differential item functioning*. Hillsdale, New Jersey: Erlbaum.

Lewin, K. (1948). Some socio-psychological differences between the United States and Germany. In G. Lewin (Ed.), *Resolving social conflicts: Selected papers on group dynamics* (pp. 1935-1946). New York: Harper.

Lord, F.M. (1980). *Applications of item response theory to practical testing problems*. Hillsdale, NJ.: Erlbaum.

Mellenbergh, G.J. (1996). Measurement precision in test score and item response models. *Psychological Methods, 1*, 293-299.

Meredith, W., & Millsap, E.E. (1992). On the misuse of manifest variables in the detection of measurement bias. *Psychometrika, 57*, 289-311.

Michie, C., & Cooke, D.J. *The structure of violent behavior: A hierarchical model*. (unpublished manuscript).

Moghaddam, F.M., Taylor, D.M., & Wright, S.C. (1992). *Social psychology in cross-cultural perspective*. New York: Freeman.

Monahan, J., Steadman, H., Appelbaum, P., Robbins, P.C., Mulvey, E.P., Silver, E., Roth, L.H., & Grisso, T. (2000). Developing a clinically useful actuarial tool for assessing violence risk. *British Journal of Psychiatry, 176*, 312-319.

Nakanishi, M. (1987). Perceptions of self-disclosure in initial interaction: A Japanese sample. *Human Communication Research, 3*, 167-190.

Nisbett, R.E., & Cohen, D. (1996). *Culture of honor: The psychology of violence in the South*. Oxford: Westview Press.

Raju, N.S., Van der Linden, W.J., & Fleer, P.F. (1995). IRT-based internal measures of differential functioning of items and tests. *Applied Psychological Measurement, 19*, 353-368.

Reise, S.P., Widaman, K.F., & Pugh, R.H. (1993). Confirmatory factor analysis and item response theory: Two approaches for exploring measurement invariance. *Psychological Bulletin, 114*, 552-566.

Salekin, R.T., Rogers, R., & Sewell, K.W. (1996). A review and meta-analysis of the Psychopathy Checklist and Psychopathy Checklist--Revised: Predictive validity of dangerousness. *Clinical Psychology: Science and Practice, 3*, 203-215

Santor, D.A., & Ramsay, J.O. (1999). Progress in the technology of measurement: Applications of item response models. *Psychological Assessment, 10*, 345-359.

Steadman, H., Mulvey, E.P., Monahan, J., Robbins, P.C., Appelbaum, P., Grisso, T., Roth, L., & Silver, E. (1998). Violence by people discharged from acute psychiatric inpatient facilities and by others in the same neighborhoods. *Archives of General Psychiatry, 55*, 393-401.

Tanzer, N.K. (1995). Cross-cultural bias in Likert-type inventories: Perfect matching factor structures and still biased? *European Journal of Psychological Assessment, 11*, 194-201.

Tengstrom, A., Grann, M., Langstrom, N., & Kullgren, G. (2000). Psychopathy (PCL-R) as a predictor of violent recidivism among criminal offenders with schizophrenia. *Law and Human Behavior, 24*, 45-58.

Van de Vijver, F.J., & Leung, K. (1997). *Methods and data analysis for cross-cultural research.* Thousand Oaks: Sage.

Van de Vijver, F.J., & Poortinga, Y.H. (1994). Methodological issues in cross-cultural studies on parental rearing behavior and psychopathology. In C. Perris, W. A. Arrindell, & E. Eisemann (Eds.), *Parental rearing and psychopathology* (pp. 173-197). Chichester, UK: Wiley.

Waller, N.G., Thompson, J.S., & Wenk, E. (2000). Using IRT to separate measurement bias from true group difference on homogenous *and* heterogenous scales: An illustration with the MMPI.. *Psychological Methods, 5*, 125-146.

Zimring, F.E., & Hawkins, G. (1997). *Crime is not the problem: Lethal violence in America.* Oxford: Oxford University Press.

III

Comparative Themes: Selected Legal and Management Issues

Multi-Problem Violent Youth
R.R. Corrado et al. (Eds.)
IOS Press, 2002

The Utility of an Integrated Multi-Service Instrument within France

Catherine Blatier and Raymond R. Corrado

In France, the potential for gathering information that is relevant to identifying risk factors for violent behavior in adolescence and young adulthood is encouraging. More than most other advanced industrial and liberal democratic countries, France has led the way in implementing systematic programs to assist children considered to be in danger. In this chapter, the policies and programs that are available to such children will be described. As in most countries, the age of the child determines which government ministry assumes primary program jurisdiction, even though, at the latter ages, such as adolescence, programs from multiple ministries routinely provide services and/or assume responsibility.

1. Pregnancy to Six Years Old: The Ministry of Health and Solidarity

The 'Protection Maternelle Infantile' (PMI) is first informed of a pregnancy by a doctor. Once this information is processed, a public health nurse initiates contact with the prospective mother. The visits between the expectant mother and the public health nurse focus on the health status of the mother and consists of a standardized screening of potential problems, including nutrition, smoking, and emotional stress. Of particular concern during this screening process are women who are very young, women aged 40 or more who are primipara, women who are expecting several children, such as twins, or those who already have several children. Once the baby is born, social workers assume the responsibility of visiting the home in order to examine if the mother has any problems with her infant. These social workers observe the child in order to identify any such problems.

Under the French system, there are three obligatory visits to the doctor that the mother must keep. The first visit occurs at the 8th day of the child's life. This is followed by a visit when the child is 9 months old and a final visit when the child turns two years old. These visits with the doctor are mandatory, and in order to ensure the parents' participation, government family subsidies may be withheld from those parents who fail to attend.

When a child is 3 years old, the school health prevention system assumes responsibility for the child's health. At this point, the mother has access to the PMI special centre. This centre is serviced by medical doctors, nursery nurses, midwifes, psychologists, specialized education workers for young children, and social workers. In consultation with the mother, a variety of action plans are devised to assist the mother based on the individual needs identified in the initial meetings. These plans can include, but are not limited to, discussion groups and information sharing sessions with other mothers. Specialists also help educate mothers in dealing with any specific issues that may arise with her baby. For instance, basic parenting skills, such as when to awaken the little child and health education, are often taught in these discussion groups. However, for those mothers identified by social workers as being "at risk" with their children, and for immigrant mothers with language and cultural barriers that inhibit their access to the PMI centers, the consultative and education services are often delivered to the home, on a one-to-one basis.

Another component of this comprehensive approach to the total health of children from pre-birth forward is the frequent and mandatory medical supervision of the child. The mother has 20 obligatory visits to a doctor from the time of the birth of her child to when the child reaches six years old. Specifically, she must have nine medical supervisions during the child's first year and three visits during the child's second year. These visits are followed by an additional two visits per year until the child's sixth birthday. It is during these consultations that exhaustive diagnostic medical and psychological-social profiles are conducted. During these visits, any initial learning disabilities are identified and remedial assistance is provided. Again, particular psychological and social services are typically designated for higher risk families, namely those either who are impoverished or those that have a history of health problems, especially mental disorders. Once a problem is identified, the parent or doctor can request assistance from the 'Centre d'Action Médico-Sociale Précoce' (C.A.M.S.P.). This centre provides specialized treatment for children. For example, this centre might provide a specific treatment for a child with sensorial, motor, or mental deficits.

A complimentary resource for young pregnant mothers, or mothers with babies less than three years of age, who need special financial and psychological support, is the 'Centre Maternel' (CM). The CM is designed to address the problems associated with financial stresses and psychological disabilities that put mothers and babies at risk. All of these special service centers, therefore, are part of a nation-wide system, including the country-side, of identifying, as early as possible, those families and children who have actual or potential problems that put the former at risk for further problems, including delinquency and violence.

In addition, as soon as conduct disorders, or related mental disorders, are identified, families can be referred to the Centre Medico-Psychologique pour Enfants (CMPE). This medical centre for children provides a wide array of therapy services ranging from traditional psychiatric and psychological approaches to procedures for correcting speaking and motor malfunctioning.

All of the resources described above involve either health or social services with the focus being on the earliest diagnosis of problems and intensive responses to future risks for additional health and mental problems among children. Another important and pervasive resource is the centralized national education system.

2. The Ministry of Education

As mentioned in the introduction, France is unusual in comparison to other advanced industrial countries, particularly Canada and the United States. Part of the distinction between France and other advanced industrial countries is based on the mandatory, universal, and primarily tax-funded nature of a complete range of services for children and adolescents. The educational system is an integral part of these services.

Parents can bring their children to preschool when they are two years old. Over one-third (35%) of attendees to preschool are supported by bursaries or scholarships. However, virtually all three years old children (99%) enter the Ecole Maternelle preschools, as do all four and five year old children. In France, enrollment in school is obligatory for all children between the ages of six and sixteen. Moreover, for a small number of parents, home schooling, utilizing correspondence learning, is an option. In either approach, during the elementary phase (grades one to five), children who are having difficulties are referred to a diagnostic and remedial service team provided by the Réseau d'Aide Specialisée Aux

Enfants en Difficulté (R.E.S.A.D). This team typically consists of a psychologist and two specialized teachers who intervene pedagogically and/or with mental health issues. The psychologist's role can be extensive depending on the seriousness of the mental disorder profile, especially if the child has major difficulties functioning in a regular school environment. Approximately five thousand students are dealt with by RESAD teams throughout schools in France.

During the child's seven-year secondary school stage, a different set of resources or programs are accessed for older children and adolescents with problems. The primary responsibility for these students now lies with "social assistants" rather than with psychologists. The social assistant's focus is on identifying both "children in danger", such as those who are abused, and "children at risk", such as those who are habitually truant. Located within each school, a nurse is also available for students with drug and alcohol problems, as well as for those students with other health problems. However, if, for instance, a youth has a conduct problem, like ADHD, he or she will usually be sent to special classes within the school to assist in learning. These students may also be treated with Ritalin, or, in extreme cases, they may be sent to a special educational establishment for children with conduct disorders.

There are also two routine medical examinations when the children are eleven years old and fifteen years old. During these routine visits, children and youth who are identified as having conduct disorders or other psychological problems can be referred to a special service, namely the "Centre Médico-Psycho-Pédagogique" (C.M.P.P.), which is provided by the Ministry of Education. Psychiatrists, psychologists, and speech or motor function therapists are available at this centre to assist in the treatment of a myriad of problems.

Once in secondary school, there are considerably fewer systematic diagnostic and treatment interventions for children than is found in the elementary school system. In part, the rationale for this reduction in diagnostic and treatment services is based on the notion that older children and adolescents are far more capable of making their needs known to parents or to the appropriate authorities. Also, other national agencies have jurisdiction in identifying and responding to children at risk, such as the "Aide Sociale a l'Enfance" (A.S.E.), or the Child Protection Board.

3. Administrative and Judicial Protection for Children in Danger

As a state service, the Child Protection Board (A.S.E.) is mandated to ensure both the protection of the child and offer services, primarily educational, to the juvenile justice system for those cases under the latter's jurisdiction. The A.S.E. has a broad age mandate to deal with minors and their families, with emancipated minors, and with majors under 21 years old who encounter social difficulties. The A.S.E. may intervene with either financial support to the families, by working with families and children to prevent the marginalization of the latter, or to facilitate social integration by socio-educative community activities. Since 1958, children 'in danger' are entrusted to the non-judicial administrative protection of the A.S.E. Often, these designated children remain at home and a social worker or a special education worker is assigned to work with the children and their family. However, where it is determined that it is necessary to separate the child from their parents, and to place the child in a special educational facility, the A.S.E. first seeks the cooperation of the parents. If the parents cannot agree among themselves, cannot reach an agreement with their child, or cannot reach an agreement with the intervention of special social and education workers, the social workers can bring the matter before the juvenile court.

There are two main criteria that would allow for direct intervention by a juvenile court judge. The first criteria for intervention by a judge is based on a positive assessment that the child is considered to be in danger if there is a conflict between the child and someone in the family. The second criteria is if it is deemed necessary to take action in the best interest of the child. Acceptance of either of these criteria is sufficient for a judicial intervention. The juvenile court judge's mandate is to deal with neglected, abused, or maltreated children under civil procedure, and with delinquents under a quasi-criminal procedure. Specifically, under the former procedure, the juvenile court judge is charged with protecting children 'whose health, safety, and moral security are in danger', or to act in cases in which 'upbringing conditions are seriously threatened' (Article 375, Civil Code). It should be kept in mind that, although the judge is under no legal obligation to do so, in most cases, the judge first attempts to obtain the approval of the parents before intervening.

The judicial role is complicated due to the fact that whether a case proceeds under judicial or administrative protection, there is no difference in how evidence is presented procedurally, or in the nature of judicial making. For example, anyone, including social workers, may bring a matter before the juvenile court. Similarly, anyone can bring a matter before the A.S.E. for a strictly administrative determination. To reiterate, in principle, "risk of danger" calls for administrative care and "state of danger" calls for judicial care. Also, if, in the course of the administrative care, the family disagrees with the intervention and attempts to impede it, the case is automatically brought before the juvenile court. This is important because French juvenile court judges are specialized judges, in that they deal with each minor during the preliminary investigation, the judgment, and continue to be in charge of the minor after the judgment is made (Blatier, 1998).

In addition, the juvenile court judge has the option, at the request of the youth, to retain the youth under the court's protective care until age 21. During this period, the judge determines what specific problems are inhibiting the youth's ability to reintegrate into the community. The judge also has the authority to devise, implement, and monitor a plan of intervention that can include, but is not limited to, foster care, special education, and training programs.

Unlike most other countries, these extraordinary powers at the investigation and judgment implementation stages allow French juvenile court judges to direct cases. It also gives judges considerable influence over both accessing state resources or programs and the involvement of parents. According to a long time critic of the more restricted role of juvenile court judges (Hackler, 1984), the French model is far superior because it is based on informal procedures and powers that maximize the quantity and quality of information the judge is able to utilize in devising, implementing, and monitoring the outcome of individual cases. In contrast, Hackler argues that the more procedurally rigid distinctions between criminal and civil matters regarding children and youth, and the limited role of judges in monitoring the implementation of their decisions in countries, such as Canada and the United States, results in children "falling between the cracks" of the numerous and uncoordinated ministries and their multitude of programs. In effect, there is no one to ensure that the appropriate resources will actually be provided to the appropriate children and youth.

4. Juvenile Court Processing of Delinquents

Since the Order of February 2, 1945, the Welfare model of juvenile justice has prevailed in France. This model focuses decision-making on rehabilitative objectives rather than on the due process objectives of the Justice model or the incapacitation objectives of the Crime Control model (Corrado, 1992). The juvenile court judge, therefore, adopts an inquisitorial role in order to obtain information that can assist the court in devising a plan that can help the youth avoid further delinquencies. In effect, the delinquent act is regarded as a sign that there is a need for intervention. The aim of the Welfare model is to understand the delinquent act, rather than to condemn it.

The judge relies on specialists, including psychologists and specialized education workers, to explain the meaning of the delinquent act. Moreover, the aim is not restricted to only understanding the nature of the act, but also to prevent a relapse or a repeat of the behavior. Nonetheless, the juvenile court judge and the juvenile criminal court are able "when circumstances and the personality of the delinquent justifies it, to condemn juveniles older than 13 years to a penal sentence" (Article 2).

Recently, an intervention was introduced whereby the prosecutor in charge of minors is supposed to process certain delinquents, rather than a juvenile court judge. This change occurred to expedite the response to minor delinquencies. Unlike the juvenile court judge, the prosecutor gives an immediate response to the act.

In contrast to the rapid decisions prosecutors make based on limited information, the judge typically requests that the special educational worker assess the family environment in which the youth was raised (Article 8). The judge then has a wide range of options at his/her disposal to respond to the problem profile of the youth and the family. These options include, but are not limited to, placing the juvenile in a children's home; making a provisional order for the juvenile; appointing a social worker to assist the youth and family; putting a youth into the care of a special educational establishment; or placing the youth in custody if necessary.

For minor offences, usually property offenses, the apprehended youth is processed by the police and taken home. Subsequently, the police report or file is sent to the public prosecutor who decides whether to proceed further. For a second offence, and/or more serious offenses, the public prosecutor decides whether to bring the case before the juvenile court judge or the "examining magistrate in charge". The latter is a public prosecutor and a judge in charge of minors who specializes in juvenile matters. This procedure also occurs if the offence has been committed with adults, or in cases of a major violent crime, such as murder, drug dealing, and burglary. When the latter option is selected, it is usually because the prosecutor wants a more punitive outcome than typically is expected from the juvenile court. As stated above, the latter court routinely directs delinquent youths to educational interventions and rarely to custodial sanctions.

During the 1990s, the juvenile court incorporated a new option, the reparation order. The reparation order is considered an educative measure because it is designed to teach the youth a sense of responsibility with respect to their actions; the capacity to take into account the meaning of the infraction, with respect to a law that aims to ensure justice within the community; and the recognition that the victim has been wronged. Usually, a social worker investigates the feasibility of a reparation order at the bequest of the juvenile court judge and then, with the consent of the youth and victim, a mutually agreeable plan to compensate the latter is undertaken. Other measures are also available to juvenile court judges.

5. Juvenile Court Outcome Options

The initial, and critical, step in the decision making process is the judge's Social Inquiry. This inquiry focuses on three main areas, namely, the problem profile of the youth's family, the extent of the youth's criminal record, and the youth's school behavior. Regarding the first information objective, a social welfare worker is directed to conduct a

Social Orientation and Investigation Measure (Investigation d'Orientation Educative). In this measure, the behavior of the child is observed in the family context. The most common outcome that results from the information provided to the juvenile court judge is Educative Action in Open Environment (Action Educative en Milieu Ouvert). This order requires that the child live with the parent(s), and that the family meet regularly with a social worker. The situation is evaluated within six months after the order is issued in order to determine whether to extend or terminate the order. The Social Inquiry Report can also lead to educative measures, such as admonishing the youth, to probation, and/or to reparations to victims. More intrusive or extensive measures are also available.

6. Placement in Special Institutions

The choice of a special institution is based on an assessment of the child's needs at several levels. The evaluation is based on the social, psychological, academic, and health-related needs of the youth. These welfare institutions, therefore, emphasize rehabilitative interventions, and include both juveniles under care and juvenile delinquents. Special education programs are also a core element of this institutions. During the first six months, the judge can direct a more extensive investigation and assess the evolution of the youth's situation. After this six month period, and the preliminary judgment, a new judgment can be given for a period not exceeding two years. Before the end of the two years, judges have to review their decisions, and cannot extend their judgments for longer than two year periods respectively.

As in all cases, prison, or custody orders, are the most intrusive option and rarely used in France. For the first 11 months of the year 2000, 556 youths under 18 were imprisoned. This figure constitutes less than 1% of all the prisoners, minors and adults, in France. Only 18 girls and 15 minors, under 16 years old, were included in this total. As mentioned above, only for the most serious and violent offenses do judges send youths to prison. Medical examinations are mandatory and psychological referrals can be ordered as well. Social workers generally do in-take screening with a focus on education assessment. Attendance at the prison education classes typically is proposed. Technical or job related training is also available. Individual counseling is provided by psychologists or psychiatrists upon request. Projective tests, most commonly, the Rorschach and Thematic Apperception Test, are utilized in assessment processing. Group therapy is an option that is also available in custody for those youths who want to participate.

7. Information Processing Issues Regarding the Utility of a Needs-Risk Assessment Instrument for Violence in France

As in most jurisdictions, the juvenile court in France is the initial state institution that can respond to violent youth with wide ranging and authoritative measures, including both rehabilitative and incarceration options. Similarly, it is this court that is in the position to identify and react to the risk-needs factors most often associated with multi-problem profile families because of the court's dual jurisdiction for "children in need of care" and juvenile delinquents. As enumerated above, the juvenile court judge has inquisitorial, educative-rehabilitative, protective, and punitive functions. In addition, the social worker is another major source of information. Typically, the cooperation of the parents maximizes the court's understanding of the youth's problems, as well as their likely genesis. Additionally, for many of the most serious cases, it is not uncommon for these youths to have been in contact with social workers since they were five years old. Finally, several medical and educational programs described previously are also likely to have been accessed by some of these youths as toddlers and children up to the age of five years. In effect, there is the potential for the juvenile court to obtain relatively complete historical records of the risk factors for violence from pre-birth to young adulthood. Not unexpectedly, this court's enormous power to gather this information, and its ability to restrict the access of parents and other relevant parties to it, has caused an intense political debate about arbitrary power.

It is in this political context that the utility of introducing a systematic and comprehensive risk-needs assessment instrument would have to be considered. It raises the specter of an even more powerful, arbitrary, and potentially capricious juvenile court. Nonetheless, attempting to gather such information already is mandated by several laws based on welfare models developed since the early 20th century to respond to impoverished, neglected, abused, and delinquent children. There is, however, an awareness that coercive or intimidating procedures and roles of state institutions, regarding families with "at risk" or delinquent children, need to be minimized.

One of the protocols that facilitates the exchange of information from the family to the juvenile court consists of the judge's attempt to engage the family in an "agreement" concerning the appropriate judicial intervention (Garapon, 1985). This court is partly distinctive because it has an explicit educative function with the family. In effect, the family is approached by the judge, social worker, and other specialists in, for the most part, an assisting capacity. As well, the judges effectively seize control of each case until an intervention plan is completed. This further builds a trust and understanding between the court and the family.

A more adversarial tone can occur for delinquency cases. Social workers do not routinely share the information they gather about the youth and the family with the family. For some critics, this secrecy is inappropriate. The family is at a disadvantage since they are not necessarily aware or understand all the information that is used as the basis for the judge's intervention plan. Until this debate is resolved, the status quo allows social workers wide ranging authority to gather virtually the entire history of a family with a 'child at risk' or a delinquent.

Typically, the Social Inquiry report is completed in three or four visits with the family. Again, the focus is usually on estimating the risk posed to the children or youths if they remain within the family. Where it is deemed necessary, experts will also be consulted to assess the specific risk factors in health, mental health, and school domains. All expert reports are signed by the consultants, while some of the social workers' reports are signed by the Director of the service in charge of the social inquiry service. The Director also reviews the information before the Social Inquiry report is submitted to the judge.

Various concerns have been raised about the current information gathering and processing in France. Most importantly, concerns have been raised about the validity of data that is largely descriptive rather than evaluative. Other concerns are based on the notion that risk estimates are derived exclusively from an interviewer's impressions. The fundamental question is to what extent do social workers and other experts bias their descriptions of risk information about the family and child to fit personal/professional ideologies? (Rieman & Angleitner, 1993; Shweder & D'Andrade, 1979; Shweder, 1981). Another issue is to understand the degree to which interpretations of impressions made several days or more after the actual interview took place are valid. (Le Poultier, 1990). It can be argued that these validity problems are lessened in proportion to the completeness, immediacy of information gathering to risk assessment decisions, and the minimization of interpretative criteria of the instrument(s) utilized in compiling the Social Inquiry report.

The multi phase needs-risk assessment instrument presented in this book is a step in the direction towards resolving some of the outstanding issues that have been discussed in this chapter. For France, this instrument would provide a valuable database to better understand the growing complex and demographically diverse clientele of families with children at risk for a variety of problems associated with violence. Equally important, for those agencies and workers responsible for providing already extensive services, such an instrument would facilitate the standardization of inquiries and the integration of the data in a manner that is absent under current practices. In turn, this data integration may result in the delivery of services that will be more effective in reducing the risk factors that likely lead to child and youth violence in France. Given the recent political trends calling for more punitive policies toward this violence, in the face of cross national research that has consistently revealed that imprisoning youth does not protect the public (Tournier, 1991; Tournier, & Portas, 1997) it is hoped that innovative policy approaches will allow France to continue its tradition of progressive responses to multi-problem families with children at risk.

References

Amiel, C., & Garapon, A. (1988). La justice des mineurs entre deux ordres juridiques: justice imposée et justice négociée. *De quel droit? De l'intérêt...aux droits de l'enfant*, Vaucresson: Centre de Recherche Interdisciplinaire.

Beauvois, J-L. (1984). *La psychologie quotidienne*. Paris: P.U.F.

Beauvois, J-L. (1990). L'acceptabilité sociale et la connaissance évaluative. *Connexions*. 56, 7-16.

Blatier, C. (2000). Locus of control, causal attributions and self-esteem: A comparison between prisoners. *International Journal of Offender Therapy and Comparative Criminology, 44,* 97-110.

Blatier, C. (1999a). Towards a constructive response to young offenders: reparation at the levels of justice and individual psychology. *Journal of Social Work Practice, 13,* 211-220.

Blatier, C. (1999b). Juvenile justice in France: The evolution of sentencing for children and minor delinquents. *British Journal of Criminology, 39,* 240-252.

Blatier, C. (1999c). *La délinquance des mineurs: L'enfant, le psychologue, le droit.* Grenoble: Presses Universitaires.

Blatier, C. (1998). The specialized jurisdiction: A better chance for minors. *International Journal of Law, Policy and the Family, 2,* 115-127.

Corrado, R. R. (1992). Introduction. In R. R. Corrado, N. Bala, R. Linden, & M. Le Blanc (Eds.), *Juvenile justice in Canada: A theoretical and analytical assessment.* Toronto: Butterworths Canada.

Garapon A. (1985). *L'Ane portant des reliques. Essai sur le rituel judiciaire.* Paris, Le Centurion.

Hackler, J. (1988). Practicing in France what Americans have preached: The response of French judges to juveniles. *Crime and Delinquency, 34,* 467-485.

Le Poultier, F. (1990). *Recherches évaluatives en travail social.* Grenoble : P.U.G.

Rieman, R., & Angleitner, A. (1993). Inferring interpersonal traits from behavior: Act prototypicality versus conceptual similarity of trait concepts. *Journal of Personality and Social Psychology, 64,* 356-364.

Shweder, R.A., & D'Andrade, R.G. (1979). Accurate refection of systematic distortion? A reply to Block, Weiss and Thorne. *Journal of Personality and Social Psychology, 37,* 1075-1084.

Shweder, R.A. (1981). Fact and artifact in trait perception: The systematic distortion hypothesis. In B. A. Maher & W. B. Maher (Eds.), *Progress in experimental personality research.* New York: Academic Press.

Tournier, P. (1991). La détention des mineurs. Observation suivie d'une cohorte d'entrants. Paris: C.E.S.D.I.P.

Tournier, P., Mary, F. L., Portas, C. (1997). Au-delà de la libération.Observation suivie d'une cohorte d'entrants en prison. Paris: C.E.S.D.I.P.

Zuckier, H.(1986). The paradigmatic and narrative modes in goal-guided inference. In R.M. Sorrentino and E.T. Higgins (Eds). *Handbook of motivation and cognition: Foundations of social behavior*, pp.465-502. New York: Guilford Press.

Multi-Problem Violent Youth
R.R. Corrado et al. (Eds.)
IOS Press, 2002

261

Principles of Formulating Forensic Psychology Reports Concerning Minor and Juvenile Perpetrators of Violent Criminal Acts in Poland

Jozef K. Gierowski and Teresa Jaskiewicz-Obydzinska

In Poland, a separate juvenile court was established in 1919. Since the very beginning, the legal code took into the consideration the psychological aspects of juvenile delinquency. Thus, the term "recognition," interpreted as the ability to understand and assess the criminal nature of an act, was introduced. This term, although being somewhat controversial, was in force in Poland until 1982 (Czerederecka, 1995, Gierowski, 1980). In 1982, the new *Juvenile Delinquency Bill* was enacted. The *Bill* has two basic aims. First, it is intended to prevent juvenile crime and the "demoralization" of youth by creating conditions conducive to the return to normal life of those who have been in conflict with the law or society. Second, it is intended to strengthen the role of the family as the educator of and caregiver for juveniles.

The <u>Bill</u> defines a "juvenile" in three different ways. Insofar as the prevention of demoralization is concerned, a juvenile is any person below age 18. (The *Bill* does not define the concept of demoralization, and it is therefore open to wide interpretation. In practice, however, it is assumed to refer to an acutely intensive and durable form of maladjustment.) With respect to criminal offenses, a juvenile is any person between the ages of 13 and 17. Finally, with respect to the provision of rehabilitation – including custodial, educational, medical, and correctional services – a juvenile is any person aged 21 or younger.

Independent of the *Bill*, the new Polish *Criminal Code* from 1997 lowered the age limit for legal responsibility for certain serious, violent offenses (e.g., terrorism, murder, aggravated assault, kidnapping, robbery) to 15 years. Legal responsibility depends on "circumstances of the case, degree of maturity of the perpetrator, his personal features and circumstances, and especially, whether correctional and educational measures undertaken before have, or have not been successful" (Górecki & Stachowiak, 1998).

In effect, the new law requires juvenile courts to consider individual features of a juvenile, his family circumstances, and educational conditions. To assist in this assessment, the court may refer the juvenile to a specialist center; if a juvenile is to be placed in a correctional institution, such examination is obligatory. The assessments are carried out primarily in Family Diagnostic and Consultative Centers at regional courts. In certain cases, especially when a juvenile has committed a serious crime, the Psychology Department of the Institute of Forensic Research in Cracow conducts the assessment. The Department examines minors and juveniles exclusively at the request of the court. For minors, an expert's opinion is requested by the Family and, for juveniles, by the Criminal Court.

The assessment focuses on such things as socialization experiences, risk for future criminality, and treatment needs related to psychological, educational, or medical problems. The assessment should also identify the psychological mechanisms believed to account for the criminal behavior.

The primary focus of the assessment is socialization, which is generally understood to mean the process of human development resulting from deliberate, intentional (educational) influences of the social surroundings, as well as from unintentional ones. The causes of incorrect socialization may include poor educational and family conditions and bio-psychical factors leading to disorders in the development of personality in the scope of its basic structures and functions. Such disorders may take two basic forms. First, young people may identify with the values and aims of a criminal subculture. In this case we use the term *antisocial personality*. It may happen, however, that for various reasons, the psychological mechanisms responsible for socialization do not develop. In such situations, biopsychological variables may play an important role (e.g., low level of intelligence, damage to central nervous system), being so detrimental to adaptation abilities that young people are not able to integrate with any social group and assume its values and norms as their own. Especially important here are the functional disorders in the cognitive, emotional, and motivational spheres resulting in incorrect evaluation of a situation, limited control of affects and drives, over-excitement, and a high level of aggression. Such disorders of regulative and integrative functions of personality are called *asocialization*. There is also a possibility that a correctly socialized person will break the law. It occurs, for example, under a high pressure of outside circumstances or as the result of psychopathological variables (e.g., psychosis).

The psychological examination of a minor or a juvenile for judicial purposes should aim to determine the causes of disorders in the process of socialization. A thorough identification and analysis of symptoms of the maladjustment exhibited by a minor or a juvenile is required here, as well as the period when such symptoms occurred, their intensity, frequency of occurrence and social noxiousness of the act. The attitude of a juvenile towards the acts he has committed should also be taken into consideration—his or her understanding of the criminality of the act, sense of guilt, and, first of all, motivation to change his or her behavior and correctional aptitude.

Such an assessment, showing the formation of a minor's psyche in its developmental aspect, could be fully useful for court purposes. What is essential here is the possibility to determine the advantage of either the environmental or biopsychological conditioning of specific personality features or behavioral disorders. It constitutes the fundamental condition concerning the planning and delivery of adequate educational, correctional, or medical interventions. Personality assessment also falls within the responsibility of the forensic psychological expert. It applies to situations when the young people examined are the perpetrators of criminal acts or when they manifest the symptoms of demoralization. The assessment of personality should not be based solely on the pathogenetic paradigm, which concentrates on seeking the conditions responsible for the faulty functioning of the personality – the broadly understood risk factors – but also take into account the salutogenetic approach. This approach is based on the search for the psychosocial conditioning for the correct functioning of personality. Following Antonovsky (1979), instead of asking only about the causes of behavioral disorders we pose the question, what factors constitute the potential resources of an individual, on the basis of which one can plan and execute psycho-correctional measures?

Figure 1 summarizes the process for examining minors and juveniles at the Institute of Forensic Research in Cracow. Psychological examinations include analysis of case files; interviews with custodians, focused on obtaining the maximum information concerning the development, rearing experiences, and environmental conditions; psychological tests; and interviews with minor aimed at determining their level of understanding of the situation in which criminal acts were committed, as well as the motives for the acts. In the assessment process, one takes into account the results of the psychiatric examination and other additional medical tests.

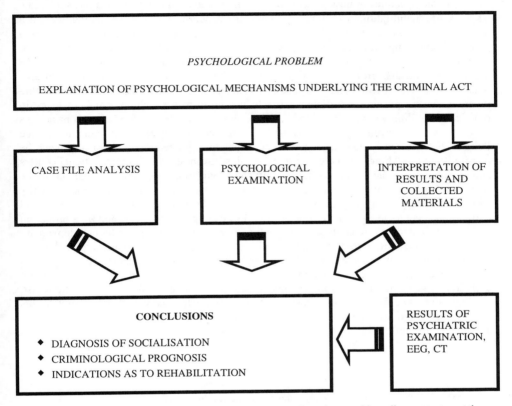

PSYCHOLOGICAL PROBLEM

EXPLANATION OF PSYCHOLOGICAL MECHANISMS UNDERLYING THE CRIMINAL ACT

CASE FILE ANALYSIS

PSYCHOLOGICAL EXAMINATION

INTERPRETATION OF RESULTS AND COLLECTED MATERIALS

CONCLUSIONS

♦ DIAGNOSIS OF SOCIALISATION
♦ CRIMINOLOGICAL PROGNOSIS
♦ INDICATIONS AS TO REHABILITATION

RESULTS OF PSYCHIATRIC EXAMINATION, EEG, CT

Figure 1: Process for preparing psychological reports concerning minors and juvenile perpetrators at the Institute of Forensic Research in Cracow.

Taking into the consideration the necessity to obtain uniform methods used in different countries to examine the conditioning of aggressive behavior in minors and juveniles, we would like to underline that it is important from the practical point of view to differentiate between interpersonal aggression due to a violent character and aggression resulting from dysfunctions of drive or affect. In the first case, aggression manifested by young people results primarily from disorders of socialization that are the consequence of incorrect processes of learning. In the second case, aggression is directly related to such biopsychological factors as temperament and character variables, damage to central nervous system, deterioration of self-control resulting from weaknesses in cognitive processes, emotional over- excitement, impulsiveness, and disorders in the drive sphere. To evaluate normal and abnormal aspects of personality, we use a variety of projective tests and self-report questionnaires. The final evaluation of aggression manifested by minors, of its origin, symptoms and intensity, is the result of the analysis of all material collected during the examination and concerns both the personality structure as well as its function.

Finally, when formulating an effective psychological opinion in cases concerning minors and juveniles, it is important to keep in mind the following problems. First, evaluators should not disregard the seemingly unimportant behavioral disorders exhibited by the youngest minors which are often a tell-tale sign of the distorted formation of their social attitudes. It precludes the application of preventative measures. Second, the process of rehabilitation is not clearly conceptualized in law. Existing legal regulations refer to most general principles of how to deal with juveniles and do not constitute a basis to create a system of rehabilitation institutions up to modern standards. Generally speaking, there is a lack of specialized centers and institutions that would cater for individual needs of a juvenile, depending on intensity and character of his lack of adaptation or incorrect social adaptation. The expert psychologist voicing an opinion cannot, therefore, indicate or choose the optimal means, from the psychological point of view, to realize indispensable rehabilitative and therapeutic purposes. Third, existing legal regulations concerning juveniles do not provide a sufficient basis to use up-to-date psychological knowledge in practical rehabilitative procedures. Indeed, they appear to contradict fundamental principles of developmental psychology. Such is the case with the recently resolved regulation by the Polish Diet, currently undergoing further legislative procedures, which legalizes he possibility to use "direct coercive measures" against juveniles placed in educational centers and reformatories. The argument that this just the legalization of a status quo can hardly be accepted. Such coercive measures are tantamount to legalized violence and may actually worsen a young person's aggressive behavior. Fourth, the quality of most rehabilitative and correctional centers leaves much to be desired. For example, the supervision by a probation officer is executed by rare contacts between the officer and a juvenile, or by people who are not sufficiently qualified to do the job. Finally, there is a lack of systematic approach to try to prevent the demoralization and to correct a faulty social adaptation. Poor material conditions of the existing institutions, the lack of model solutions and abilities to carry out correctional tasks in inter-disciplinary teams lead to the situation in which a psychologist, carrying out specific diagnostic tasks is not sufficiently motivated to formulate his conclusions in the form of clear therapeutic and rehabilitative tasks, having a well-defined aim and a program.

References

Antonovsky, A. (1979). *Health, stress and coping.* San Francisco: Jossey-Bass.

Czerederecka, A. (1995). The participation of a psychologist in civil cases concerning a child and a family. *Problems of Forensic Sciences, 32*, 123-136.

Gierowski, J. K. (1980). Rozeznanie nieletniego – problemy diagnostyczne i kompetencyjne. *Psychiatria Polska, 2*, 173-178.

Górecki, P., & Stachowiak, S. (1998). *Ustawa o postępowaniu w sprawach nieletnich.* Komentarz. Zakamycze.

Multi-Problem Violent Youth
R.R. Corrado et al. (Eds.)
IOS Press, 2002

265

Violent Offenders with Mental Health Problems: The Italian Youth Study[1]

Luca Iani, Gaetano De Leo and Antonella Ciurlia

Italian research on adolescents has typically focused on multi-problem families rather than multi-problem youth. However, even though the construct of multi-problem youth has been only occasionally addressed by Italian researchers, we believe that both Social Services and the Youth Justice System routinely are faced with cases involving multi-problem diagnostic and treatment issues. Yet no validated instrument for the assessment of violence and multi-problem risks for youth, at least within the Justice System. Such an instrument not only could fill a fundamental gap in the research field, but also allow the Social Services and the Justice Services to collect the integrated basic information. This would facilitate the comparison of case management data needed by the multiple agencies responsible for specific types of problems for a single youth. However, what would the function and the usefulness of this instrument be, from a practical point of view, and when should it be administered? For example, when the risk assessment tool is administered after the crime has already been committed, it provides insights about the juvenile's needs that can be used to plan a more appropriate and specific treatment. The intake stage of the juvenile by the Social or Justice Services would provide important diagnostic evidence. Nonetheless, if our aim is to reduce the risk of serious juvenile delinquency, it is necessary to collect data well before a major crime is committed. This would require the screening of at risk multi-problem children and youth. The screening instrument would provide the diagnostic data needed to plan intervention projects aimed at the development of social and personal skills that would increase protective factors against future violent offending (Guerra, 1998). The challenge of developing an appropriate screening instrument that can be used in Italy and elsewhere is enormous: "The group of individuals with multiple problems cannot be characterized as being serious delinquents, nor can these multiple problems be used to identify the group of serious delinquents" (Huizinga & Jakob-Chien, 1998). In effect, it is necessary to utilize an instrument that is based on the research demonstrating a relationship between particular problem profiles and violence.

1. Definitions

The concept of youth violence is not a simple one (Reppucci, Woolard, & Fried, 1999). Researchers from diverse disciplines have often considered different aspects and constructs, not always from the same conceptual domain. The most common conceptualization is based on direct criminal offense indicators, i.e., the commission of one or more criminal acts such as homicide, aggravated assault, robbery (or armed robbery), kidnapping, voluntary manslaughter, rape and arson (Loeber, Farrington, & Waschbusch, 1998). Tedeschi and Felson (1994) adopted a more abstract conceptual definition based on the construct of *coercive action*, which is "an action taken with the intent to impose harm on another person or to force compliance." Focusing on the social explanation of action and on the actor's motivations, intentions and expectations rather than on drives, pattern of arousal or other internal sources of behavior, Tedeschi and Felson employ a social-interactionist approach, which focuses on the dynamic relationship between antagonists. There are even broader theoretically based concepts that are relevant to understanding how violence can be described. If we consider the point of view of the actor as a decision-maker, then values, expectations, scripts, definitions of the situation, and assessment of alternatives in the choice of a goal-directed action should also be taken into account.

The choice of a conceptualization of violence is extremely important given the inherent complexity of the multi-problem construct. It is not uncommon for theorists to include subtypes of violent offenses. In effect, the explanation of violence is not separated from its conceptualization. For example Toch and Adams (1994) refer to the concept of "disturbed violent offenders." Unfortunately, they give concepts such as disturbed, multi-problem and mentally ill a similar meaning, thereby risking conceptual confusion by including, under a single violent concept, overlapping of populations which do not necessarily share the same clinical and diagnostic characteristics. Within the disturbed violent offender concept, the authors identified three different types: delinquents with histories of mental health problems, delinquents with histories of substance abuse and mental health problems, and delinquents with histories of substance abuse.

From our perspective, it is important to differentiate the concept of violent youth from the risk/problem factors often associated with it. The approach is evident in the Denver Youth Survey of multi-problem violent youth (Huizinga, Esbensen, & Weiher, 1991, 1996). In this study, four typologies of problems relevant for juveniles delinquency were derived: school problems, substance abuse problems, mental health problems and victimization problems. Besides the above differences in conceptualizations, it is necessary to describe how each juvenile system processes multi-problem youth and identifies some as violent.

2. The Italian Juvenile Justice System Functioning

In Italy, few juvenile offenders are imprisoned. This reflects the Welfare Model philosophy of juvenile law which seeks to minimize the involvement of juvenile offenders in the legal system. The focus instead is on rehabilitative resources directed at the ongoing developmental processes of each youth.

Another important philosophical change occurred in 1988, when the Juvenile Criminal Proceeding Code came into force. Juvenile courts were encouraged to introduce forms of mediation between the youth and the victim of the crime, according to a mediation model that promotes the youth's sense of responsibility for the crime committed. This attitude change, in combination with specific developmentally focused rehabilitative interventions, are seen as an efficacious response to multi-problem youth and the reduction of future violent offending.

3. Intake Screening Process

The initial processing of juveniles occurs at the First Welcome Centers (Centri di Prima Accoglienza, C.P.A.). A youth typically has the first interview with a psychologist while waiting for trial. However, because of the brief stay of most juveniles in the C.P.A. (usually from 2 to 4 days), it is difficult to identify the multi-problem profiles. This intake limitation emphasizes the need for an integrated multi-agency approach to identify these profiles. The problem is compounded further because in Italy there are no differentiated treatments for juveniles who committed serious violent crimes compared to those who committed other kind of crimes. Yet, there is mounting evidence that the need to make treatment distinctions is important given several fundamental changes occurring within Italy and its juvenile justice system.

4. The Research on Juvenile Offenders with Mental Health Problems

The Central Office for Juvenile Justice (Ufficio Centrale della Giustizia Minorile, (U.C.G.M.[2]) is responsible for examining policy issues. The U.C.G.M. monitors juvenile services and conducts related research. One of the most important issues that arose in the 1990s was the changing profile juvenile offenders and the types of crimes. Most critically, the ethnic profile had changed fundamentally as result of recent immigration from the Balkans, especially Albania and Bosnia and North Africa.[3] Drug trafficking and use appear to have risen concomitantly.[4] Finally, a controversial issue involved the apparent increasing association of young offenders with criminal adult organizations. The UCGM initiated a research project to examine these purported trends.

In the first phase of the research, 66 Juvenile Services centers throughout Italy were asked to provide data on the ethnic profiles of juvenile offenders in their regions and whether there was any association with drug trafficking and use with criminal organizations. The second research phase consisted of a review of the diagnostic criteria and treatment approaches adopted by the various Juvenile Service teams for offenders with severe behavioral problems. A questionnaire was designed to screen all juvenile intakes during 1999. The first objective was to identify the profile of psychological problems and the diagnostic methods utilized by service professionals. The second objective was to profile a wide range of other relevant offender characteristics including age of onset of offending, drug abuse, family problems and possible connections with criminal organizations.

As mentioned above, there were 42,107 juveniles referred to Juvenile Court in 1998. Similar to other western European and North American countries, most of these youth (81.8%) were between 14 and 17 (Farrington, 1986). Department of Justice reported that nearly half (20,106) of these youths were referred to Juvenile Services. Of the 66 Justice Service Centers that participated in the research project, a majority had more than 100 referred youths and one third had over 200.

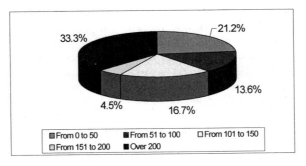

Figure 1: Total number of juvenile offender intakes

5. Juvenile Offenders with Mental Health Problems

Of the 20,106 juveniles referred to Justice Services, 636 were diagnosed with a mental health disorder (see Table 1). Slightly less than half of these mentally disordered offenders committed their first official offence between 15 and 16 while a small percentage

(4.4%) offended before they were 13. At the other age extreme, approximately one-fifth (21.2%) had their first reported offence between 17 and 18.

Table 1: Age of onset of offending of juveniles with mental health problems

9/10 yrs-old	3%
11/12 yrs-old	1.4%
13/14 yrs-old	18.75
15/16 yrs-old	44.2%
17/18 yrs-old	21.2%

This total figure of 636 mentally ordered juveniles is important, since spread across 66 Juvenile Justice Service Centers, it does not constitute an onerous burden for any single center. In effect, the case management requirements of most centers in dealing with these youths who are in need of substantial treatment planning and, often, on-going diagnostic reviews is unlikely to strain a center's resource capacity. Of course, this assumes that each center has minimum diagnostic and treatment resources. Nonetheless, virtually three quarters of all the centers in the research project had fewer than 16 mentally disordered offenders in 1999 (see Figure 2).

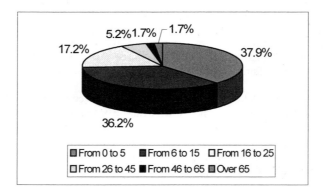

Figure 2: Offenders with psychopathological diagnosis took in charge

Despite the relatively small number of mentally disordered offenders, it is important to describe their disorders (see Table 2). A variety of methods were employed to obtain these data. Most youths were interviewed (87.9%) while half were subjected to various psycho-diagnostic tests as well including those based DSM IV criteria (25.9%). Most interviews were conducted by a Juvenile Justice Service psychologist (65%) while half of the cases also included a psychiatrist from the local health agency. For another approximately one-third (37.9%) of the youths, a psychologist was involved. Interestingly, despite the presence of certain symptoms, the most frequent category (37.7%) was the absence of a specific diagnosis. This finding is somewhat perplexing. It is possible to conjecture that there is a general reluctance among these mental health professionals to label youths with potentially damaging mental disorders. It also is likely that there are fundamental methodological difficulties or validity problems in making a diagnosis with this adolescent sample which might cause some hesitation in contrast to an adult client sample. For the youngest juveniles, especially, the presenting of symptoms is confounded by maturation or developmental issues. It is these concerns that might be better addressed by the risk-needs instrument outlined in the second part of this book. Most importantly, the longer the period of observation of symptoms, i.e. from early childhood, the more likely a mental health professional would have the confidence in making a diagnosis during adolescence.

Table 2: Types of psychopathological diagnosis formulated

37.7	Absent
15.1	Borderline Personality Disorder
8.8	Personality Disorder
7.7	Conduct Disorder
4.7	Depression
4.2	Mild Mental Retardation
2.2	Anxiety
2.2	Depressive Anxiety
2.2	Emotional-Relational Immaturity
2.0	Narcissistic Disorder
13.2	Other

Where a diagnosis occurred, the most common is the Borderline Personality Disorder (15.1%). This is some what surprising since Conduct Disorder is more typically associated with serious delinquencies (Loeber & Farrington, 1998). Again, given the large percentage of cases without a diagnosis, the data presented in Table 2 must be interpreted with considerable caution. However, the profile of symptoms presented in Table 3 is consistent with delinquency research since impulsivity is evident in nearly 60% of the diagnoses followed by emotional instabilities (51.7%). Along with these two traits, the remainder of the symptoms reveal a "classic" profile of serious multi psychiatric-problem youths.

Table 3: Symptoms displayed by juveniles in charge of the Justice Services

Symptoms (on 636 subjects)	
Impulsivity	57.9%
Instability of interpersonal relations, disturbed self-imagine, unstable mood, and strong impulsivity	51.7%
Emotional instability due to a marked mood reactivity	42.6%
Distorted identity: enduring distorted self-image and self-perception	38.8%
Unmotivated and harsh rage or difficulties in controlling rage	38.8%
Unstable and intense interpersonal relations characterized by an alternation of over-idealization and depreciation	35.1%
Persistent sense of emptiness	27.7%
Desperate efforts to avoid a real or imagined abandon	26.7%
Suicidal or self-mutilating behaviors	18.6%
Paranoid ideation, or severe dissociation symptoms related to stress	15.6%
Flattened emotions, absent speech, apathy (at least for a month)	7.4%
Delirium (at least for a month)	4.9%
Disorganized or catatonic behavior (at least for a month)	4.7%
Cognitive retardation	4.7%
Incapability of expressing emotions	4.1%
Hallucination (at least for a month)	3.6%
Disorganized speech (at least for a month)	3.6%
Acting out	1.3%
Other	37.7%

6. Violent Offenders with Mental Health Problems

Since there are no available data on the number of juveniles who committed violent crimes out of the total reported offenders in year 1999, 1998 data will be used. Of the 42,107 reported offenders, 20% (8,422) committed violent crimes. However, the most common crime committed by mentally disordered youths involved property offences (56.4). Only 23.7% (151 youth) of offences were violent.[5] The remainder consisted of drug offenses (13.5%), crimes against institutions (1.7%), and 4.6% with missing offense information.

This low frequency of violent offending in general and for mentally disordered youth in particular (.75% of all reported offenders), is extremely important in terms of policy development regarding the feasibility and utility employing an age-integrated risk-needs management instrument in Italy. In practical terms, it would likely be used routinely only for a small number of cases---for the multi-psychiatric and violent youths. In turn, it would allow for maximizing resources allocated to helping these youths avoid further violence.

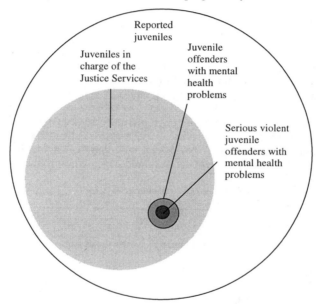

Figure 3: Incidence of the category of violent offenders with mental health problems within the total number of reported offenders took in charge.

Another major policy issue involves youth with mental health problems who also use drugs and alcohol. Substance abuse is an obviously serious aggravating problem for these youths. One-third (33.4%, n= 213) reported no drug use so they were excluded from the research sample (N=636) when considering youths' drugs use, so that only 423 youth are included in these analyses. As indicated in Figure 4, while slightly more than half of the sample (52.5%) did use cannabis, less than a third (31.4%) used it regularly. Far fewer youths with psychiatric problems ingested hard drugs such as cocaine and opiates. Again, only a small number of these youths used these drugs either regularly (7.7%) or were addicted (9.9%). Alcohol use was surprisingly low since only, approximately, one quarter (26.3%) of the youths stated they consumed it.

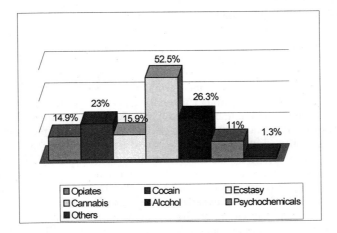

Figure 4: Types of drugs used by juveniles with a psychopathological diagnosis in charge of the Justice Services

Table 4: Modality of drug use among juvenile offenders with psychopathological diagnosis

Modality of drug use	Occasional Use	Regular Use	Addiction
Opiates	4.6%	3.0%	6.4%
Cocaine	12.9%	4.7%	3.5%
Ecstasy	9.7%	4.2%	1.1%
Cannabis	17.3%	31.4%	2%
Alcohol	11.8%	11.5%	1.3%
Psychochemicals w/o medical supervision	4.6%	4.1%	1.3%
Others	-	-	-

Finally, several social contextual concerns were examined for mentally disordered youths. Most of them (63.1%) had experienced serious intra-familial conflicts while less than half (39.9%) stated their families had been through serious socio-economic problems. A very small proportion (13.2%) had any involvement with criminal organizations. This problem profile confirms the existence of a complex connection between delinquency, mental health and socio-economic factors.

7. Conclusions and Recommendations

The policy conclusions based on the data presented above are tentative because of the validity limitations of this initial research project. An important concern is the absence of several key dimensions utilized in major research projects such as the Denver Youth Survey (Elliot, Huizinga, & Menard, 1989). By restricting the focus to primarily mental health, it was not possible to examine related critical factors such as peer groups, school, developmental stages, neighborhood and employment (Loeber & Farrington, 1998). As well, data was gathered only from diagnostic interviews. It is necessary to obtain collateral data from parents by utilizing instruments such as Achenbach's Child Behavior Checklist (Achenback & Edelbrock, 1983). Juvenile Court reported crimes also substantially underestimate the actual number of crimes our sample of youths committed. In effect, without self-reported data, it is not possible to have confidence in the extent of a youth's criminal involvement (Dunford & Elliot, 1984). Finally, our small sample size of youth with psychiatric problems precluded any multivariate statistical analysis. Without this, it is difficult to ascertain patterns among the various factors associated with delinquency and mental health, and in turn, policy concerns such as the purported increase in immigrant youths with substance abuse and family problems.

Despite these methodological caveats, several policy inferences can be made from this initial research. Most importantly, there is a definite need to gather more valid data on the youths who are being sent through the Italian juvenile justice and related systems. Without such data, media and political controversies, based too often on stereotypes, might inappropriately influence policy and program development. For example, it appears that the multi-problem of profile of most youths in this study are not as severe as in other countries such the United States especially the prevalence of alcohol and drug abuse, violence and mental disorders. Nonetheless, there are enormous information gaps about when psychiatric and delinquency problems are detected and responded to by the various social service, educational, mental health and juvenile justice agencies in Italy. This concern is compounded by the changing ethnic profile of youths in urban centers in the last decade. Language and cultural issues have and will make it increasingly difficult to identify and assist multi-problem youths and their families. It is in this evolving context that an age integrated risk-needs management instrument could be useful.

Notes

[1] Acknowledgements. We thank Dr. Iannace and Dr. Brauzzi from the U.C.G.M. for letting us use their data on juvenile offenders with mental health problems.
[2] UCGM coordinates those Juvenile Justice Services' programs involving both risk prevention and treatment for juveniles and professional training.
[3] According to data from the National Statistic Institute (ISTAT), the number of foreigner juveniles reported to the Juvenile Court increased from 7928 in 1991 to 10.926 in 1998, over a total amount of 42.107 reported juveniles.
[4] According to ISTAT data, the number of reported juveniles for this type of crime increased from 2.181 in 1991 to 4145 in 1998.
[5] The category of violent crimes adopted in this study is different from the one used in other studies, for it does not include crimes such as robbery, kidnapping and arson. The Italian official records include these crimes in the general category of crimes against property.

References

Achenbach, T.M., & Edelbrock, C.S. (1983). *Manual for the child behavior checklist and revised child behavior profile*. University of Vermont, Burlington.
Dunford, F.W., & Elliott, D.S. (1984). Identifying career offenders using self-reported data. *Journal of Research in Crime and Delinquency, 21,* 57-86.
Farrington, D.P. (1986). Age and crime. In M. Tonry & N. Morris (Eds.), *Crime and justice: An annual review of research* (Vol. 7, pp. 189-250). Chicago: University of Chicago Press.
Guerra, N. (1998). Serious and violent juvenile offenders: Gaps in knowledge and research priorities. In R. Loeber & D.P. Farrington (Eds.), *Serious & violent juvenile offenders: Risk factors and successful interventions* (pp. 389-404). Thousand Oaks, CA: Sage.
Huizinga, D., Esbensen, F.A., & Weiher, A.W. (1991). Are there multiple paths to delinquency? *Journal of Criminal Law and Criminology, 82,* 83-118.
Huizinga, D., Esbensen, F.A., & Weiher, A.W. (1996). The impact of arrest of subsequent delinquent behavior. In R. Loeber, D. Huizinga, & T.P. Thornberry (Eds.), *Program of research on the causes and correlates of delinquency: Annual report 1995-1996*. Washington D.C.
Huizinga, D., & Jakob-Chien, C. (1998) The contemporaneous co-occurrence of serious and violent juvenile offending and other problem behaviors. In R. Loeber & D.P. Farrington (Eds.), *Serious & violent*

juvenile offenders: Risk factors and successful interventions. (pp. 47-67). Thousand Oaks, CA: Sage.

Loeber, R., Farrington, D.P., & Waschbusch, D.A. (1998). Serious and violent juvenile offenders: In R. Loeber and D.P. Farrington (Eds.), *Serious & violent juvenile offenders: Risk factors and successful interventions* (pp. 13-29). Thousand Oaks, CA: Sage.

Reppucci, N.D., Woolard, J.L., & Fried, C.S. (1999). Social, community and preventive interventions. *Annual Review of Psychology, 50,* 387-418.

Tate, D.C., Reppucci, N.D., & Mulvey, E.P. (1995). Violent juvenile delinquents: Treatment effectiveness and implications for future action. *American Psychologist, 50,* 777-781.

Tedeschi, J.T. & Felson, R.B. (1994). *Violence, aggression, & coercive actions.* Washington, DC: American Psychological Association.

Toch, H., & Adams, K. (1994). *The disturbed violent offender.* Washington, DC: American Psychological Association.

Multi-Problem Violent Youth
R.R. Corrado et al. (Eds.)
IOS Press, 2002

Adolescent Murderers: A Genoa Sample

Giovanni B. Traverso and Monica Bianchi

There have been few Italian empirical studies concerning homicides committed by minors. Throughout the world, in the last 25 years, children and juveniles have drawn much attention from experts, professionals, the public at large and the mass media as well, above all when they are victims of crime (especially victims of physical and/or sexual abuse). On the contrary, when they are studied as the perpetrators of an offense, they are usually considered to be vandals, property damagers, shoplifters, drug addicts, or at most as car thieves, burglars or robbers, which are all delinquent/criminal behaviors that generally connote a low level of physical interpersonal violence.

Although both the official Italian statistics and Italian self-report studies about juvenile delinquency (Gatti et al., 1994; Traverso, Esposito, Leone, & Ciappi, 1994) demonstrate that juvenile violent crime, and in particular homicide, is a fairly rare phenomenon (for the period 1961-1993, the offense of homicide, in terms of crimes known to the Police, shows a mean annual rate per 100,000 population of 0.09, that is less than 1 crime every 1,000,000 population), a growing international literature has recently sounded an acute alarm concerning the increase in violent behavior in youth (Fox, 1996; Snyder, Sickmund, & Poe-Yamagata, 1996; Snyder et al., 1995; Loeber, Farrington, & Waschbusch, 1998). This has stimulated the interest of researchers looking to acquire new insights into and understanding of the origins of criminal careers (risk factors) in juveniles, in search of explanations for serious and violent conduct by juvenile offenders, and for answers to the problem in terms of effective intervention or prevention programs as well (see, for all, the work done by the Office of Juvenile Justice and Delinquency Prevention's Study Group on Serious and Violent Juvenile Offenders, 1999; Loeber & Farrington, 1998).

Given these, combined with the scarcity of Italian research on homicides committed by minors (apart from the raw data furnished by the National Office of Statistics, the only information available in Italy in the specific field is contained in a volume published by Bandini Bandini, Gatti, and Traverso, 1983, and in a research conducted by Portigliatti-Barbos & Scatolero, 1985), and following the example of foreign previous similar research (including Bender, 1959; Walsh-Brennan, 1975; Benedek & Cornell, 1989; Labelle, Bradford, Bourget, Jones, & Carmichael, 1991; Bailey, 1996), we were induced to carry out the present study with the objective to deepen our understanding of the phenomenology of juvenile homicide, as well as to try to specify the possible importance and role of those clinical, developmental and environmental variables which might be useful to the creation of significant typologies, useful to the setup of effective prevention and rehabilitation programs.

1. The Sampling and Methodology of the Research

This research was carried out from data files of the Juvenile Court of Genoa (*Tribunale per i Minorenni*), whose jurisdiction includes not only the entire Liguria Region but the city and surrounding district of Massa Carrara too, and the Adult Criminal Court of Genoa (*Corte di Assise d'Appello*), where the information needed for this work was gathered once the authorization to consult the dossiers and sentences was obtained.

The data gathered covers a period of 36 years, from 1960 to 1995, and includes cases of voluntary homicide, attempted homicide and infanticide committed by minors between 14 and 17 years of age (<18 years of age). The cases numbered 30, with 39 minors implicated as perpetrators, 37 males and 2 females. The victims were 38, 27 males and 11 females. The material gathered included first of all personal data about the perpetrator of the crime, such as sex, age at the moment of the crime, place of birth, place of residence, scholastic level, type of work, any previous criminal activity.

Particular attention was given to finding out some important information about the family environment (like the composition and structure of the family and its socioeconomic conditions), including the presence in the family members of any type of psychopathology or chronic maladjustment (such as substance abuse, specific psychiatric pathologies, the use of violent and abusive behavior, or finally previous criminal records). The place where the crime was carried out, which weapon was used, possible complicity and the correlation of the crime under study with the perpetration of other crimes were all considered variables which could contribute to the study of the phenomenology of the crime. The relationship between the offender and the victim, and the characteristics of the victim himself/herself were considered very important variables, which permit the drawing of some interesting hypotheses about the typology of homicide committed by minors in Genoa.

The consultation of the medical, psychological and psychiatric reports permitted on one hand a qualitative deepening of the individual and social factors related to the explosion of the violence, and on the other allowed a better understanding of the expert's role in the process of assessing criminal responsibility issues (in Italian jurisprudential terms, the so-called "maturity" of the juvenile), and his/her eventual dangerousness, defined as the probability of reoffending.

2. The Sociodemographic Characteristics of the Perpetrators

Table 1 indicates that homicide, both attempted and completed, is committed nearly always by males (94.9%) in their later juvenile years (79.5%), born especially in Liguria or in the south of Italy (added together, more than 70% of the subjects in our sample was born in these areas), and residing mostly in Genoa or the Liguria Region (64.1%).

Most of the teens in our sample came from fairly large families of a low socioeconomic level, which were often broken up for various problems such as death, abandon or separation of the parents, and more than half of them (51.4%) had been institutionalized at least once for varying lengths of time. Half of the subjects (50.0%) completed only elementary school, with resulting difficulty in finding work (47.2% were unemployed). Despite these above-mentioned negative socio-demographic characteristics, only 30% of the subjects in our sample had previous criminal records. The crimes were usually committed by two or three individuals acting together but never structured in real gangs (with the exception of one case which involved a band of Palestinian terrorists); in some cases the small group was made up by minors and adults in different combinations.

The comparative analysis of the characteristics of the subjects who completed a homicide with those who were tried for attempted homicide, shows some important and statistically significant differences in the distribution of several variables including sex (within a typically male phenomenon, females are more likely to kill their victim) and age: younger subjects tend to belong more likely to the category of attempted murderers. There is also a statistically significant difference between homicide and attempted homicide cases in the area of origin of the perpetrator, since the murderers were most often from Genoa, Liguria or abroad, while attempted murders seem to be more common among teens coming from south Italy. What is more, the murderers generally came from a higher socioeconomic level and were less likely to have a previous criminal record than the teen accused of attempted murder.

3. The Characteristics of the Victims and of the Crime

The homicidal events most often take place in the streets or in open areas (47.4%), or in private homes (31.6%), and involve primarily a male victim with an age usually higher (89.5%) than that of the perpetrator (see Table 2). The weapon used to carry out the crime is in just over half the cases a gun or a knife, whereas the use of a blunt instrument or other weapon is rarer.

Comparing murder and attempted murder, there are statistically significant differences as regards the age of the victim, since a homicide is more probable when the victim is younger or has the same age of the perpetrator. The victim of a homicide is more likely to be unemployed, and the preferred weapon in a homicide, compared with an attempted homicide, is generally a blunt instrument or asphyxia, whereas the firearm or knife is generally the prerogative of an attempted homicide. Lastly, streets and open or public places are more likely the preferred areas for attempted homicides.

4. The Relationship Between Perpetrator and Victim

The analysis of the perpetrator-victim relationship (Table 3) is of the greatest importance in this research because it allows us to indicate three types of homicide (homicide within the family, homicide between acquaintances and homicide between strangers, our modal category). In fact, 72.5% of the homicidal acts occurred between strangers, and within this typology, the great majority were of the subtype of aggression with intent to kill during the commission of another criminal act, usually robbery. This fact significantly differentiates juvenile homicide from adult homicide, since the latter involves people with closer interpersonal links in a much higher percentage of case (Bandini, Gatti, & Traverso, 1983).

Table 1: Socio-demographic characteristics of juvenile offenders charged with voluntary homicide and attempted homicide (Genoa, 1960-1995).

Socio-demographic Characteristics	Voluntary Homicide		Attempted Homicide		Total	
Sex						
M	22	(91.7)	15	(100.0)	37	(94.9)
F	2	(8.3)		---	2	(5.1)**
Total	24	(61.5)	15	(100.0)	39	(100.0)
Age						
14-15	4	(16.6)	4	(26.7)	8	(20.5)**
16-17	20	(83.4)	11	(73.3)	31	(79.5)
Total	24	(61.5)	15	(38.5)	39	(100.0)
Place of birth						
Genoa and Liguria	10	(41.7)	4	(26.7)	14	(35.9)**
Northern Regions	4	(16.7)	3	(20.0)	7	(17.9)
Central Regions	2	(8.2)	1	(6.7)	3	(7.7)*
Southern Regions/Islands	4	(16.7)	6	(40.0)	10	(25.6)**
Abroad	4	(16.7)	1	(6.7)	5	(12.8)**
Total	24	(61.5)	15	(38.5)	39	(100.0)

Place of residence						
Genoa and Liguria	15	(62.5)	10	(66.7)	25	(64.1)
Northern Regions	2	(8.3)	2	(13.3)	4	(10.3)**
Central Regions	2	(8.3)	2	(13.3)	4	(10.3)**
Southern Regions/Islands	1	(4.2)	1	(6.7)	2	(5.1)**
Abroad	4	(16.7)		---	4	(10.3)**
Total	24	(61.5)	15	(38.5)	39	(100.0)
School Grades						
Some Elementary School		---	2	(13.3)	2	(5.6)**
Elementary Diploma	8	(38.1)	10	(66.7)	18	(50.0)**
Junior High School Completed	9	(42.9)	2	(13.3)	11	(30.6)**
Some Secondary School	4	(19.0)	1	(6.7)	5	(13.8)**
Total (N.R.=3)	21	(58.3)	15	(41.7)	36	(100.0)
Institutionalization						
None	13	(65.0)	4	(26.7)	17	(48.6)**
< 1 year	1	(5.0)	3	(19.9)	4	(11.4)**
1 to 3 years	4	(20.0)	4	(26.7)	8	(22.9)*
> 3 years	2	(10.0)	4	(26.7)	6	(17.1)**
Total (N.R.=4)	20	(57.1)	15	(42.9)	35	(100.0)
Occupational Status						
Unemployed	10	(47.6)	7	(46.7)	17	(47.2)
Unskilled or semiskilled	5	(23.8)	7	(46.6)	12	(33.4)**
Student	4	(19.1)	1	(6.7)	5	(13.9)**
Other	2	(9.5)		---	2	(5.5)**
Total (N.R.=3)	21	(58.3)	15	(41.7)	36	(100.0)
Number of brothers						
None	6	(30.0)	2	(15.4)	8	(24.2)**
1 brother	4	(20.0)	4	(30.8)	8	(24.2)*
2 or more brothers	10	(50.0)	7	(53.8)	17	(51.6)
Total (N.R.=6)	20	(60.6)	13	(39.4)	33	(100.0)
Family and marital conflict						
Family harmony	11	(55.0)	7	(53.8)	18	(54.5)
Family breakup (divorce, etc.)	9	(45.0)	6	(46.2)	15	(45.5)
Total (N.R.=6)	20	(60.6)	13	(39.4)	33	(100.0)
Socio-Economic Status						
Low	14	(66.7)	14	(93.3)	28	(77.8)*
Medium	6	(28.6)	1	(6.7)	7	(19.4)**
High	1	(4.7)		---	1	(2.8)**
Total (N.R.=3)	21	(58.3)	15	(41.7)	36	(100.0)
Prior Criminal Records						
Yes	3	(15.0)	7	(53.8)	10	(30.3)**
No	17	(85.0)	6	(46.2)	23	(69.7)**
Total (N.R.=6)	20	(60.6)	13	(39.4)	33	(100.0)
Complicity						
No	8	(33.3)	3	(20.0)	11	(28.2)**
Yes	16	(66.6)	6	(80.0)	28	(71.8)
Total	24	(61.5)	15	(38.5)	39	(100.0)

* p <.05, ** p< .01

The association between completed homicide and the first two types of relationship (homicide within the family and homicide between acquaintances) is very strong and statistically significant.

5. Qualitative Considerations About Minors who Commit Homicide

Qualitative data about the most important family and personal characteristics of the subjects being studied, as well as the prevalent motivational components and modus operandi, have been obtained by a close reading of the medico-legal psychiatric/psychological experts' reports ordered by the Courts, as well as from the work carried out by the social workers responsible for the area. The results show the following:

a) 43% of the minors had serious social dysfunctions in their family. In the vast majority of the cases the father, already arrested and/or convicted, was overtly violent with the other family members;

b) moreover, in 43% of the cases, there was an open neuropsychiatric pathology in the family of minors who perpetrated homicide, usually epilepsy or alcohol abuse/dependence (the father in all the cases suffers by the latter pathology);

c) in the vast majority of the cases (25 subjects over 30, that is 83.3%), the minors suffered from various neuropsychiatric problems, many of which involved real organic disturbs, such as permanent consequences of cranial traumatisms, epilepsy (demonstrated by bio-electric cerebral alterations), innate cerebropathic oligophrenia (mild mental retardation), organic signs of substance abuse (usually alcohol) and personality disturbs;

d) the data demonstrate that 28% of the subjects presented all three above-mentioned negative conditions (social dysfuctions in the family, neuropsychiatric pathology in one or more family members, personal neuropsychiatric disturbs) contemporaneously, while 51% presented at least two, and 72% at least one. Only 10% of the subjects were exempt from the problems referenced above. Lastly, in a small percentage of the cases not assessed by experts (about 18%), it was impossible to know the anamnesis of the family and of the perpetrator of the crime.

Table 2: Some victims' and offenses' characteristics by type of crime (Genoa, 1960-1995).

Victims' and offenses' Characteristics	Voluntary Homicide		Attempted Homicide		Total	
Victim's sex						
M	15	(71.4)	12	(70.6)	27	(71.1)
F	6	(28.6)	5	(29.4)	11	(28.9)
Total	21	(55.3)	17	(44.7)	38	(100.0)
Victim's age						
Younger than offender	2	(9.5)	1	(5.9)	3	(7.8)**
O/V of the same age	1	(4.8)	---		1	(2.6)**
Older than offender	18	(85.7)	16	(94.1)	34	(89.5)
Total	21	(55.3)	17	(44.7)	38	(100.0)
Victim's occupational status						
Unemployed	5	(29.3)	2	(12.5)	7	(21.2)**
Employed	8	(47.1)	9	(56.2)	17	(51.5)
Other (retired, students, etc.)	4	(23.6)	5	(31.3)	9	(27.3)*
Total (N.R.=5)	17	(51.5)	16	(48.5)	33	(100.0)
Weapon used						
Gun	5	(23.8)	6	(35.3)	11	(28.9)**
Knife	5	(23.8)	6	(35.3)	11	(28.9)**
Blunt instrument	4	(19.0)	2	(11.8)	6	(15.8)**
Asphyxia	4	(19.0)	---		4	(10.6)**
Other	3	(14.4)	3	(17.6)	6	(15.8)
Total	21	(55.3)	17	(44.7)	38	(100.0)
Crime location						
Offender's/victim's house	11	(52.3)	1	(5.9)	12	(31.6)**
Bar, restaurant, etc.	3	(14.3)	4	(23.5)	7	(18.4)**
Street, open site	6	(28.6)	12	(70.6)	18	(47.4)**
Other	1	(4.8)	---		1	(2.6)**
Total	21	(55.3)	17	(44.7)	38	(100.0)

* p <.05, ** p< .01

As for motivation, about half the cases (46.2%) took place during, or immediately after, the commission of another crime, usually robbery or theft. About a fourth (26.7%) were motivated by violent arguments or angry reactions between relatives or anyway between people linked by a strong emotional bond. Another 21% of the crimes were due to vengeance or score settling, and in the remaining two cases the motivations were participation in a terroristic action and due to honor (infanticide).

Consistent with the above information, there were secondary charges alongside the principal charges in a high number of cases. These secondary charges included destroying the cadaver, unlawful carrying of arms, driving without a license, resisting arrest and making false declarations.

Based on the information outlined above, it was possible to individuate, and consistent with international literature (Cornell, Benedek, & Benedek, 1989), two major groups of motivations for homicide. The first is openly delinquent (a group of 25 subjects which, like Cornell et al., 1989, we have called the "crime" group), while the second has more openly conflictual aspects (a group of 14 subjects, which we have named the "conflict" group).

6. Towards a Typology of the Adolescent Murderer: A Multivariate Analysis

We analyzed the data using a multivariate quantitative statistical analysis. This approach allowed us to verify the possible presence of significantly statistical differences between the subjects under study divided in two groups based on a) the prevalent motivation in the crime committed ("conflict" and "crime" groups), and b) the two different charges ("homicide" and "attempted homicide" groups).

Table 3: Offender/Victim Relationship in homicide/attempted homicide cases, Genoa, 1960-1995.

Relationship Offender/Victim	Voluntary Homicide		Attempted Homicide		Total	
Son/Daughter (O), Parent (V)	2	(7.6)	---		2	(3.9)
Parent (O), Son/Daughter (V)	1	(3.8)	---		1	(2.0)
Other relative	5	(19.2)	---		5	(9.8)
Total type I (*)	8	(30.8)	---		8	(15.7)
Friend	1	(3.8)	---		1	(2.0)
Lover	---		1	(4.0)	1	(2.0)
Acquaintance	4	(15.6)	---		4	(7.8)
Total type II (**)	5	(19.2)	1	(4.0)	6	(11.8)
Stranger (O), Policeman (V)	---		9	(36.0)	9	(17.6)
Stranger (during the perpetration of another offence)	7	(26.9)	12	(48.0)	19	(37.3)
Stranger	6	(23.1)	3	(12.0)	9	(17.6)
Total type III (***)	13	(50.0)	24	(96.0)	37	(72.5)
Total Relationship	26	(51.0)	25	(49.0)	51	(100.0)

$\chi2 = 26.599$, d.f.=8, p<.001
(*) Type I: Homicide within the family
(**) Type II: Homicide between acquaintances
(***) Type III: Homicide between strangers

In order to verify the hypotheses mentioned above, we used multiple regression associated with a discriminant analysis, which allowed us to verify the degree of accuracy in the classification by considering a set of independent variables of the subjects within the separate groups. In order to use the regression technique, all variables were dichotomized.

Based on the strength of the relationship between the variables shown in the correlation matrix, the multivariate analysis mentioned above was carried out with the following results.

7. A Classification Based on Motivation (The "Conflict/Crime" Model)

The analysis of the variance produced a statistically significant coefficient ($F=160.278$, $p<.001$, $R^2=.959$) which indicates the general significance of the model. All of the independent variables of the model had statistically significant coefficients. This can be interpreted by noting that the group for which the prevalent motivation was identified as a "conflict", when compared to the group for which the homicide or attempted homicide were carried out during a crime or directly connected to another crime, appeared to be significantly correlated with 1) a higher socioeconomic conditions, 2) a lower involvement in prior criminal activity, 3) a higher percentage of crimes committed inside the family or among people with close and binding acquaintanceship, 4) a higher percentage in the use of firearms, 5) a more probable outcome of attempted homicide. The "crime" group obviously had the opposite characteristics.

The results of the discriminant analysis, used to verify the correct classification of the subjects in the groups "conflict" and "crime", allows us first of all to confirm the statistical significance of the model used (Wilks' lambda=0.497, $F=6.691$, gl=5, $N=39$, $p<.001$) which permits us to correctly classify 85% of the 39 cases. In detail, there is a correct classification of 10 subjects belonging to the "conflict" group and 23 belonging to the "crime" group, with, respectively, 4 errors for the first group and 2 errors for the second group.

8. A Classification Based on the Specific Charge (The "Homicide/Attempted Homicide" Model)

The analysis of the variance produced a statistically significant coefficient ($F=106.503$, $p<.001$, $R^2=.924$) which indicates the general significance of the model. All of the variables of the model are statistically significant. This can be interpreted by noting that the group judged for "homicide", as compared to the group judged for "attempted homicide", appears to be significantly correlated with 1) a lower involvement in prior criminal activity, 2) a higher percentage of crimes committed inside the family or among people with close and binding acquaintanceship, 3) a higher percentage in the use of firearms, 4) the commission of a homicide carried out in direct connection with the perpetration of another crime.

The results of the discriminant analysis, used to verify the correct classification of the subjects in the groups "homicide" and "attempted homicide", allows us first of all to confirm the statistical significance of the model used (Wilks's lambda=0.689, $F=3.830$, gl=4, $N=39$, $p<.01$) which permitted us to correctly classify 74.3% of the 39 cases. In detail, there is a correct classification of 18 subjects belonging to the "homicide" group and 11 belonging to the "attempted homicide" group, with, respectively, 6 errors for the first group and 4 errors for the second group.

9. Conclusion

This research allows us to define the main socio-demographic characteristics of the juvenile murderers of our sample, the typology of the victim, the relationship offender-victim, and finally some features of the crime itself and modus operandi. The comparative analysis between homicide and attempted homicide cases shows important significant differences concerning the two types of crime.

This research permits also the elaboration of two multivariate models which explain a high level of variability in the sample with reference to the prevalent motivations and to the specific charges. The most significant variables are socioeconomic status, prior criminal record, relationship between the perpetrator and the victim and the weapon used. Following these models, the discriminant analysis confirmed, descriptively, the possibility of constructing a typology of juvenile homicide based on motivation ("conflict" and "crime" groups) and on the specific charges ("homicide" and "attempted homicide" groups).

These results, which seem confirmed by some US studies about the same matter (Cornell et al., 1989), suggest the existence of important differences in the different groups described and, if they should be confirmed by further research, could constitute the foundation for the identification of different models in the development of violent behavior. On a more general theoretical level, our data seems to conform to and be explained by an integrated theoretical approach like, for example, the one proposed by Walgrave (1992) which attempts an integration of the theories of control with elements taken from cultural theories of deviance (Ferracuti & Wolfgang, 1966; Shaw, & McKay, 1942; Sutherland & Cressey, 1924), and from the theory of labeling (Becker, 1966; Lemert, 1972).

If we add to this complex scheme the possible intervention, as our data seems to demonstrate, of factors linked to neuropsychiatric (even if not psychotic) disturbs, the prospective possibility of a truly "integrated" approach in the explanation of the phenomenon we have studied, appears possible. An approach, in need of further validation, in which we would consider the different factors (biological, psychological/psychiatric, familial, sociocultural and those stemming from labeling and stigmatization) intertwining and correlating in such a characteristic and specific way as to produce that rare and aberrant event which is the case of an adolescent who murders another human being.

References

Bailey, S. (1996). Adolescents who murder. *Journal of Adolescence, 19,* 19-39.

Bandini, T., Gatti, U., & Traverso, G.B. (1983). *Omicidio e controllo sociale.* Milano: Franco Angeli.

Becker, H.S. (1966). *Outsiders.* New York: The Free Press.

Bender, L. (1959). Children and adolescents who have killed. *American Journal of Psychiatry, 116,* 510-513.

Benedek, E.P., & Cornell, D.G. (Eds.) (1989). *Juvenile homicide.* Washington, D.C.: American Psychiatric Press.

Cornell, D.G., Benedek, E.P., & Benedek, D.M. (1989). A typology of juvenile offenders. In E. P. Benedek & D.G. Cornell, (Eds.), *Juvenile homicide* (pp. 59-84). Washington, D.C.: American Psychiatric Press.

Ferracuti, F., & Wolfgang, M.E. (1966). *Il comportamento violento. Moderni aspetti criminologici.* Milano: Giuffrè Editore.

Fox, J.A. (1996). *Trends in juvenile violence. A report to the United States attorney general on current and future rates of juvenile offending.* Washington, D.C.: Bureau of Justice Statistics, Department of Justice.

Gatti, U., Fossa, G., Lusetti, E., Marugo, M.I., Russo, G., & Traverso, G.B. (1994). Self-repoted delinquency in three Italian cities. In J. Junger-Tas, G-J. Terlouw, & M.W. Klein (Eds.), *Delinquent behavior among young people in the western world. First results of the international self-report delinquency study* (pp. 267-278). Amsterdam: Kugler.

Hirschi, T. (1969). *Causes of delinquency.* Los Angeles: University of California Press.

Labelle, A., Bradford, J.M., Bourget, D., Jones, B., & Carmichael, M. (1991). Adolescent murderers. *Canadian Journal of Psychiatry, 36,* 583-587.

Lemert, E.H. (1972). *Human deviance, social problems, and social control.* Englewood Cliffs: Prentice- Hall.

Loeber, R. & Farrington, D.P. (Eds.) (1998). *Serious & violent juvenile offenders. Risk factors and successful intervention.* Thousand Oaks: Sage Publication, Inc.

Loeber, R., Farrington, D.P., & Waschbusch, D.A. (1998). Serious and violent juvenile offenders. In R. Loeber & D.P. Farrington (Eds.), *Serious & violent juvenile offenders: Risk factors and successful interventions* (pp. 13-29). Thousand Oaks: Sage.

Portigliatti Barbos, M. & Scatolero, D. (1985). L'omicida minorenne. In G. Canepa (Ed.), *Fenomenologia dell'omicidio* (pp. 221-341). Milano: Giuffrè Editore.

Shaw, C.R. & McKay, H.D. (1942). *Juvenile delinquency and urban areas.* Chicago: University of Chicago Press.

Snyder, H.N., Sickmund, M., & Poe-Yamagata, E. (1995). *Juvenile offenders and victims: A national report.* Washington, D.C.: Office of Juvenile Justice and Delinquency Prevention.

Snyder, H.N., Sickmund, M., & Poe-Yamagata, E. (1996). *Juvenile offenders and victims: 1996 update on violence.* Washington, D.C.: Office of Juvenile Justice and Delinquency Prevention.

Sutherland, E.D. & Cressey, D.R. (1924). *Principles of criminology.* Philadelphia: Lippincott.

Traverso, G.B., Esposito, R., Leone, G., & Ciappi, S. (1994). I risultati di uno studio pilota sulla delinquenza giovanile con la tecnica dell'autorilevazione. *Rassegna Italiana di Criminologia, 3,* 397-410.

Walgrave, L. (1992). *Délinquance systématizée des jeunes et vulnerabilité societale.* Géneve: Meridiens.

Walsh-Brennan, K.S. (1975). Children who have murdered. *Medical Legal Journal, 43,* 20-24.

Multi-Problem Violent Youth
R.R. Corrado et al. (Eds.)
IOS Press, 2002

283

Juvenile Aggression: A Risk Factors Model of Instrumental and Reactive Violence and Ensuing Protective Intervention Programs[*]

Frans Willem Winkel, Eric Blaauw and Ad Kerkhof

As detailed in several recent reviews (Loeber & Farrington, 1998; Weitekamp & Kerner, 1992; see also the Farrington, and Reppucci, Fried, & Schmidt chapters in this book), there is a large and growing empirical literature focused on risk factors for juvenile delinquency and aggression. According to some, this literature tends to be markedly atheoretical, and merely provides ever increasing lists of risk factors, that are not integrated in a general theoretical framework (e.g., Guerra, 1998). Others (e.g., Rogers, 2000, p. 597) have suggested that risk focused studies are "unabashedly one-sided"; they emphasize risk factors to the partial or total exclusion of protective factors (Loesel & Bender, in press). In view of these criticisms we will first present a theoretical model of juvenile instrumental and reactive aggression. Next, a number of intervention programs derived from this model, all based on the scared s'traight principle[1], will be reviewed.

1. The PDO-model: *Proximal*, *Distal*, and *Opportunity* Risks

Within the framework of the PDO-model–the basics of which are represented in Figure 1–the emergence of juvenile aggressive behavior is assumed to be the outcome of a complex interaction between proximal, distal, and opportunity risk factors, which provide starting points for developing effective intervention-programs. Figure 1 combines two explanatory models of delinquent and aggressive behavior, discussed in more detail elsewhere (Winkel, 1993, 1996; Winkel & Baldry, 1997; Winkel & Vrij, 1996). The upper right part of the figure outlines a basic model of instrumental aggression, while in the lower right part the focus is on reactive aggressive behavior. Both types of aggressive behavior are assumed to be mediated by similar types of inter- and intra-personal distal risk factors, while they differ mainly in terms of proximal risk, or the main motive directly underlying these behaviors. Instrumental aggression is "rationally" driven, while reactive aggression is "emotionally" driven behavior.

A number of empirically derived propositions underlying Figure 1 can be summarized in a few points. First, aggression is a *learned* response. Second, *opportunities* for learning are available in various domains, including the home environment (observing inter-parental conflict, and aggressive conflict tactics; observing or personally experiencing various forms of abuse), contacts with aggressive friends (peer-mediated aggression), and the consumption of media violence (mass media-mediated aggression). Third, *exposure* to aggressive models is variable, and is associated with the number of opportunities present, their frequency, and duration. Accumulated exposure (e.g., over domains) exerts a stronger effect on aggressive inclinations. Fourth, learning takes place through (a) direct *observation* of the behavior of others, and (b) direct or vicarious *reinforcement* (e.g., observing that other's aggressive behavior is rewarded). Both learning mechanisms also involve the *transmission* of aggression condoning norms and values. Fifth, the emergence of aggression is facilitated by *personal predispositions*, such as high impulsivity, absence of perspective taking skills, a low threshold for perceiving and experiencing aversiveness (e.g., over-attribution of hostile intent) and deficits in controlling impulses. Accumulated risk, e.g., in terms of psychopathy, exerts a stronger influence. Sixth, aggressive inclinations do not translate into behavior if there is no (perceived) *opportunity*, for example, if the person does not feel competent to engage in the behavior or if the target is considered inappropriate. In these conditions displacement may occur. Seventh, given the presence of (intra- and interpersonal) distal risk factors, reactive aggression specifically occurs in response to experiencing an *aversive event* of any sort (frustration, high temperatures, boredom, deprivation, provocation, air pollution, etc.) producing negative affect, and strong physiological arousal, particularly in the absence of cognitive mechanisms to control or channel arousal (e.g., anger control cognitions). Eighth, given such distal risk factors, instrumental aggression results from "rational choices" and becomes more likely if the 'attitude towards the pertinent behavior' (e.g., as operationalized in 'reasoned action theory'; Ajzen & Fishbein, 1980) is more positive.

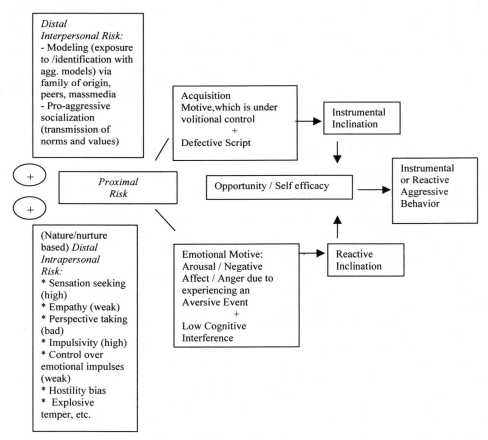

Figure 1: Figure (proximal risk)–ground (distal risks) forcefield underlying instrumental and reactive aggressive behavior

1.1 Interactions

To avoid an overly simplistic impression we would like to note that the modeling effects are moderated by a host of other variables. General moderators (Comstock & Paik, 1991; Comstock & Strausburger, 1990) of the aggressive impact of exposure to mass media aggressive models include efficacy (whether the behavior seen on screen rewarded or punished), normativeness (the degree to which violence is justified or lacks negative consequences for the perpetrator), pertinence (e.g., the relevance of the media scene or situation to the viewer's social context), and viewer susceptibility (e.g., whether the viewer easily becomes aroused or frustrated). Findings from laboratory experimental studies, reviewed by Donnerstein and Linz (1995), suggest that aggressive behavior is more likely to follow violent media depictions if (1) those who act aggressively are rewarded and not punished, (2) the violent behavior is seen as justified, (3) cues in the portrayed violence have a similarity to those in real life, (4) there is a similarity between the aggressor and the viewer, (5) the viewer imagines being in the aggressor's place, (6) behavior that is motivated to inflict harm or injury is portrayed, (7) the effects of violence are lowered, e.g., there is no pain, remorse, or sorrow, (8) violence is portrayed more realistically, (9) violence is not subjected to critical commentary, (10) the violent portrayals seem to please the viewer, (11) the behavior includes physical in addition to verbal abuse, and (11) the viewer is left in a state of arousal or frustration after viewing violence.

In a similar fashion, moderators of peer-mediated modeling (see Huesman, 1994) include: the adolescent's attachment to the delinquent peer group (e.g., the extent to which he feels emotionally close to these peers), the amount of peer-contact (e.g., the amount of time spent together), and the group's level of aggressive involvement (e.g., the amount of anti-social talk or aggressive conversations). Moderators of the effects of observing interparental violence were recently reviewed by Carlson (2000) and by Mohr, Noone, Lutz, Fantuzzo, and Perry (2000).

Hidden interaction effects, resulting from the interplay of social and biological transmission-processes of predispositions, may also lie behind the distal intrapersonal risk factors, as depicted in Figure 1. For example Bars, Heyrend, Simpson, and Munger (2001) recently analyzed the association between abnormal electroencephalographic (EEG) findings—including evoked potentials—and a specific type of explosive behavior in children and adolescents. An evoked potential is a regular pattern of electrical activity recorded from neural tissue that is evoked by a controlled stimulus. In pattern-reversal visual evoked potential studies, a subject is shown a rapidly alternating pattern, and the subject's EEG data are analyzed with respect to things such as amplitude and latency. Findings indicated that subjects who exhibited explosive aggression were significantly more likely to produce high-amplitude P 100 wave forms in such studies than subject not exhibiting explosive behavior. Bars et al. (2001) suggest that a subgroup of persons exhibiting explosive behavior may thus have a predisposition that is an innate characteristic of their central nervous system.

1.2 "Frustration" and Defective Scripts

The "reactive part" of the model depicted in Figure 1 may be considered an updated version of frustration-aggression theory, developed by Dollard, Doob, Miller, Mowrer, and Sears (1939). Parallel to current suggestions offered in the experimental social psychological literature, the concept of frustration was expanded to include any kind of aversive event producing negative affect. Moreover, the (originally assumed) automatic relation between frustration and aggression was replaced by a conditional linking of these two concepts, through incorporating distal intra- and interpersonal risk factors. A more elaborate version of the model was used to design a training program for conductors in the public transport system, providing guidelines to prevent and control reactive aggression by passengers. As the focus of this program is on adults, it will not be considered here (see Winkel, 1996).

The instrumental part of the model was further elaborated in terms of a "defective scripts theory" of juvenile delinquency. Some of its more basic concepts include desensitization, defective scripts, light versus heavy anchoring, generalized desensitization:

Desensitization. Emotional blunting is suggested as a hallmark of juvenile offending. Offenders tend to be desensitized. Feelings of shame, guilt, or regret do not emanate from behavioral performance; while considerations of future behavior show no signs of anticipated regret, or guilt. Desensitization is assumed to be consequent to a defective or crime-conducive behavioral script.

Defective script. Defective scripts show a lack of confluence of behavior and behavioral consequences; the interconnections between the pertinent behavior and its damaging or victimizing consequences tend be loose or absent. Law-abidingness is characterized by an 'identity' or 'Bandura-script': reprehensible conduct = detrimental consequences = victim. Defective scripts consist of biased perceptions of the behavior, its consequences, and the victim resulting from that behavior.

Light and heavy anchoring. Desensitization is lightly anchored if the pertinent biases merely result from ignorance. Desensitization is heavily anchored if these biases are rooted in processes of active cognitive re-structuring. In the social-psychological literature these processes are referred to as the process of moral justification, downward or palliative comparison, euphemistic labeling (regarding behavior), the process of neutralization, of denial and diffusion of responsibility, and of rationalization (regarding consequences), and the processes of moral exclusion, victim blaming, and dehumanization (regarding victim).

Generalized desensitization. Heavy anchoring is apt to be found in persistent offenders. Persistent offending is, at least partly, due to a process of generalization of desensitization from stepping-stone or gateway behaviors, via troublesome, daring, and oppositional behavior to more serious forms of criminal behavior ("behavioral aggravation"; Le Blanc & Loeber, 1993).

In terms of interventions the bottom line of this theoretical conception is to base programs on the "scared s'traight" principle. The major focus of such programming is *S*ensitization through *C*ognitive and *A*ffective *R*eparation of *D*efective *S*cripts.

2. Intervention and Prevention Programs

Wasserman and Miller (1998) define prevention approaches in terms of universal (primary prevention), selected (secondary prevention), and indicated programs (tertiary prevention). Universal programs are applied to an entire population of children and adolescents, such as a classroom, a neighborhood, or a school. Selected programs target high risk children who may already show some level of antisocial behavior. Indicated programs treat children already showing clear signs of delinquent or antisocial behavior. As illustrated below, all such programs could be based on the "scared s'traight" principle.

Our first illustration relates to a program entailing a teacher manual and a booklet, written in age-appropriate language, targeting children in the age range of 8 to 12 years. The booklet consists of nine stories depicting inappropriate behaviors that "actually should not happen" (stepping stone behaviors, such as throwing pine cones at automobiles from a viaduct, stealing a watch, and vandalism). All stories present behavioral interactions between a perpetrator and a victim, and are structured in such a way that the behavioral episode first evolves exclusively from the perspective of the perpetrator, and then from the perspective of the victim. Stories are preceded by a "multiple-perspectives" introduction, suggesting that material objects have an inside, and an outside, a front and a reverse side, and that an argument with a friend entails the other's perspective and your own perspective.

One of the scenarios was selected for further study, both in the Netherlands and Italy (Baldry & Winkel, 2001). Utilizing illustrative material from the booklets this scenario was presented by an actress, who was fluent in Dutch and Italian. Presentations, which included the main independent variable, were video-taped. The main focus of the studies was to examine the main and interaction (due to background factors, such as empathy) effects of these videos on the extent to which subjects associated positive and negative emotions with personally engaging in the pertinent behavior (and other behaviors, to test for potential carry-over effects), and on their awareness of damaging consequences resulting from that behavior for the victim. The Italian study was conducted at a 'middle school' in Rome, and consisted of a sample of 32 boys and 35 girls, with a mean age of 11.3 years. Subjects were randomly assigned to video-presentations relating to (1) a victim-implicit narrative, (2) a victim-implicit narrative, combined with a multiple perspectives instruction, and (3) a victim explicit narrative. Analyses revealed significant type of video main effects on positive emotions associated with the behavior, and on the likelihood of psychological distress and fear responses from the victim. A victim explicit narrative the most successfully reduced positive emotions, and raised subjects' awareness of victimizing consequences. Prior to video exposure, Bandura's Moral Disengagement Scale (Bandura, Barbaranelli, Caprara, & Pastorelli, 2001), a general indicator of script defectiveness, was administered. Subjects exhibiting high versus low disengagement associated significantly more positive emotions with engaging in the behavior. On this variable, moreover, a significant type of video by disengagement (low/high) interaction emerged, suggesting that the victim explicit narrative was particularly effective in reducing positive emotions in subjects exhibiting high disengagement. All subjects were re-approached, as part of another study, three months later. This study included items to assess subjects' intentions to personally engage in the pertinent behavior. Findings revealed that subjects previously exposed to the victim explicit narrative reported significantly lower behavioral intentions than subjects exposed to the other two experimental conditions. Utilizing a slightly different experimental design, including a no video exposure control group, and a victim implicit and explicit condition, the Italian findings were partly replicated in Amsterdam. Subjects exposed to a victim explicit narrative were for example significantly more aware of potential victimizing consequences for the driver than subjects exposed to the implicit condition. Quite unexpectedly, however, in comparison to controls boomerang effects emerged in the latter subjects. These subjects associated significantly more positive emotions, and significantly less negative emotions with throwing pine cones at cars passing by than control subjects. The size of these boomerang effects was substantially reduced, and even nullified on negative emotions, by a victim explicit narrative. The Dutch findings suggest that discussing inappropriate steppingstone behaviors in classroom settings may easily enhance the formation of defective scripts, if these discussions do not explicitly focus on its damaging and victimizing consequences. More generally, the findings underline the continued need to evaluate school focused intervention-programs.

The prevention of various forms of sexual intimidation among young adolescents was the major focus of a massive mass media campaign. The campaign included more extensive brochures and documentaries, specifically designed for use in school settings. To evaluate the impact of one of these documentaries Winkel and De Kleuver (1997) conducted an experiment among 198 male and female pupils with a mean age of 16 years. The main aim of the experiment was to examine the effects of a perpetrator-focused (highlighting the negative consequences for perpetrators engaging in various forms of sexual intimidation) and a victim focused (stressing negative consequences for victims) video on 5 dependent measures, including: (1) the evaluation of macho behavior in male-female interactions, (2) sexual intimidation myths acceptance, (3) the conditional acceptance of coerced sex, (4) the perceived likelihood of negative consequences for perpetrators, and (5) the perceived likelihood of psychological harm for victims.

Both videos were designed in a similar fashion; for example, the order in which factual information was presented was the same. All videos provided information on the prevalence and 'legal definitions' of sexual harassment, sexual assault, and rape. Moreover the two videos consisted of fragments of conversations of several boys and girls, while they are interviewed in (same sex) pairs. During these interviews personal dating-experiences are discussed. The boys for example note that boys, at least in some peer-groups, gain respect by telling that they managed to overcome initial resistances of a girl through verbal or physical force; that sexual assaults tend to be related to the victim's own behavior, and that they tend to talk about girls in a rather 'macho' manner, when they are in groups, and tend be more modest, when they are alone. The interviews with the girls focus on their disliking boys' macho behavior, and that they sometimes feel very frightened, when they are confronted with groups of boys late at night, after visiting pubs. The bottom line of these various interviews is that girls actually mean no, if they say 'no', and that boys much more easily tend to sexualize personal encounters, in which girls only try to be friendly. The central focus of the two videos is a series of detailed interviews with real victims and perpetrators of sexual harassment, sexual assault, and rape. These interviews form the basis of the experimental manipulation; differences between videos typically emerged here. These interviews relate to three different scenarios:

1. A group of boys is spying on a group of girls who are dressing up at a sports center. Fragments of conversations among the boys are heard, including statements "I would like to tie her up, and screw her" and bets, such as "If you grab her tits, we'll pay you 2.50 guilders each." The boys are messing around, obviously waiting for the girls to come out. When a girl is leaving the center one of the boys (Hans) grabs her, touches her breasts and kisses her. While a motor cyclist is passing by, noticing the incident, the girl is able to get free and escape the scene. While fleeing, Hans is still yelling threats at her. At night, Hans is visited by the police.

2. *Linda*, who works at a supermarket, is regularly bothered by her boss. While refilling the shelves he is standing too close behind her, telling that she is special and so forth. For a total period of 6 years she doesn't dare to tell anyone about what is going on. After her boss actually raped her in the back of the store, she was very afraid of males and boys; she felt filthy and ashamed. Eventually her boyfriend finds out about it, after observing her odd responses to mass media sexual violence. He encourages her to report to the incident to the police. Her boss gets convicted, but she has lost her job in the meantime.

3. In a disco *Peter* gets acquainted with a girl and asks her if she wants to come out with him. Being outside, while fondling, he starts talking about sexual intercourse and the need to use a condom not to contract AIDS. Trying to undo her pants, she tells him that she doesn't want this. He then gets furious, and forces her to the ground, saying "If you say A, you have to say B also." She is trying everything possible to resist him, holds her legs tight together, and begs him to let her go. He threatens her with a sharp object, and when people are passing by he covers her mouth with his hands. While some other persons are passing by she gets free and runs away. Later on, Peter is arrested by the police.

Both videos utilized scenario 1. The victim-focused video also included the full Linda-scenario, with added fragments of an interview with her boyfriend, discussing the psychological impact of living with a raped partner, and of finding out about her negative life experiences. Scenario 3 was also included, but only from the viewpoint of Jeannette, the girl who was sexually assaulted by Peter. This video thus strongly focused on the negative psychological consequences of sexual victimization. Also discussed were, post traumatic responses, such as fear of males, feelings of fear of being touched, the reluctance to talk openly about the incident, and feelings of self-blame, guilt and shame.

The perpetrator-focused video used a minor part of the Linda-scenario (2), providing factual information about what had happened. All fragments related to psychological distress were left out. Besides Scenario 1 this video especially considered the Peter-scenario in full detail. Moreover, a police officer throws light on the criminal justice aspects of sexual intimidation (discussing the procedures used during interviewing suspects, the risk of having to stay at the police station overnight, etc.), while images of a jail are shown, where an arrested person is locked in. This video thus heavily concentrated on the negative consequences for the perpetrator, including the various formal responses from the criminal justice system, and negative reactions from the social environment. These social reactions also emerged in Peter's interview: after his arrest he felt forced to move to another village, and to go to another school (due to fear of being recognized and stigmatized in his old village).

Analyses revealed significant main effects due to type of condition (controls, perpetrator and victim focused video) on myths acceptance and on the conditional acceptance of forced sex. The most favorable values on both myths and conditional acceptance were found in subjects exposed to a victim-focused video. Particularly in boys, self-reports of various forms of sexual intimidation were significantly associated positively with myths acceptance, and negatively with the perceived likelihood of victims suffering psychological distress. Various main effects were modified by significant interactions. Significant condition by gender effects emerged on the evaluation of macho behavior, on myths acceptance, and on conditional acceptance of forced sex. On all three variables a similar pattern emerged. For male subjects, the central target of the videos, the perpetrator-focused message appeared to backfire. Boys exposed to this message evaluated macho behavior in interacting with girls more positively. Instead of weakening myths about sexual intimidation, this message seemed to strengthen them. Moreover, the message appeared to enhance boys' conditional acceptance of coerced sex. In boys, the perpetrator-focused message thus resulted in various unwanted outcomes that did not emanate from a victim-focused strategy. Thus, if the aim is to repair defective scripts the victim-focused strategy appears to present a more viable option.

The "scared s'traight" principle also provided the basis for designing a new alternative sanctions program for juvenile offenders in 1988. The Focus on Victims project was developed by the Netherlands Victim Support in cooperation with the Arnhem Agency for Alternative Sanctions, which forms part of the Council for Child Care and Protection Services. The program started as an experiment in 4 jurisdictions, and was expanded to all jurisdictions in 1990. Utilizing a variety of means, including specially created videos, in depth group discussions, home work assignments, and role playing sessions, the basic program highlights the social, psychological, and material impact of crime on victims. An extended version introduced in 2001, partly based on Novaco's work (1999), also targets skills to recognize anger symptoms, to control and manage anger responses, and skills to resist peer pressure to engage in delinquent acts. Cases are referred by public prosecutors and juvenile judges. Participants are typically males, in the 14–18 age range, who committed various acts of violence. The program is conducted in small groups (4 to 6 persons), and consisted of 781 (137 females and 643 males) in 1997, and 799 (144 females and 655 males) clients in 1998 (Report 1996 - 1998).

Groenhuijsen and Winkel (1994) and Winkel (1993) evaluated the program's efficacy in changing perceptions of crime seriousness, of damaging consequences resulting from crime, and victimization awareness. Measures were administered prior to and post participation in the program. Analyses revealed significant main effects on all pertinent variables, and thus suggested that learning in the desired direction took place due to program exposure. Twenty six percent of the participants re-entered the criminal justice system (e.g., were re-arrested by the police) within a period of at least 3 up 15 months following the program. Keeping in mind the study's methodological limitations, such as the absence of a control group and non-random assignment of clients, the program thus appears to have a beneficial behavioral impact on a majoritiy of participants. Further analyses revealed a sigificant interaction, suggesting that the program's impact on perceptions of crime seriousness was moderated by subject status, in terms of being or not being a recidivist.

An obvious argument to explain why a minority of participants has fallen back into their old habits is to suggest that they did not learn from the program. The interaction clearly does not support this argument. Perceptual changes in crime seriousness typically emerged in recidivists. The interaction effect underlines the continued necessity to consolidate repaired scripts, and to actively create opportunities for their "survival." However, one might also argue that the currently used measures are simply too transparent, and can easily be used to fake improvement. To further investigate this possibility a number of studies are currently conducted in which defective scripts are measured in a subtler manner through using modified Stroop-test procedures.

3. Discussion

Behavioral manifestations of juvenile instrumental and reactive aggression were assumed to be mediated by an interacting force field of distal, proximal, and opportunity risk factors. The model discussed here offers a variety of starting-points for developing interventions aimed at controlling aggressive responses. An obvious implication of the model is that only comprehensive programming, targeting a series of risk factors, can be expected to result in strong, long lasting, behavioral effects. Programs based on the "scared s'traight" principle may contribute to achieving this goal. Our review of intervention programs suggests that this principle can be applied in different contexts, including programs targeting young school children in the 8–12 age range, young adolescents in the 14–17 age range, and young delinquents in the 13–18 age range. Findings moreover revealed that programs not fully utilizing this principle, such as victim implicit and perpetrator focused strategies, may easily backfire. Instead of controlling, such programs may eventually stimulate the emergence of aggressive behavior.

Notes

* Author Note. Preparation of this chapter was supported by a grant from the Achmea Foundation Victim and Society (Stichting Achmea Slachtoffer en Samenleving).
1 The acronym before the apostrophe stands for: Sensitization via Cognitive and Affective Reparation of Defective Scripts.

References

Azjen, I., & Fishbein, M. (1980). *Understanding attitudes and predicting social behavior*. Englewood Cliffs, NJ.: Prentice Hall.
Bandura, A., Barbaranelli, C., Caprara, G.V., & Pastorelli, C. (2001). Sociocognitive self-regulatory mechanisms governing transgressive behavior. *Journal of Personality and Social Psychology, 80*, 1, 125 –135.
Baldry, A., & Winkel, F.W. (2001). Early prevention of delinquency. In G. B. Traverso & L. Bagnoli (Eds.), *Psychology and law in a changing world: New worldwide trends in theory, practice and research*

(pp. 35 – 51). London: Routledge.

Bars, D.R., Heyrend, F., Simpson, C., & Munger, J. (2001). Use of visual evoked-potential studies and EEG data to classify aggressive, explosive behaviors of youths. *Psychiatric Services, 52,* 81–87.

Carlson, B.E. (2000). Children exposed to intimate partner violence: Research findings and implications for interventions. *Trauma, Violence, & Abuse: A Review Journal, 1,* 321– 343.

Comstock, G., & Strausburger, V.C. (1990). Deceptive appearances: Television violence and aggressive behavior. *Journal of Adolescent Health Care, 11,* 31–44.

Comstock, G., & Paik, H. (1991). *Television and the American child.* San Diego: Academic Press.

Donnerstein, E., & Linz, D. (1995). *The media.* In J.Q. Wilson, & J. Peterselia (Eds.), *Crime* (pp. 237 – 267). San Francisco: Institute for Contemporary Studies.

Dollard, J., Doob, L.W., Miller, N.E., Mowrer, O.H., & Sears, R.R. (1939). *Frustration and aggression.* New Haven: Yale University Press.

Groenhuijsen, M., & Winkel, F.W. (1994). The 'focusing on victims program' as a new substitute penal sanction for youthful offenders: A criminal justice and a social psychological analysis. In G. F. Kirchhoff, E. Kosovski, & H. J. Schneider (Eds.), *International debates of victimology* (pp. 306–329). Moenchen-gladbach: WSVN-Publishers.

Guerra, N. (1998). Serious and violent juvenile offenders: Gaps in knowledge and research priorities. In R. Loeber & D. P. Farrington (Eds.), *Serious and violent juvenile offenders: Risk factors and successful interventions* (pp.367 – 389). Thousand Oaks: Sage

Huesman, L.R. (1994). (Ed.). *Aggressive behavior: Current perspectives.* New York: Plenum.

Le Blanc, M., & Loeber, R. (1993). Precursors, causes, and the development of criminal offending. In D. F. Hay & A. Angold (Eds.), *Precursors and causes in development and psychopathology* (pp. 233– 265). Chichester: Wiley.

Loeber, R., & Farrington, D.P. (1998). *Serious and violent juvenile offenders. Risk factors and successful interventions.* Thousand Oaks: Sage.

Loesel, F., & Bender, D. (in press). Protective factors and resilience. In J. Coid & D. Farrington (Eds.), *Prevention of adult criminality.* Cambridge: Cambridge University Press.

Mohr, W.K., Noone Lutz, M.J., Fantuzzo, J.W., & Perry, M.A. (2000). Children exposed to family violence: A review of empirical research from a developmental-ecological perspective. *Trauma, Violence, & Abuse: A review journal, 1,* 264–284.

Novaco, R. (1999). *Assessment and treatment of anger.* Workshop. EAPL/AP-LS Psychology and Law Conference. Trinity College, Dublin, July 5-6. Southampton: University of Southampton Press.

Report 1996 – 1998. (1999). *Meer dan de feiten op een rij.* Utrecht: Stichting Slachtoffer in Beeld.

Rogers, R. (2000). The uncritical acceptance of risk assessment in forensic practice. *Law and Human Behavior, 24,* 595–606.

Wasserman, G.A., & Miller, L.S. (1998). The prevention of serious and violent juvenile offending. In R. Loeber & D. P. Farrington (Eds.), *Serious and violent juvenile offenders: Risk factors and successful interventions* (pp. 197-248). Thousand Oaks: Sage

Weitekamp, E.G.M., & Kerner, H.J. (1992). (Eds.). *Crossnational longitudinal research on human development and criminal behavior.* Dordrecht, NL: Kluwer Academic.

Winkel, F.W. (1993). Opvattingen van jeugdige delinquenten over crimineel gedrag: 2 studies naar 'gunstige definities' en gerichtheid op criminaliteit. *Tijdschrift voor Ontwikkelingspsychologie, 20, 2,* 151-170.

Winkel, F.W. (1996). A propositional theory of reactive violence: Some implications for controlling aggression against personnel in the mass transit system. In G. Stephenson & N. Clark (Eds.), *Issues in criminological and legal psychology* (Vo. 26, pp. 84-95). Leicester, UK: British Psychological Society Press.

Winkel, F.W., & Baldry, A.C. (1997). An application of the scared s'traight principle in early intervention pro-gramming: Three studies on activating the other's perspective in pre-adolescents' perceptions of a stepping-stone behavior. In G.M. Stephenson & N. Clark, (Eds.), *Procedures in criminal justice: Contemporary psychological issues* (pp. 3-14). Leicester, UK: British Psychological Society.

Winkel, F.W., & De Kleuver, E. (1997). Communication aimed at changing cognitions about sexual intimidation: comparing the impact of a perpetrator-focused versus a victim-focused persuasive strategy. *Journal of Interpersonal Violence, 12,* 513 - 530.

Winkel, F.W., & Vrij, A. (1996). Het bestrijden van criminaliteit. In B. Klandermans & E. Seydel (Eds.), *Over-tuigen en Activeren: Publieksbeïnvloeding in theorie en praktijk* (pp. 180 - 303). Assen, NL: Van Gorcum.

IV

Risk/Needs Management Instrument: A Conceptual Framework

Multi-Problem Violent Youth
R.R. Corrado et al. (Eds.)
IOS Press, 2002

295

An Introduction to the Risk/Needs Case Management Instrument for Children and Youth at Risk for Violence: The Cracow Instrument

Raymond R. Corrado

One of the major goals of the workshop held in Cracow was the development of an instrument for the assessment and management of risk and need factors related to youth violence. In the next chapter, we present the framework for this instrument, which is currently known as the *Cracow Instrument*. It was developed by our Simon Fraser University team, guided by the presentations and discussions that took place at the Cracow meeting.

1. Key Concepts

From the outset, our primary focus has been on risk for *youth violence*, which we define as the actual, attempted, or threatened physical harm of another person perpetrated by a child or adolescent. However, research indicates that youth who commit serious violence also tend to commit other forms of serious (i.e., repeated and persistent) delinquency. As well, many risk and needs factors associated with youth violence also are associated with serious delinquency. Both youth violence and other forms of serious delinquency are similar insofar as they involve major violations of explicit social norms, such as "externalizing" behaviors that involve major violations of others' rights. This means that any risk/needs instrument for youth violence lacks some degree of specificity in that it is also relevant to the assessment and management of those at risk for serious delinquency. Cohort research confirms that there are several similar risk factors within the multi-problem profiles that characterize violent versus non-violent youth (Loeber & Farrington, 1998).

Importantly, this instrument is not intended to, and is unlikely to be useful for, assessing and managing risk for less serious violations of social norms, such as truancy, simple substance abuse, or minor property crimes, or risk for poor mental health adjustment, such as "internalizing" behaviors like suicide, anxiety, or depression. Instead, this instrument targets the kinds of violence that results in children and adolescents being classified as serious violent juvenile or young offenders. While minor violent acts may be indicators of more serious violent potential for a few youth, these behaviors are usually assumed to be within norm-aggressive behavior. In contrast, a physical attack that causes significant damage and/or the violation of a victim's body, such as a sexual assault, is the type of violence that is the object of risk and needs interventions.

Accordingly, our instrument focuses on assessment and management. We conceptualize *assessment* as the process of evaluating individuals and the environments in which they live to understand their risk for engaging in violence. Because research indicates youth violence is a complex and multi-determined phenomenon, and because people responsible for case management come from diverse educational and professional backgrounds, we decided that any assessment instrument should be comprehensive in nature and consider information about the person obtained using a variety of methods, such as self reports, peer reports, official records, and physical examinations. We also decided that it was impossible, given the current state of the science literature, to construct an instrument based solely on empirical research; practical, legal, and policy considerations were deemed to be equally important and relevant.

In contrast to assessment, *management* is the process of intervening in a case to reduce the risk of violence. It is a general term that subsumes many different kinds of interventions intended to contain or reduce risk. Legal interventions place restrictions on freedoms to limit the opportunities for the person to engage in serious delinquency or violence. Examples include arrest, diversion from the justice system, incarceration, community supervision, court-ordered treatment, and removal from the family home. Habilitative interventions address deficits in biological, psychological, or interpersonal functioning that may be causally related to serious delinquency or violence. Examples here include pharmacological treatment, family therapy, educational training, and provision of employment or housing. Some management strategies are purely preventive in nature, whereas others are reactive and initiated only when problems become apparent. We decided that to facilitate comprehensive and integrated case management, the assessment instrument should be written in plain language, so the findings would be comprehensible to diverse groups of service providers, easily translated across cultures, and easily integrated into existing or planned computer databases.

In order to facilitate the adoption and effective implementation of such a multi-stage developmental data instrument by all relevant government agencies, the instrument must serve a multi-purpose function. For example, chapters 20-23 of Section I discuss the cases of France, Italy, and Poland. In these countries, there are several ministries and agencies that are responsible for identifying families at-risk for multi-problems associated with potentially violent children and youth. In effect, each ministerial agency would benefit from using our instrument.

Since violent youth are typically under the legal mandate of juvenile courts, it is usually the Ministry of Justice, or some equivalent governmental department within federal systems, like the United States, that has the necessary resources to deal with case management decisions. Not surprisingly, most of the data management instruments, therefore, involve risk prediction for adolescents between 12-18 and young adults up to the age of 21. However, as described in Chapter 14, there are several recent clinical assessment instruments that utilize broad databases to assist case managers in making intervention decisions for adolescent offenders beyond violence risk prediction objectives. These interventions typically consist of treatment programs specifically targeting a particular profile of health, mental health, family, and educational needs. The province of Quebec, for example, employs LeBlanc's MASPAQ instrument to address multi-problem youths, including those at-risk for violence. Similarly, Washington State's juvenile justice system has a more narrowly focused instrument to assess continually the risk-needs profile of the adolescents in its system (see Chapter 17).

Historically, the next trend in risk-needs instrument development focuses on children between 6-12 years of age. In the province of Ontario, and the state of Minnesota, there are two instruments being validated for eventual system-wide use to reduce the likelihood of future serious and violent behaviors. These instruments are described in Chapters 15 and 16. Again, the policy objective is to identify those children and their families in need of intervention programs to treat specific multi-problem profiles. As with the adult risk-needs instruments, the adolescents and children instruments are being employed cross-nationally with encouraging results.

Not surprisingly, the developmental periods where risk-needs instruments for violence are the least developed are the pre-natal and early childhood stages. Typically, cohort research has been conducted on children eight years old and older. It is only recently that this cohort research has begun to examine younger samples of children (Trembly, Masse, Kurtz, & Vitaro, 1997). Nonetheless, most risk-needs instruments include information involving certain limited early childhood risk factors, usually involving parents and abuse issues. In constructing our instrument, we reviewed a wide range of recent research in order to build the contextual foundation for the initial age domain of our instrument.

2. Structure of the Instrument

We decided that the instrument should reflect the various aspects of individual functioning that have been implicated as playing a role in youth violence, including functioning in the biological, psychological, and interpersonal domains. This decision reflects our view that there are multiple causal pathways to youth violence, and also our view that intervention in a given case must be based on the factors that are both present and believed to be causally relevant for that individual.

Broad domains of functioning are useful for organizing and guiding assessments in general terms. However, information about specific risk and need factors within each domain is necessary for effective communication and for making decisions about which intervention strategies to use in a given case. We decided that the risk and need factors should be defined at a basic level, which is the primary focus of communication. We attempted to develop clear, principled linguistic definitions of the concepts underlying each risk and need factor. We then attempted to specify various indicators—that is, concrete operationalizations or measures—of each risk and need factor. We expected that there would be good consensus among various researchers and service providers concerning the identification of relevant functional domains, substantial but lower agreement about the specific risk and need factors within those domains, and relatively little agreement about the indicators for each risk and need factor.

We also decided that the instrument should reflect functioning in these domains over time. This decision reflected our belief that a developmental perspective is crucial to understanding the causes of youth violence. From a practical perspective, this would also facilitate the assessment and tracking of changes in functioning across key developmental stages. After reviewing the scientific and professional literature, and after considerable discussion, we decided to organize it according to four major developmental stages: pre-natal, lasting from conception to birth; early childhood, lasting from birth through age 5; late childhood, lasting from ages 6 through 12; and adolescence, lasting from ages 13 through 18, the beginning of early adulthood.

These age domains were selected primarily based on our review of the literature on risk factors for violence suggesting that each of these domains were associated with different patterns of risk factors. An additional consideration was the type of agency or person responsible for obtaining the risk factor information. It is important to keep in mind that our age domains are designed, in part, to provide case management information for a variety of policy objectives, in addition to risk-needs management.

The first age domain consists of indicators that have been identified in the relatively recent literature on violence involving areas such as genetic predispositions, mother's health, and birth complications. The second age domain focuses on attachment themes, parenting and abuse issues, and learning and educational problems. The third age domain includes family, school, and peer constructs. The fourth age domain is where, historically, most of agency responsibilities and resources have been expended on intervening with youth at-risk for violence.

While there are several risk prediction instruments that are employed for this last age domain, there are few case management instruments. Most importantly, there are no instruments that directly integrate the prenatal–perinatal age domain and the infancy–early childhood age domain into a cumulative case management risk-needs for violence dataset. Indeed, it is only recently that risk-needs case management instruments concerned with reducing delinquent and violent behaviors have begun to be developed.

It is assumed that our instrument will be used in conjunction with, or complimentary to, existing risk-needs instruments for adolescents. Table 1 provides a comparison of how our instrument dataset compliments other risk-needs instruments. As the Table makes clear, there is substantial consistency among the various instruments, although our instrument is much broader. We believe that it makes good sense to view the instruments as complementary. In other words, we recommend using the new instrument in conjunction with, rather than as a replacement for, more established and well-validated instruments, where they exist. As explained in Chapter 1, the theoretical genesis of our instrument is derived from the key theme that multiple pathways characterize the different sequence of variables that describe and explain how some youths become serious and violent young offenders. The different pathways that distinguish early-onset, or early-starter, offenders and violence versus late onset, or late-starters, are of particular importance, as are the related constructs of "life-course persistent" versus "adolescent limited" offenders (Moffitt, 1993). These constructs are reviewed in Chapter 3 by Farrington, who also discusses the critical link between multiple risk factors for multi-problem boys. These constructs and theoretical themes are elaborated further in Chapters 2 and 4. Similarly, throughout all of the chapters in Section I on Risk Factors, it is evident that violent adolescents are more often characterized by problems beginning at the earliest age domains than non-violent adolescents.

Table 1: Concordance Between Conceptual Framework and Current Risk Assessments for Young People

	Conceptual Framework	EARLB-20	SAVRY	YLS/CMI
ENVIRONMENTAL	Prenatal/ Perinatal Complications			
	Obstetrical Complications	?		
	Maternal Substance Use @ Pregnancy			
	Living Conditions			
	Exposure to Toxins	?		
	Community Disorganization	?	?	
	Family Socio-economic Status	X		
	Residential Mobility			
	Exposure to Violence	?	?	
	Peers			
	Peer Socialization	X	X	X
	School			
	School Environment			

INDIVIDUAL	Biological			
	Birth Deficiencies	?		
	Parental History of Mental Illness	?		
	Executive Dysfunction	?	?	?
	Chronic Underarousal			
	Abnormal Biochemical Activity			
	Psychological			
	Cognitive Delays/Disorders	X	X	?
	Personality Traits/Disorders	?	X	?
	Other Mental Illnesses			
	Antisocial Attitudes	X	X	X
	Poor Coping Ability	X	X	?
	Functional			
	School Functioning	X	X	X
FAMILY	Parental Characteristics			
	Teenage Pregnancy			
	Maternal Coping Ability			
	Parental Antisocial Practices/Attitudes	X	X	
	Parental Education and IQ			
	Family Dynamics			
	Familial Supports	X		
	Family Conflict / Domestic Violence	X	X	
	Family Structure/ Single-Parent Family		X	
	Parent- Child Relationship			
	Ineffective Parenting	X	X	X
	Early Caregiver Disruption	X	X	
	Parent/Child Attachment			?
INTERVENTIONS	Previous Interventions			
	Accessibility to Interventions			
	Familial Responsivity to Intervention	X		
	Youth Responsivity to Intervention	X	?	
EXTERNALIZING BEHAVIOR	General Behavioral Problems	X	X	X
	Violence/Aggression	X	X	X
	General Offending		X	X
	Substance Use	?	X	X

Note.

X = captures the risk factor in all or most respects, ? = captures the risk factor in part

EARL-20B – Early Assessment List for Boys: Version 2 (Augimeri et al., 2001)

SAVRY – Structured Assessment of Violence Risk in Youth (Version 1) (Bartell et al., 1999)

YLS/CMI – Youth Level of Service/Case Management Inventory (Hoge & Andrews, 1999)

While relatively few in number and overwhelmingly males, these early onset or life-course persistent offenders are the primary policy target for our instrument's case management objectives. In Chapter 4, Lösel discusses the importance of intervention strategies at different developmental stages. It is evident as well in the chapters in Section II that identifying the most complete range of risk factors is the most important case management approach to providing constructive interventions in juvenile justice systems, schools, and social work agencies.

Clearly, the identification of risk factors at the earliest age domains is still very tentative. The relationship between genetic predispositions, parental risks, and birth complications to at-risk behavior are all based on innovative research, yet much of the research is speculative. However, the research presented in chapters 5, 7, 10, and 11 provides sufficient initial supporting evidence for the inclusion of a broad array of risk indicators at the earliest age domains. As mentioned above, the assumption is that our instrument's risk indicators will be continually adjusted as the research into these constructs continues and expands.

As LeBlanc explains in Chapter 14, it is vitally important that any risk-needs instrument be theoretically based. We believe that most of the chapters in Sections I and II provide a strong theoretical underpinning for our instrument. There are key constructs that link all of the age domains. The constructs provide the rationale for the indicators that we are currently matching to every construct in the instrument. Again, as discussed in the introductory chapter, all of the age domains and related constructs were reviewed jointly by all of the participants at the two workshops in Vancouver and Cracow. Modifications and additions were then made subsequently by our Simon Fraser University research team.

3. Using the New Instrument

Most researchers or service agencies work in specific contexts, and their assessment and management activities, therefore, are focused on specific subsets of the risk factors included in the new instrument. Even people working in similar contexts tend to define risk factors in dissimilar ways, often by operationalizing them in terms of different indicators. Finally, legal and professional traditions limit the way in which people talk about and deliver services to youths. In our view, this gives the false impression that "nobody agrees" about how best to assess and manage youth violence. Our hope is that the new instrument will serve, most fundamentally, as a kind of dictionary or lexicon that will permit clear communication about terms, concepts, and ideas relevant to youth violence. This should be true regardless of whether the communication is between service providers working in different agencies, between researchers working in different countries, or between policymakers and researchers or service providers. The instrument's structure is intended to balance brevity, simplicity, and flexibility, on the one hand, and comprehensiveness, complexity, and consistency on the other.

Still, clear communication is not enough. The key goal of the instrument is to guide the assessment and management of risk and need factors related to youth violence. We hope this goal will be accomplished in two major ways. First, the new instrument should be a comprehensive, multi-disciplinary assessment and management tool. It is likely that some risk or need factors in some domains are, on average, more strongly predictive of violence (all else being equal) than others. It is also likely that one could construct very brief actuarial instruments, comprising only a handful of items that predict violence as well as the findings of a comprehensive, time-consuming assessment instrument. However, it is our view that useful decisions about interventions with an individual person must be based on something more than stereotypical beliefs about what is useful on average.

Second, the new instrument should promote basic and evaluative research. Basic research focuses on clarifying the importance and nature of risk and need factors. The new instrument should encourage research on (putative) factors that have received relatively little attention. It should also help researchers to synthesize research findings, insofar as it presents a common language for talking about risk and need factors. It should also stimulate research comparing the various indicators that are or could be used to measure risk and need factors. Evaluative research examines the effectiveness of various management strategies. The new instrument should encourage research on a broad range of interventions, as well as assisting the identification and control of potential confounding or nuisance factors.

We do not assume that most researchers or service providers have the interest or the resources to use the new instrument in its full form. Obviously, this would require considerable time and money. Nonetheless, the instrument should still be useful in helping these people to rethink and revise their assessment procedures, and in recognizing the limitations of their existing assessment procedures.

4. Conclusion

We view the Cracow instrument as a work in progress. Efforts at SFU continue in an attempt to finalize its structure and content. Upon completion, we will submit the instrument to various people, including all those who attended the ARW in Cracow, for review. The instrument will then be revised, taking into account the comments and suggestions we receive. The revised instrument will be used as the basis for international multi-site research, as well as in many specific research projects and by various agencies. Feedback from scientists and service providers will be crucial in making future revisions to the instrument. Only through such a collaborative process can we hope to make meaningful advances in the prevention and management of youth violence.

References

Augimeri, L. K., Koegl, C. J., Webster, C. D., & Levene, K.S. (2001). *Early Assessment Risk List for Boys: EARL-20B (*Version 2*)*. Toronto, ON: Earlscourt Child and Family Centre.

Bartell, P., Borum, R., & Forth, A. (1999). *SAVRY: Structured Assessment of Violence Risk in Youth* (Version 1). Consultation Edition.

Hoge, R.D., & Andrews, D.A. (1999). *The youth level of service/case management inventory and manual (revised)*. Ottawa: Department of Psychology, Carleton University.

Loeber, R., & Farrington, D.P. (1998). *Serious and violent juvenile offenders: Risk factors and successful interventions*. Thousand Oaks, CA: Sage.

Moffitt, T. E. (1993). Adolescence-limited and life course-persistent antisocial behavior: A developmental taxonomy. *Psychological Review, 100*, 674-701.

Tremblay, R.E., Masse, L.C., Kurtz, L., & Vitaro, F. (1997). From childhood physical aggression to adolescent maladjustment: The Montreal Prevention Experiment. In R.D. Peters & R.J. McMahon (Eds.), *Childhood disorders, substance abuse, & delinquency: Prevention and early intervention approaches* (pp. 1-62). Thousand Oaks, CA: Sage.

Multi-Problem Violent Youth
R.R. Corrado et al. (Eds.)
IOS Press, 2002

A Preliminary Conceptual Framework for the Prevention and Management of Multi-Problem Youth

Candice Odgers, Gina M. Vincent and Raymond R. Corrado[*]

Research has demonstrated a number of factors that put children and youth at risk for involvement in future violence. Admittedly, there is no absolute consensus regarding the most salient risk factors for youth violence or the most direct pathway to involvement in violent offending among adolescents. There is, however, an increasing amount of support for the development of empirically based risk and case management tools in order to assist in the response, prevention, and prediction of youth violence (Augimeri, Koegl, Webster, & Levene, 2001; Bartel, Borum, & Forth, 1999).

Recent literature reviews (Hawkins et al., 1998; Reppucci, Fried, & Schmidt, this book; U.S. Department of Health and Human Services, 2001) and meta-analyses (Lipsey & Derzon, 1998) have provided valuable contributions to the state of research and knowledge in this area. A systematic translation of this body of research into practice, however, is still pending. Therefore, the primary objective of this chapter is to provide a comprehensive summary of the factors associated with youth violence by outlining a proposed conceptual framework for a set of instruments designed to assist practitioners and researchers in responding to youth violence and related anti-social behaviors.

As illustrated in Table 1, the factors that were selected for inclusion in this conceptual framework either have demonstrated an empirical association with youth violence, or have an expected clinical or theoretical contribution to this relationship. We have conceptualized the influence of these factors on youth violence and other antisocial outcomes from a developmental perspective. To date, developmental concerns have been largely neglected in this area, despite the widespread recognition of the potential for developmental approaches to assist in answering fundamental questions relating to the development, prediction, and treatment of violence among youth. As Shaw and Winslow (1997) note, in order to successfully identify pathways to violence, the transactions that occur between children and their environments must be plotted developmentally because risk factors are not necessarily stable across the lifespan. In accordance with this view, the eventual structure of the presented conceptual framework will divide factors by the relevant developmental stage. In this chapter, we first explain the structure of the risk factors included in this conceptual framework. We then provide the framework with a brief explanation and justification of each risk factor. Finally, we conclude with some details regarding a future scoring system and future directions.

Table 1: Risk Factors and Basis For Inclusion

Conceptual Framework		Empirical	Theoretical/ Clinical
ENVIRONMENTAL	Prenatal/Perinatal Complications		
	Obstetrical Complications	X	
	Maternal Substance Use During Pregnancy	X	
	Living Conditions		
	Exposure to Toxins	X	
	Community Disorganization	X	
	Family Socio-economic Status	X	
	Residential Mobility		X
	Exposure to Violence	X	
	Peers		
	Peer Socialization	X	
	School		
	School Environment		X
INDIVIDUAL	Biological		
	Birth Deficiencies	X	
	Parental History of Mental Illness	X	
	Executive Dysfunction	X	
	Chronic Underarousal	X	
	Abnormal Biochemical Activity	X	
	Psychological		
	Cognitive Delays/Disorders	X	
	Personality Traits/Disorders	X	
	Other Mental Illnesses		X
	Antisocial Attitudes		X
	Poor Coping Ability		X
	School Functioning	X	
FAMILY	Parental Characteristics		
	Teenage Pregnancy	X	
	Maternal/ Parental Coping Ability		X
	Parental Antisocial Practices/ Attitudes	X	
	Parental Education and IQ	X	
	Family Dynamics		
	Familial Supports		X
	Family Conflict / Domestic Violence		X
	Family Structure/ Single-Parent Family	X	
	Parent- Child Relationship		
	Ineffective Parenting	X	
	Early Caregiver Disruption	X	
	Parent/Child Attachment		X
INTERVENTIONS	Previous Interventions		X
	Accessibility to Interventions		X
	Familial Responsivity to Intervention		X
	Child/Youth Responsivity to Intervention		X
EXTERNALIZING BEHAVIOR	General Behavioral Problems	X	
	Violence/Aggression	X	
	General Offending	X	
	Substance Use	X	

1. Proposed Structure

Throughout the youth violence literature, a standard set of domains for grouping risk factors has emerged. Although there is some variation across reviews (Herrenkohl et al., 2000; Reppucci et al., this book; U.S. Department of Health and Human Services, 2001), in general, predictors of youth violence have been grouped based on environmental, individual, and family criteria. Given that this division of factors is both theoretically and conceptually straightforward, the majority of the factors included in this review have been grouped according to these three domains.

Simply put, the environmental domain includes social and early medical risk factors that are external to the individual. The individual domain pertains to biological, psychological, and functional attributes of the person. The family domain pertains to risk factors specific to family dynamics. Factors within the family domain have received considerable attention by researchers, yet the exact nature of the influence of the family on youth violence and related antisocial behaviors still remains somewhat convoluted. One of the difficulties in detangling the influence of family factors is the separation of the effects of inherited traits and biological pre-dispositions from the effects of related social learning processes. For this reason, we define the "family" as the caregiving family (a dynamic variable in many cases) throughout the family domain, and attempted to incorporate risk factors seen as being genetically influenced into the biological section of the individual domain. A more unique dimension of this conceptual framework is the inclusion of our last two domains, namely interventions and outcome behaviors. The fourth domain takes into account the potential for various types of interventions to change the developmental course toward youth violence and related antisocial activities. In addition, this domain provides a guide for professionals to plan case management and intervention strategies. The final domain, externalizing behaviors, includes factors that may ordinarily be classified under the individual domain. We made the decision, however, to create a separate domain for these factors given they both represent the behaviors that this review is targeting, and have been consistently identified as the strongest predictors of serious and violent offending in adulthood. In other words, the behaviors included within this domain can often be classified as both outcomes, in terms of the type of behavior that we are attempting to identify, treat, and predict; as well as significant predictors for future violence at the next developmental stage.

2. Environmental Domain

"Children who are unfortunate enough to have both biological and social deficits are theorized to be at the highest risk for persistent antisocial behavior" (Brennan & Mednick, 1997, p. 269). The environmental domain includes biological and social risk factors that are a product of the individual's environment. According to Moffitt's (1993a) developmental biosocial theory, medical risk factors that occur early in the lifespan are theorized to have the greatest impact on serious antisocial outcomes. Since researchers have begun to scrutinize this theory, they have discovered that medical complications during the early stages of life are more likely to have detrimental affects on an individual's later antisocial behavior when combined with other environmental deficits. For example, Hodgins, Kratzer, and McNeil (this book) noted that although some studies have found an association between obstetrical complications (a biological risk factor) and later offending, this correlation largely has been the result of an interaction between these complications and factors such as ineffective parenting and socially disadvantaged environments.

The environmental risk factors included in this section have been selected based on either their empirical association with antisocial behavior and violent offending in adolescence or their clinical impact on antisocial youth. Although it is difficult to disentangle the interactive effect of environmental and individual level contributors on youth violence, there are, nonetheless, a number of environmental factors that have demonstrated an independent relationship with violent behavior among youth. The environmental factors have been incorporated under the following four sections: I) Prenatal/Perinatal Complications, II) Living Conditions, III) Peer Influences, and IIII) School.

2.1 Prenatal/Perinatal Complications

Pre/Perinatal complications involve fixed risk factors that are hypothesized to lead to disorders by causing brain damage (Brennan & Mednick, 1997). Brennan and Mednick summarized the findings from studies investigating pre and perinatal insults into the three following points. First, they have additive negative effects, as children with more insults during these stages had the more deviant outcomes. Second, most infants who suffered perinatal complications did not exhibit deviant outcomes. Third, those children who did exhibit antisocial outcomes following these complications were more likely to also have experienced adverse environmental circumstances. Thus, it appears the interaction of medical and social risk factors has a synergistic effect. The specific manifestations of these medical risks (e.g., damage to the central nervous system, birth deficiencies) are addressed under the individual domain.

Obstetrical Complications. Obstetrical complications are theorized to affect offending behavior by causing damage to areas of the fetal brain associated with impulsivity (Litt, 1972). This specific link has not been tested, however, there is evidence that neurological dysfunction is noticeable through middle childhood following perinatal complications (Hertzig, 1982).

For the most part, investigations into the relationship between obstetrical complications and serious and violent offending have produced mixed findings, largely as a result of methodological issues (see Hodgins et al., this book, for a review). The Danish Perinatal Project (Raine, Brennan, & Mednick, 1994) found that the presence of delivery complications, in combination with early childhood rejection (defined as public institutional care of the infant, an attempt to abort the fetus, and an unwanted pregnancy), was predictive of violent criminal offending at ages 17-19. It should be noted, however, that males with birth complications alone were not significantly different in their rate of violent offending than males without birth complications or rejection. In 1997, these researchers extended this project to age 34 and found the interaction was predictive of violent crime only in males whose first violent offense was prior to age 18. Few studies have differentiated and compared pregnancy (prenatal) complications from delivery (perinatal) complications. Hodgins et al. (this book) found pregnancy complications, as opposed to delivery or neonatal complications, combined with poor parenting in the early years of life, slightly increased the risk of offending and more than doubled the risk of violent offending. Conversely, Kandel and Mednick (1991) reported the opposite to be the case.

Eventual scoring of this factor will be intentionally over-inclusive given that we still know little about which type of complication, at which stage, creates the greatest impact on later serious and violent offending. Thus, obstetrical complications are defined broadly in accordance with McNeil (1988) to include deviations from the normal course of events during *pregnancy*, *delivery*, and the early *neonatal* period. Examples of obstetrical complications are poor maternal physical health, forceps extraction, breech delivery, and preeclampsia.

Maternal Substance Use During Pregnancy. Today, the neurological damage caused by prenatal exposure to substances such as alcohol (see Conry & Fast, 2000, for a review) and crack, and its link to increased susceptibility to criminal offending is well known.

However, more recent research has pointed to the significant, adverse effects of exposure to cigarette smoke. As summarized in Hodgins et al. (this book), studies have found a strong association between smoking and offending even after controlling for a number of spurious factors. "Maternal smoking during pregnancy specifically increases the risk of early onset and persistent offending which includes violence, among males" (Hodgins et al., this book). Conversely, Gibson and Tibbetts (1998) found the interaction between maternal smoking and low Apgar scores significantly predicted later offending in their sample of 832 African-American boys; however, in isolation, neither factor was strongly related to subsequent offending. Then again, Gibson and Tibbetts considered mothers to be "smokers" even if they had smoked only one cigarette during pregnancy.

As common sense may dictate, women who smoke while pregnant are more likely than women who do not smoke while pregnant to be characterized by other factors related to future antisocial behavior in their offspring (see Hodgins et al., this book, for a summary). For example, they are more likely to be young, use alcohol and drugs during pregnancy, and have an antisocial male partner. Each risk factor that is present in addition to maternal cigarette use increases the child's risk for persistent offending. For example, smoking during pregnancy was found in Rasanen et al.'s (1999) sample to increase the risk of criminality by 2.4 times. However, when the mother was also younger than 20, the odds of subsequent offending in the offspring increased to 10.8 times.

2.2 Living Conditions

Exposure to Toxins. Environmental toxins most commonly identified as having a negative affect on the fetal brain have been lead, cadmium, and manganese. Toxic levels of minerals, such as lead, may be related to neurological dysfunction. Despite this suspicion, little research has addressed the relationship between early exposure to these toxins and subsequent violence (Hodgins et al., this book). Several studies have discovered a correlation between low-level lead toxicity and other risk factors for violence and/or offending in children and adolescents; such as low IQ scores and poor school achievement (Needleman et al., 1979). In a more recent prospective study, bone lead levels, measured in males at ages 7 and 11, were found to be related to aggression, delinquency, and somatic complaints at age 11 (Needleman et al., 1996). Another study found high levels of lead and cadmium predicted "acting-out" behavior in 80 elementary school children (Marlowe et al., 1985). These findings have been consistent despite controls for socioeconomic and parental factors. Further, this association has been tested in many stages of development. One longitudinal study found umbilical cord blood lead levels to be related to IQ scores at age 2, and blood lead levels at age 2 to be related to IQ at age 10 (Bellinger, Leviton, Waternaux, Needleman, & Rabinowitz, 1987).

Brennan and Mednick (1997) noted that few studies have investigated the relationship between mineral toxicity and adult antisocial behavior. Moreover, other than a few examples (e.g., Needleman et al., 1996; Needleman, Schell, Bellinger, Leviton, & Allred, 1990), studies examining the relationship between lead, delinquency and other risk factors have been cross-sectional. Therefore, we know little about the stability of the ostensible impact of mineral toxicity on antisocial behavior.

Community Disorganization. The influence of community disorganization on involvement in violent behavior varies significantly across developmental periods. Specifically, living in a disorganized area has its largest correlation with violence during adolescence, while it tends to operate indirectly through its influence on family variables during earlier developmental stages (US Department of Health and Human Services, 2001).

During adolescence, measures of community disorganization, such as the presence of neighborhood violence (Durant et. al, 1994; Singer et al., 1995; Widom, 1989);

accessibility to firearms and drugs (Maguin et al., 1995); the number of neighborhood adults who are involved in antisocial behaviors and violence (Elliot, 1989; Maguin et al., 1995); high levels of social disorganization (Osgood & Chambers, 2000; Sampson & Lauritsen, 2994); exposure to prejudice (McCord & Ensminger, 1995); and, the absence of structured activities (Bursik & Grasmick, 1993) have all been related to increased involvement of youth in violence (Brewer, Hawkins, Catalano, & Neckermen, 1995). The eventual scoring of this item will consider indicators of the overall quality of the neighborhood and community, such as the availability of drugs and firearms in the neighborhood, neighborhood adults involved in crime, gang activity, and an absence or lack of prosocial, structured, and supervised activities.

Family Socio-economic Status (SES). The existence of a relationship between poverty and youth violence has been well established (Elliott, Huzinga, & Mernard, 1989; Farrington, 1989). However, family SES remains a complicated variable to study and quantify due to its interaction with other environmental and family factors. As Reppucci et al. (this book) note, the effect that poverty has on later violence is highly dependent on the influence of a multitude of other variables. For example, family SES has been related to measures of community disorganization and exposure to violence (US Department of Health and Human Services, 2001), and is believed to be strongly related to other negative parental and family outcomes, such as family break up and conflict, parental stress, and damaged parent-child relations (including abuse and neglect). Finally, research has demonstrated that the potential effects of poverty are amplified when combined with high levels of community disorganization (Sampson & Lauritsen, 1994), namely prejudice and structural inequalities (Hill et al., 1994).

Given the interactive nature of family SES with other key risk factors for violent offending, there is likely substantial overlap between SES and community disorganization. The separation of family SES and community disorganization will allow for an examination of the unique contribution of each item. On a cautionary note, although it may be tempting to combine other family factors with measures of family SES, it is still advisable to keep these items separate given the overwhelming evidence that they continue to operate as distinct factors (Rutter, Giller, & Hagell, 1998). The eventual scoring of this item will also consider the amount of financial strain on the caregiving family. Example indicators may include low parental income, low status job(s), and the use of welfare programs by the family.

Residential Mobility. Although longitudinal research has demonstrated a relatively small relationship between residential mobility and later violent offending (Manguin et al., 1995), cohort studies of juvenile offenders tend to report high levels of residential mobility among their samples (Corrado, Odgers, & Cohen, 2001), and instability with respect to residential placements continues to operate as an important case management consideration when dealing with high risk and violent youth. High levels of residential mobility also may influence other variables such as school functioning, peer acceptance, and the level of family strain, particularly if the moves are the result of a negative event (i.e., parental job loss, presence of domestic abuse) or if the youth is being moved around frequently in the absence of a support network. The number of residential moves may also serve as a proxy indicator for the presence of a maladaptive environment, poor family dynamics, and/ or negative behaviors being exhibited by the youth. Therefore, the influence of this factor will not be represented solely by the number of transitions made by the child or youth.

It should also be noted that, while the number of residential moves has been found to only exert a short-term effect on behavior (Manguin et al., 1995), other dimensions of the event, such as the age of the child and the type of precipitating factors that lead up to the child leaving the home, have demonstrated more lasting effects. For example, the earlier the age at which a child leaves home the stronger the prediction of later self report (Farrington,

1989) and official violence (Farrington, 1989; McCord & Ensminger, 1995; Wadsworth, 1978). The eventual scoring of residential mobility will consider the overall stability of the individual's living situation. For example, movement of the youth between several foster placements and/or institutions, versus movement from community to community with the family will create higher stress for the child/youth.

Exposure to Violence. Research has shown that exposure to violence in the infancy stage has been associated with a variety of emotional and behavioral problems, such as excessive irritability, emotional distress, and delayed language development (Gabarino, 1997). Research has also linked exposure to violence in middle childhood and adolescence with involvement in aggressive behavior (Paschall, 1996).

With respect to exposure to violence in the media, reports suggest that children and adolescents who have been exposed to the greatest amount of media violence are more likely to engage in later aggressive and delinquent behavior (Osofsky, 1999). However, research in this area has been methodologically limited. Specifically, the majority has focused solely on the effects of short-term media exposure and has been largely conducted within laboratory settings; while the effects of alternative forms of media, such as video and computer games, remains largely unknown. Despite the limitations of the research to date, the general consensus is that exposure to violent forms of media exerts a low to moderate effect on the involvement of children and youth in physical forms of aggression and violence (Reiss & Roth, 1993).

The eventual scoring of this item will take into account the overall amount of consumption/exposure to violence outside of the individual's home (i.e., in school or in the community). Variables such as exposure to media violence and prejudice, as well as indicators of the level of conflict within the school and neighborhood, also will be included.

2.3 Peers

Peer Socialization. The influence of peers on antisocial and violent behavior varies considerably across developmental stages. Specifically, the presence of antisocial peers constitutes the highest risk factor during the adolescence phase. Nonetheless, having weak social ties to conventional peers and associating with antisocial or delinquent children in early and middle childhood appears to have a small impact on aggressive behavior in childhood (Farrington, 1989). As common sense would dictate, peer influence increases as the child begins school and begins to have prolonged exposure to other children.

Peer rejection has also been implicated in the later expression of aggression among children (see US Department of Health, 2001, for a review). For example, children and adolescents who report having weak social ties and experiencing rejection by conventional peers have been found to be at an elevated risk for violence (Farrington, 1989; Lipsey & Derzon, 1998; Saner & Ellickson, 1996). The fact that a large proportion of crimes are committed with peers in adolescence further reinforces the need to examine the influence of peers (Zimring, 1998). Finally, gang association is a dimension of peer socialization that has demonstrated an independent effect on violence in adolescence (Battin, Hill, Abbot, Catalano, & Hawkins, 1997; Thornberry, 1998). This item will eventually be rated in accordance with the level of exposure and/or bonding to antisocial or delinquent peers, as well as aspects of unpopularity and/or victimization.

2.4 School

School Environment. Research suggests that certain schools foster a culture of violence (Lorion, 1998) and that students who attend these schools are at an elevated risk for both victimization and association with delinquent peers involved in serious and violent offending. For example, Farrington (1989) has reported that boys who attended schools with high delinquency rates (at age 11) reported more violent behavior in adolescence. The absence of structured and supervised activities within the school also contributes to the overall level of attachment the youth has to prosocial activities and institutions. Although studies have provided mixed support for the importance of bonding to school (Elliott, 1994), in general, the development of a positive bond to school has been viewed as a protective factor against involvement in youth crime (Catalano & Hawkins, 1996; Hirschi, 1969).

Another important dimension of the school environment is the availability of resources and community support. Not surprisingly, schools within disorganized communities may suffer from an absence of community supports and resources. Schools located in socially disorganized areas tend to have higher rates of violence (Laub & Lauristen, 1998), although this is not always the case. Therefore, the level of disorganization within the school should not be inferred solely from characteristics of the broader community. Instead, the eventual scoring of this item will attempt to make an independent assessment of the overall quality of the school environment based on indicators such as; the availability of resources and structured activities, adequacy of supervision (teacher to student ratios), and the level of community commitment to the school and its activities.

3. Individual Domain

Both biological and psychological characteristics of children and adolescents interact to increase vulnerability to negative social and environmental influences. Alternatively, as is theorized to be the case in some adult psychopaths, in extreme cases biological and psychological characteristics may transcend positive social and environmental influences, resulting in persistent antisocial behavior (Hare, 1993). In the individual domain, we refer to individual differences among children and adolescents that interact with environmental risk factors to determine later antisocial outcomes.

There is good evidence for a biological predisposition toward antisocial behavior and violent offending. It is possible that neurological damage, a biological risk marker, is also responsible for the psychological characteristics covered in this domain. Raine (1997), for example, theorized that damage to the prefrontal cortex results in cognitive, personality, and arousal deficits, which in turn create aggressive behavior. Nonetheless, there remains substantial difficulty in detangling the nature/nurture controversy. As will be apparent in this review, psycho-biological research continues to result in inconsistent findings in many areas. As Linnoila (1997) suggested, this is likely due to the study of the brain and behavior being hampered by the "relative crudity of the available tools" (p. 336) and other methodological limitations. Thus, to the reader, it may appear that there is considerable overlap in the risk factors discussed within this domain. Despite potential overlap, given current universal limitations in our knowledge base, we consider the factors delineated here to represent different pathways to offending and violence. It is the hope that this framework will provide a uniform structure for studying biological components and their interactions with psychological, environmental, and social factors. The individual characteristics domain is categorized into the following three sections: Biological, Psychological, and Functional.

3.1 Biological

This section pertains to objective and inferred biological deficiencies. In contrast to biological risk factors covered in the environmental section, here we refer to the phenotypic or potential genotypic expression of these risk markers.

Birth Deficiencies. Birth deficiencies refer to physical abnormalities that appear in the perinatal stage. This factor is differentiated from obstetrical complications given that complications do not always result in tangible deficiencies. This item instead calls for an objective examination of the infant's physical abnormalities. Low birth weight, for example, has been implicated in subsequent neurological dysfunction (e.g., Fitzhardinge & Steven, 1972; McCormick, 1985) that may be related to later disruptive behaviors. Alternatively, some studies have indicated that a higher than average body size at birth is associated with violence and aggression. Hodgins et al. (this book) for example, found that their group of male offenders who started offending prior to age 18 were heavier than average at birth. Others have found that a high body weight prior to age 1 was associated with violent offending as an adult (Rasanen, Hakko, Jarvelin, & Tiihonen, 1999).

Minor physical anomalies (MPAs), small aberrations in external physical characteristics, are other deficiencies noticeable at birth that have been linked to violent criminal arrests (e.g., Kandel, Brennan, Mednick, & Michelsen, 1988) and attentional and hyperactive problems (e.g., Firestone & Peters, 1983). The presence of many MPAs is thought to represent neurological impairments that may predispose individuals to behavioral problems. Eventual scoring of this item will include deficiencies appearing at birth such as anoxia, low apgar scores, abnormally low or high birth weight, and disabilities or chronic illnesses identified during the perinatal stage. MPAs; including low-seated ears, furrowed tongues, long third toes, and single transverse palmar creases also will be considered.

Parental History of Mental Illness. The heritability of antisocial behaviors and criminality has been well documented, though the exact genetic epidemiology is less understood (see Carey & Goldman, 1997, for a review). Given the contributions of twin and adoption studies, there is sound justification for a genetic influence on behavior, accounting for approximately 40% of the variance in behavior (Plomin, Owen, & McGuffin, 1994). Antisocial and criminal behaviors are no exception (see Carey & Goldman, 1997). In a study of 5,182 adult male adoptees, for example, Moffitt (1987) discovered that 15.09% had criminal conviction records. For those adoptees with at least one biological parent who had a history of psychiatric hospitalization, the conviction rate rose to 19.25% compared to 14.45% of those without a hospitalized biological parent. This association was particularly strong when the parent was hospitalized for a personality or substance abuse disorder, as opposed to other mental illnesses. Parental mental illness appears to have an additive effect with parental criminal history. The 1% of the adoption sample that had a history of both parental criminality and parental hospitalization accounted for 12.2% of the total sample's convictions.

Carey and Goldman (1997) noted, studies examining the heritability of violent behavior have been less consistent than those examining broader antisocial behaviors (e.g., convictions, self-reported delinquency, hostility scales); but this is largely due to the very limited number of studies addressing violence specifically. Eventual scoring of this item will require an examination of the biological parents' history. Mental illness is defined broadly; including major mental illnesses, personality disorders (most importantly antisocial personality disorder), and substance abuse disorders.

Executive Dysfunction. Deficiencies in the brain's self-control functions have been linked to juvenile delinquency and adult criminal behavior (see Henry & Moffitt, 1997; Pontius, this book, for reviews). Self-control, or "executive" functions include sustaining attention and concentration, abstract reasoning, concept formation, planning, goal-setting, initiation of purposive sequences of behavior, self-monitoring, and self-awareness. Consequently, the individual with serious executive dysfunctions is expected to be impulsive and hyperactive. Executive functions are governed by the brain's frontal lobe (e.g., Pontius, 1972). Locations of frontal deficits are important, as Raine and Venebles (1992) argued that location dictates the nature of the manifested psychopathology. For example, impairments in the orbitofrontal region result in impulsive behaviors (e.g., Roussey & Toupin, 2000). The etiology of frontal lobe deficits is not always conclusive, but could be the result of disruptions to the fetal brain in the pre/perinatal stages caused by maternal drug abuse during pregnancy, obstetrical complications, birth deficiencies, etc. (Moffitt, 1993a). Deficits may also be the result of a disturbance in the maturation process of the prefrontal cortex, which extends beyond puberty (Pontius, 1972).

As Henry and Moffitt (1997) reviewed, the association between executive dysfunction and disruptive behaviors has been tested in incarcerated adolescents, inner-city boys, and non-criminal adolescents; with some of the most prominent dysfunction being evident in adolescent boys exhibiting both CD and ADHD symptoms (e.g., Moffitt & Silva, 1988). A number of outcomes of executive dysfunctions have been identified, including aggression, impulsivity, early onset Conduct Disorder, and persistent violence and criminality in adults.

Two complimentary but unique measures of executive dysfunction are typically found in the research investigating antisocial behavior. Neuropsychological tests (e.g., Wisconsin Card Sort Test, Porteus Maze Test, Trails B) provide information as to how efficiently the brain controls behavior. Alternatively, neuroimaging techniques (e.g., magnetic resonance imaging, positron emission tomography) provide information as to the location of the brain's structural or functional abnormality, but have been used only with adults thus far.

Chronic Underarousal. Psychophysiological differences between offenders, psychopaths, and noncriminals have been well documented over the last 50 years (e.g., Ellingson, 1954; Lykken, 1957). Generally, studies have indicated that offenders have abnormal electroencephalograms (EEGs), and lower resting skin conductance levels and heart rates than other individuals (see Raine, 1997, for a review). One of the more influential explanations of these psychophysiological differences is that antisocial individuals are *chronically underaroused*. Raine (1997) summarized three theoretical interpretations of the underarousal found in antisocial people. First, the fearlessness theory dictates that low levels of arousal are indicators of low levels of fear. Individuals lacking fear would be unlikely to inhibit antisocial and violent behavior. Second, the stimulation-seeking theory (Eysenck, 1964) proposes that underaroused individuals seek stimulation (e.g., drug use, criminal activity, promiscuity, and risk-taking activities) to increase their arousal to more normal levels. Finally, Raine (1997) suggested "underarousal may predispose to a disinhibited temperament that in itself may act as one early predispositional factor for criminal and violent behavior" (p. 291). As Raine concluded, disinhibited temperaments seem to be relatively stable across time and may be identified, in part, by lower heart rates and skin conductance levels.

Research on the psychophysiological basis of antisocial behavior has not been conducted as extensively in children and adolescents as in adult samples. Nonetheless, a few prospective studies have uncovered associations analogous to the adult literature, with one finding EEGs predicted adult thievery in children from as young as 1 year of age (Peterson, Matousek, Mednick, Volavka, & Polloch, 1982). Another prospective study, using three measures of psychophysiological arousal on 15 year-old normal boys (Raine, 1990), found all three measures independently predicted criminal behavior at age 24. Measures of arousal correctly classified 74.7% of subjects as criminal or noncriminal. Similarly, evidence of high autonomic activity in adolescent offenders who desist from crime suggests that high arousal may in fact act as a protective factor (Raine, 1997).

The most prevalent measures of psychophysiological arousal have been heart rate, skin conductance, and EEGs. Current research as suggested the use of the startle blink reflex as another indicator (Patrick, Bradley, & Lang, 1993). Ideally, ratings of this item should be based on more than one psychophysiological measure.

Abnormal Biochemical Activity. The link between neurotransmitters and several aspects of behavior is well known, yet still not wholly understood. Serotonin, dopamine, and norepinephrine in particular, have been implicated in the inhibition (or disinhibition) of impulsivity, hyperactivity, and aggression (e.g., af Klinteberg, in press; Berman, Kavoussi, & Coccaro, 1997); a few of the major disruptive behavior problems linked to later serious and violent offending.

Findings pertaining to the association between neurotransmitters and disruptive behaviors and aggression have been mixed; likely due to different methods of investigation, measurement, and sampling. Low levels of serotonin are thought to predispose individuals to engage in normally suppressed behaviors (Depue & Spoont, 1986). For example, low levels of serotonin (as measured by monoamine oxidase activity in blood) have been linked to motor disinhibition (responses indicative of impulsivity) in normal adolescents (af Klinteberg et al., 1991) and psychopathy and aggression in adult criminals (Belfrage, Lidberg, & Oreland, 1992). Generally, serotonin activity has been linked, both experimentally and nonexperimentally, to impulsive violence in humans (see Berman et al., 1997, for a review). Dopamine also appears to have an inverse relationship with impulsivity and aggression (i.e., recidivistic violence), but, according to Berman et al. (1997), this may be due to its positive correlation with serotonin. Blood sample measures of norephinephrine functioning in children have indicated a negative relationship between norephinephrine activity and conduct disorder symptoms (Bowden, Deutsch, & Swanson, 1987). Conversely, this relationship appears to be opposite in adults, though this may be due to methodological differences in the type of measurement taken.

As summarized by Berman et al. (1997), studies investigating the link between biochemicals and human aggression have used one of three methods to measure neurotransmitter activity: 1) central neurochemical measures (e.g., metabolites in cerebrospinal fluid), 2) pharmacochallenge measures (e.g., observation of hormonal or behavioral responses to drugs), and 3) peripheral measures (e.g., blood or urine samples). Berman et al. suggested that peripheral measures are the optimal approach for children given they are less invasive; however, such measures also have produced the most conflicting results. Neurotransmitter activity, being influenced by environmental factors and early experiences, is not static. It is possible the association with aggression will differ as a product of the developmental stage, sex, and presence of psychiatric disorders in the individual.

3.2 Psychological

Cognitive Delays/Disorders. Moffitt (1993b) proposed that neuropsychological deficits (such as language deficits) contribute directly and indirectly to antisocial behavior, mainly in life-course persistent offenders. Like executive dysfunction, cognitive disorders are the result of cerebral impairment. Unlike executive dysfunctions, however, cognitive disorders are governed by the left cerebral hemisphere (see Benson & Zaidel, 1985). Thus, cognitive deficits are a separate pathway to antisocial behavior, either as a direct result of the brain's impairment, or as an indirect result of the child's social and functional abilities.

Cognitive disorders include delayed achievement of developmental milestones, language deficits, learning disabilities, below average IQ, and attention deficits (ADD/ADHD). Low IQ is a robust factor in the development of antisocial behavior and delinquency when assessed prospectively (e.g., Lynam, Moffitt, & Stouthamer-Loeber, 1993). Verbal deficits in particular are also strongly associated with self-reported delinquency (see Henry & Moffitt, 1997; Moffitt, 1993b, for reviews). In both cases, strong correlations hold despite controls for other related factors, such as SES and race. Not surprisingly, language deficits have a relatively high comorbidity with ADHD. Signs of these deficits can emerge in the toddler stage and comorbid cases are highly likely to have histories of physical aggression that remain stable from age 3 to 15 (Moffitt, 1990). The importance of ADD/ADHD in relation to aggression and conduct problems can not go unnoticed (see Loeber, 1990). Moffitt and Silva (1988), for example, found that near 60% of individuals with a childhood diagnosis of ADD were delinquent by age 13. Eventual scoring of this item will require appropriate neuropsychological and diagnostic assessments. Individuals with diagnostic learning impairments and ADHD will be rated with the highest severity.

Personality Traits/Disorders. Personality traits and disorders that contain the main components of anger, impulsivity, hostility, and a shallow affect or lack of empathy predispose individuals to violence and criminal behavior (Hare & Hart, 1993). This is particularly evident in adults with personality disorder diagnoses; such as antisocial, borderline, and histrionic personalities (see Kropp, Hart, Webster, & Eaves, 1995; Webster et al., 1997). Though there are cautions when diagnosing individuals prior to age 18 (see American Psychiatric Association, 1994), epidemiological studies imply that traits of antisocial personality disorder can be identified as young as age 8 (Robins, 1978). Similarly, as reviewed by Vincent and Hart (this book), there is good evidence that psychopathy-related traits, another significant predictor of violence and criminality, can be identified in young children and adolescents.

In the infancy stage, we focus our attention instead on difficult temperaments (i.e., fussiness, irritability). The influence of temperament on later aggression and behavior problems may be either direct, through its link to later oppositional personalities (Graham, Rutter, & George, 1973), or indirect, through its effects on the attitudes of caregivers (Bates, 1980).

Other Mental Illnesses. Mental illness presents a moderate risk-inflating factor for violence in adults (Monahan, 1992). In a large birth cohort, Hodgins (1992) determined that men with a major mental disorder, after controlling for other related factors, were 2 ½ times more likely than men with no disorder to have committed a criminal offense. This association was twice as strong in women.

It is unlikely that the link between mental illnesses and violence in children and adolescents deviate dramatically from findings in adult populations. Though most mental illnesses that impact affect, cognition, and behavior have an onset in late adolescence or early adulthood, they carry the same symptom expression in childhood cases (American Psychiatric Association, 1994). Aggression and impulsivity are common manifestations of organic illnesses, particularly personality changes due to head injuries, which may occur at any developmental stage (American Psychiatric Association, 1994). Anxiety disorders, many of which commonly begin in childhood, have a fairly high comorbidity with conduct problems in children (Frick, Lilienfeld, Ellis, Loney, & Silverthorn, 1999). Eventual scoring of this item will require consideration of both the severity of the mental illness in regards to violence, and the presence of symptoms versus a diagnosis.

Antisocial Attitudes. In the adult violence literature, Andrews and Bonta (1995) noted that procriminal attitudes relate to criminal and violent behavior. In regards to the attitudes of children, Dodge and Schwartz (1997) purport that patterns of processing social stimuli across situations have been strongly predictive of aggressive behavior. Further, they concluded the manner with which an individual processes information mediates the effects of their experiences (e.g., physical abuse) on later conduct problems. Dishonesty, deviant attitudes, and permissiveness toward violence in early childhood and adolescence are some of the attitudinal factors that predict self-reported violence in adulthood (see Hawkins et al.,

1998, for a review).

Here we refer to procriminal or deviant attitudes that have a likelihood of eventuating in serious or violent offending. This includes such attitudes and beliefs as negative opinions of authority, untrustworthiness and dishonesty, impaired moral reasoning, failure to accept responsibility for own actions, and acceptance of interpersonal violence. Given there is not a one-to-one correlation between attitude and behavior, the rating of this item will require an exploration of the child or adolescent's thought processes and beliefs as opposed to behavioral observations.

Poor Coping Ability. Augimeri et al. (2001) noted the importance of a child's coping ability on antisocial outcomes, describing it as a measure of resiliency. In any given developmental study, there is a significant proportion of individuals who encounter adverse circumstances (i.e., impoverished environments, parental neglect) yet, do not develop disruptive nor self-harming behaviors (e.g., Werner & Smith, 1982; Widom, 1997). Though some have focused on external protective factors to account for this disparity (e.g., Garmezy, 1993), others have highlighted individual differences in coping strategies, which may be seen as dynamic adaptive processes that interact with the particular situation (Rutter, 1987).

We define poor coping ability broadly as maladaptive methods of handling stress in the home, school, and community. Maladaptive coping mechanisms are not limited to aggression and antisocial practices, but instead include depression, anxiety, drug use, and self-harming behaviors. Alternatively, individual attributes identified in children characterized as resilient include high self-esteem and self-efficacy, good problem-solving ability, self-discipline, high maturity, compassion, and adaptability (Born, Chevalier, & Humblet, 1997; Rutter, 1985). Given coping abilities may be context-specific, eventual scoring of this item will also require an account of the amount of stress present in the individual's life.

3.3 Functional

School Functioning. Research has consistently pointed to poor school functioning and academic failure as important risk factors in the development of youth violence and antisocial conduct (Denno, 1990; Lipsey & Derzon, 1998; Maguin & Loeber, 1996). Results from longitudinal studies indicate that youth with high truancy rates were more likely to engage in violence as adolescents and adults (Farrington, 1989). Manguin et al. (1995) have reported an association between the number of school transitions experienced by youth and later involvement in violence. Dropping out of school in early/mid adolescence has also been shown to be predictive of later violence (Farrington, 1989), while the formation of a positive bond to school has been identified as a protective factor against crime (Catalano & Hawkins, 1996; Hirschi, 1969). Finally, there is some evidence to suggest that individuals who demonstrate extremely high functioning in school (who have either begun school early and/ or skipped grades) may also be at an elevated risk for certain types of maladjustment and antisocial behavior.

The eventual scoring of this item will include measures of poor school achievement and performance, and low commitment and bonding to school.

4. Family Domain

Overall, family factors have received the majority of the attention in juvenile delinquency and youth violence research. They are also generally recognized as having the largest impact on the involvement of youth in aggressive and violent behavior (McGuire,

1997). Violent offenders tend to come from families that are characterized by stressful family environments, antisocial and poorly educated parents, and disruptive familial relationships.

Although the majority of family factors have not demonstrated large effect sizes with respect to self report and official violence (Hawkins et al., 1998; Lipsey & Derzon, 1998), there is strong theoretical support for weighting family factors heavily in case management and intervention decisions. As mentioned in the introduction, one of the primary obstacles when conceptualizing the role of family factors on youth violence is the separation of inherited versus learned traits. As such, caution has been exercised in order to ensure that the impact of each factor in the family domain is the product of a socialization or learning process, as opposed to impacting the development of violence through inherited traits and biological predispositions. Factors within the family domain have been grouped under the following three sections: Parental Characteristics, Family Dynamics, and Parent – Child Relationship.

4.1 Parental Characteristics

Teenage Pregnancy. Teenage childbearing has been found to predict many undesirable outcomes for the offspring, including delinquency (Morash & Rucker, 1989) and aggression (Baker & Mednick, 1984). However, this does not appear to be a one-to-one relationship, given mothers who give birth in their teenage years also are characterized by adverse childhood histories and disadvantaged social backgrounds that may put their offspring at risk for later offending, conduct problems, and poor educational outcomes (e.g., Fergusson & Woodward, 1999; Woodward & Fergusson, 1999). Perhaps most importantly, there is an increased likelihood that young mothers will not be equipped to provide adequate care, support, and supervision for their children given the characteristics that tend to be associated with becoming a parent in adolescence (i.e., a lack of financial and emotional supports and underdeveloped and/or inadequate parenting skills). The association between young maternal age and later offending may also be the result of an increased likelihood of being raised in a single-parent family, given the presence of the biological father may act as a protective factor against many adverse outcomes (Morash & Rucker, 1989).

A study by Fergusson and Woodward (1999) added clarity to this debate by examining the ability of teenage pregnancy to predict delinquent behavior in offspring at age 18 after accounting for a number of intervening factors in a large New Zealand birth cohort (the Christchurch Health and Developmental Study). Generally, decreased maternal age was associated with repeat offending and convictions in offspring, as was less nurturant child-rearing practices, higher risk of exposure to childhood physical punishment and family instability, and higher rates of parental adjustment problems. After including these intervening factors, the association between decreased maternal age and repeated offending behavior in offspring dissipated. The authors suggested that the increased risk of delinquency following teenage childbearing is mediated by child-rearing practices and family environment. Others have discovered an additive effect of teenage pregnancy on later serious and violent offending when combining it with other identified risk factors (Rasanen et al., 1999).

Maternal/ Parental Coping Ability. The effects of external stressors on a child's adjustment (for a review see Dubow et al, 1997) can be significantly impacted by the ability of parents or primary caregivers to exercise effective problem solving and coping strategies (Hawkins et al., 1998). Parental coping ability relates to the capacity of the biological mother and, in the later stages of development, other primary caregivers, to successfully deal with life stressors and child care issues. The scoring of this item will take into account the role of factors such as, parental mental and/or physical illness, unemployment, employment stress, level of social competence and problem solving abilities, and the caregivers overall ability to deal with stress. In general, the assessment of this item will be made based on the ability of the parent or caregiver to effectively navigate through daily living circumstances and obstacles. Although it is important to assess the absolute level of stress that the parent is exposed to, the most important aspect of this item is the parent's ability to respond to, and effectively function within, their environment.

Parental Antisocial Practices/ Attitudes. Studies suggest that parents who hold antisocial attitudes and engage in antisocial practices represent an environmental risk factor (Moffitt, 1997). That is, the influence of a parent's antisocial attitudes and beliefs on their child's behavior is primarily the result of how they *learn* violent behavior through observation, as opposed to being an inherited trait.

Parental criminality is one of the primary antisocial practices that has been related to later involvement in serious and violent offending. For example, a Danish study reported that men between the ages of 18-23 with criminal fathers were 3.8 times more likely to be involved in violence relative to a sample of men with noncriminal fathers (Baker & Mednick, 1984). The Cambridge study indicated that boys whose parents had been arrested prior to their 10th birthday were 2.2 times more likely to be involved in violence as compared to boys with noncriminal parents (Farrington, 1989). Other antisocial practices such as parental alcoholism and mental illness have demonstrated mixed results in studies examining their link to later violence (McCord, 1979; Moffitt, 1987).

There are only a limited number of studies that have investigated the link between parental antisocial attitudes and violent behavior among children (Hawkins et al., 1998). One of these studies, however, has provided relatively strong support for the predictive utility of parental tolerance towards violence at age 10 and self report violence at age 18 (Manguin, 1995). There is also a small body of research that has linked other adolescent health and behavior problems, such as substance use, with parental antisocial attitudes (Peterson, Hawkings, Abbott, & Catalano, 1994; US Department of Health, 2001). The eventual scoring of this item will consider the attitudes and behaviors of parents that have been related to aggression and criminality. For example, indicators will include measures of: parental criminality (official and unofficial), antisocial behavior (substance use, aggression, etc), and favorable attitudes toward antisocial practices.

Parental Education and IQ. Parental education is closely related to both the socio-economic status and income level of the family. The general consensus is that low parental IQ and educational level influences the development of violent behavior indirectly through other variables, such as the inability of low IQ parents to assist their children with schoolwork and the increased probability that they will possess poor decision making and coping abilities (US Department of Health and Human Services, 2001).

4.2 Family Dynamics

Familial Supports. Children who exhibit fewer antisocial behaviors, despite the presence of stressors and other risk factors for violence, have been shown to have higher levels of familial support (Dubow, Edwards, & Ippolito, 1997). The presence of familial supports and resources has also been linked to improved parenting styles in socially-isolated mothers (Walhler, 1980).

The rating of this item will be dependent upon the extent to which the supports available to the family constitute a positive influence. These supports may include familial assistance (Werner & Smith, 1992), community resources, such as teachers (Jenkins & Keating, 1998), day care, and mental health resources, as well as other forms of financial, social, and/or emotional assistance. As indicated in the EARL-20B (Augimeri et al., 2001) the most important aspect of this item is the level of perceived social support and relationship satisfaction, as opposed to a tally of the actual number of supports. In addition, it is important to look beyond the number of supports and evaluate the degree to which each source of support positively contributes to the family's overall functioning.

Family Conflict/Presence of Domestic Violence. Children who are exposed to high levels of family conflict and violence have also been shown to be at an increased risk for becoming involved in aggressive and violent acts. For example, Farrington (1989) found that exposure to marital conflict was predictive of later violent convictions between the age of 10 and 32 and of self report violence at the ages of 16-18 and 32. Elliot (1994) also reported finding a strong relationship between exposure to conflict between parents and self-report violence in adolescence. While Manguin et al. (1995) found that family conflict at ages 14 and 16 was predictive of self report violence at the age of 18. Finally, there is evidence that children from violent families show a greater frequency of both internalizing and externalizing behaviors as compared to children raised in nonviolent families (Bell & Jenkins, 1992).

With respect to direct experiences of violence, it is assumed that children who have been abused or neglected are more likely to become violent, although there are surprisingly few studies that have provided strong empirical support for this assertion (Reiss & Roth, 1993). Meta-analyses examining the relationship between the experience of childhood abuse and later violence (see Hawkins, et al., 1998) have reported relatively low correlations. Nonetheless, there is still a general consensus that abuse and neglect experiences contribute to the development of aggressive and violent behaviors. These types of victimization and neglect experiences have also been examined with respect to their influence on mental health and developmental processes that are related more directly to youth violence (Widom, 1989). For instance, abuse and neglect have been strongly associated with mental health problems, substance abuse, and poor school performance (Dembo et. al, 1992; Esbensen & Huizinga, 1991; Silverman, et al., 1996; Smith & Thornberry, 1995) that in turn, have been linked directly with youth violence.

Smith and Thornberry (1995) also reported that adolescents with a history of abuse and neglect were more violent according to self-reported data. This relationship also held when variables such as gender, race, family structure, family mobility, and socio-economic status were controlled for. Notably, these authors also reported that neglect had the highest correlation with later violence when compared to physical and sexual abuse. For research purposes, therefore, investigators may want to differentiate between the types of victimization (i.e., physical abuse, sexual abuse, and neglect), as well as between the witnessing of family conflict (i.e., spousal assault) and being exposed directly to family violence.

Family Structure/ Single-Parent Family. This item refers to the overall structure of the family and the distribution of resources among family members. One of the primary considerations when operationalizing family structure is whether the family has been headed by a single parent. Longitudinal research has indicated that having a two-parent status at age 13 differentiated youth who had a conviction by the age of 18 with those youth who did not have a conviction by this age (Henry, Caspi, Moffitt, & Silva, 1996). A recent study Canadian study (Lipman, Boyle, Dooley, & Offord, 1998) also found evidence that children being raised in a single parent family are at a higher risk for involvement in serious and violent offending behaviors. Similarly, Henry, Avshalom, Moffit, and Silva (1996) reported that being a member of a single parent family at age 13 was predictive of convictions for violence by the age of 18. The eventual scoring of this item will primarily be concerned with the distribution of resources among family members and the presence of a single-family structure.

Sibling Delinquency. The presence of delinquent siblings in middle childhood has been associated with later convictions for violence (Farrington, 1989). Results from Farrington's longitudinal study of London boys demonstrated that sibling delinquency contributes an effect beyond parental criminality to both self report and official offending. Specifically, Farrington found that having a delinquent sibling at age 10 increased the likelihood of being convicted for a violent offence. The influence that siblings have on the development of aggressive and violent behavior has also been found to vary by gender and by age. For example, in William's (1994) study of 193 females and 194 males, found that sibling delinquency at age 14 was more strongly associated with self report violence among girls than boys. Using data from the same prospective study, Maguin et al. (1995) reported that the relationship between sibling delinquency and engagement in violence by the subject was stronger when the measure of sibling delinquency was taken later in development. In other words, the effect of having a delinquent sibling appears to be strongest during the adolescence phase.

4.3 Parent/Child Relationship

Ineffective Parenting. Youth who engage in violence have often been subjected to a wide range of ineffective parenting practices (Reiss & Roth, 1993). Specifically, violent offenders tend to have been subjected to one or more of the following parenting styles: poor levels of supervision, an absence of routine monitoring (Gorman-Smith, Tolan, Zelli, & Huesmann, 1996; Patterson, Ried, & Dishion, 1992), a lack of parental involvement (Williams, 1994), inappropriate parental expectations (Augimeri et al., 2001), early rejection or neglect, inconsistent, lax (Weiss, Dodge, Bates, & Pettit, 1992), harsh (Eron, Huesmann & Zelli, 1991), or permissive methods of parenting (Haapasalo & Trembly, 1994; Wells & Rankin, 1988). It is hypothesized that ineffective parenting is often the result of a reciprocal relationship whereby the interaction between the child and the parent creates a stressed or otherwise dysfunctional relationship (Vuchinich, Bank, & Patterson, 1992).

There is a large body of research that has linked ineffective parenting practices with violent behavior. For example, results from McCord's (1979) 20 year follow up study, demonstrated that poor parental supervision and aggressive discipline predicted crimes against a person well into their forties. One explanation for this relationship is that children and youth who are treated aggressively may be more likely to view this type of behavior as acceptable, whereas those who receive no supervision or inconsistent guidelines may be unaware of the parameters of acceptable behavior (Department of Health and Human Services, 2000). There is also evidence that a lack of parent-child interaction may increase the youth's likelihood for future violence, specifically when parent-child communication and parental involvement is absent during adolescence (Williams, 1994).

With respect to the protective aspect of parenting styles, Deater-Deckard et al. (1998) found that the presence of maternal warmth served to buffer the relationship between harsh discipline and aggression. The extent to which parents engage in effective problem solving and provide positive forms of encouragement are also factors that have been emphasized when evaluating the quality of parenting style (Ramsey & Walker, 1998). The eventual scoring of this item, therefore, will differentiate between the various types of ineffective parenting processes as well as consider the presence of protective aspects of the parenting style.

Early Caregiver Disruption. Early separation from parents has demonstrated predictive utility with respect to later violence (Farrington, 1991; Harris, Rice & Quinsey, 1993; Hawkins et. al., 1998). Specifically, Farrington (1991) reported that children who were separated from a parent prior to the age of 10 were more likely to be involved in violence as compared to children who had not been separated from a parent prior to this age. Separation from parents before the age of 16 has also been shown to predict adult violence (Harris, Rice & Quinsey, 1993).

The precise impact of early caregiver disruption on the development of aggression and violence still remains largely unknown. As common sense would dictate, a number of other problems tend to be present in situations where young children are separated from their parents. For example, parents who abuse substances, are unable to care for their children financially, or possess poor coping strategies and parenting skills may be more likely to be separated from their children. The eventual scoring of this item will take into account the number of changes in caregivers, abandonment by a parent (infancy), and the lack of continuity in caregiving. Another important consideration in rating this item will be the level of support around the child following their separation from their primary caregiver(s).

Parent/Child Attachment. During the infancy stage, both the lack of maternal warmth and the formation of an insecure attachment to the infant's primary caregiver (Erickson, Sroufe, & Egeland, 1985) has been associated with violent behavior within adolescence. Similarly, parental indifference and rejection have been associated with increased levels of aggression and violence in childhood and adolescence (Farrington, 1991). While in childhood and adolescence, a strong attachment to parents, which has been conceptualized as a possible protective factor, has been found to place the individual at an elevated risk for involvement in violence if the parents hold antisocial attitudes and beliefs and/or engage in antisocial activities (Hawkins, et al; 1998). The eventual scoring of this item will take into account both the quality and strength of the parent-child bond, as well as the characteristics of the parental figure that the child is bonding to.

5. Intervention Domain

Intervention is an essential component of the prevention and amelioration of serious and violent juvenile offending. The intent of this intervention domain is to guide the assessor's decisions for planning appropriate interventions for the youth and, where applicable, the youth's family. Here the assessor is also required to forecast how well individuals will respond to proposed intervention plans.

According to Mulvey (1984), assessment of an individual's amenability to treatment for violence requires an understanding of the individual's personal characteristics, the context, and the treatments that are available. Thus, this domain calls for an assessment of the success of interventions already received, interventions currently needed, the availability and quality of those interventions, and the family and individual's likely receptiveness to the proposed intervention plan. The developmental stage of the youth is also important in designing the most effective treatment plans (Reppucci et al., this book). For example, with respect to prevention, the best predictor of adolescent delinquency is early conduct problems. The sooner these symptoms can be ameliorated, the better the outcome will be. By about age 15, although it is easier to identify antisocial youth, they may also become more resistant to change. Therefore, the intervention plan needs to be dynamic. This domain consists of the following four variables: I) Previous Interventions, II) Accessibility to Interventions, III) Familial Responsivity to Intervention, and IV) Child Treatability.

Previous Interventions. This item will require the professional to generate a list of interventions the youth or family has received or attempted previously. The intent of this item is to a) provide an indication of the youth's (or family's) acquired skills, b) aid in decisions regarding the youth's current level of risk (e.g., many successful previous interventions will have diminished the youth's current risk; whereas many unsuccessful, or uncompleted, previous interventions may be a sign of a higher need), and c) identify needs to aid in the development of future intervention and risk management planning.

This item will require an inventory of interventions that have been *attempted and/or completed* prior to the time of the assessment. Interventions for young people at risk for serious and violent offending have been organized in many ways (e.g., Catalano, Arthur, Hawkins, Berglund, & Olson, 1998; Lipsey & Wilson, 1998; Wasserman & Miller, 1998). The categorization provided is by no means exhaustive and is kept general in order to be useful across a variety of contexts and cultures. In addition to the particular intervention undertaken, the assessor should make note of whether the treatment was attempted or completed, and if completed, whether it had a successful impact. The broad interventions are categorized as follows: a) Hospital/medical-Based Interventions, b) Parent and Family-Based Interventions, c) Youth-Based Interventions, d) School-Based Interventions, e) Peer-Based Interventions, f) Criminal Justice Interventions, and g) Community-Based Interventions.

Accessibility to Interventions. This factor involves consideration of the *availability* of interventions, *feasibility* of the proposed intervention plan for the particular individual and targeted risk factors, and the overall *quality* of those interventions that are available. This item calls for forecasting the accessibility and quality of needed interventions at the time of assessment.

Assessors should consider the degree to which the proposed treatment is appropriate for the individual. As Reppucci et al. (this book) suggested, the youth's motivations for violence, and situational antecedents to violence are useful in determining what interventions are likely to be successful in decreasing their violence. For example, if the youth's violence stems largely from exposure to domestic violence in the home, it is unlikely that mere one-on-one psychotherapy with the youth will be effective. Rather, the intervention would need to focus on the whole family. Further, as noted by Wasserman and Miller (1998), "antisocial behavior emerges from the convergence of multiple risk factors; designating any single factor as a target for change is unlikely to be a successful intervention strategy" (p. 246). Assessors should also pay special consideration to the level of integration of services at the community level. If the appropriate service providers lack the requisite resources, or are unwilling to provide resources to the specific individual, interventions are unlikely to be successful.

Familial Responsivity to Intervention. In a recent report of the U. S. Surgeon General, parent skill-training programs and marital and family clinical therapy were recognized to be among the most effective strategies for the prevention and treatment of youth violence, as indicated by a medium to large effect size and sustainable effects (U.S. Department of Health and Human Services, 2001). The noted importance of family treatment is not surprising given that many aspects of the family, such as family conflict and ineffective parenting, are linked to violence and delinquency in children (see the Family Domain). This factor requires a forecasting of how receptive and responsive the caregiving family will be to proposed interventions. Parenting courses and other programs aimed at the family will be an essential component of intervention in many cases. Alternatively, whether the family maintains an active interest in the youth's treatment is also of relevance. To some degree, this assessment will require a review of the family's *past* and *present* commitment to intervention strategies. Indicators of responsivity would include, but are not limited to: encouraging the youth to participate in programs, corresponding with service providers, engaging in treatment plans, willingness to accept their child's problems, and level of disclosure. Possible indicators of a low likelihood of responsiveness or a poor outcome from treatment include parental depression, single-parent status, stress, low socio-economic status (Webster-Stratton & Hammond, 1990), clinical problems, mistrust of the therapist, and low social support (Cunningham & Henggeler, 1999).

Child/Youth Responsivity to Intervention. As mentioned previously, treatment effectiveness is largely dependent on individual characteristics, one of the most important being receptiveness to treatment (Augimeri et al., 2001). Kazdin (1987) suggested that youths may be identified as less amenable to an intervention "on the basis of characteristics of the sample and hypotheses about the interface of treatment and these characteristics" (p. 198). For example, individuals who are functioning well in areas related to antisocial behavior, such as school performance, may be more receptive to treatment than those who are experiencing more academic dysfunction. Kruh and Brodsky (1997) state that determining "goodness of fit" between the youth and possible treatment options is critical to the evaluation of their amenability.

Adams (1970) found that youths who were judged to be amenable to treatment were significantly less likely to recidivate following random assignment into treatment, than those deemed non-amenable. Youth considered to be amenable were those who were more intelligent, verbal, anxious, insightful, aware of their problems, and interested in change. One factor that has been found to have a *negative* relationship with amenability is the severity of the child's antisocial conduct at intake (Kazdin, Mazurick, & Siegel, 1994).

6. Externalizing Behaviors

Specific externalizing behaviors, such as aggression and general offending, at an early age, are the strongest predictors of later serious and violent delinquency (see Lipsey & Derzon, 1998). Although externalizing behaviors are individual factors, and thus, may be appropriate for the youth domain, we decided they were deserving of their own category because (a) they are also the behaviors this instrument is designed to prevent, treat, or manage; and (b) they consistently have been identified as the strongest predictors of serious and violent offending in adulthood.

The younger antisocial externalizing behaviors are exhibited, the more likely they are to be linked to later serious and violent offending. Many researchers have emphasized the relative predictive power of age of onset in identifying who will become the more serious and persistent violent offenders (e.g., Loeber, 1990; Moffitt, 1993; Tolan & Thomas, 1995). During the adolescent stage, however, it is not particularly uncommon for individuals to engage in some of the behaviors listed under this domain. For example, in their meta-analysis, Lipsey and Derzon (1998) found that substance use during early childhood (defined as age 6 to 11) had a relatively large effect size in the prediction of later violent offending, whereas substance use during early adolescence (defined as age 12 to 14) had a very low effect size. We have divided this domain the four following key risk factors: General Behavioral Problems, Violence/Aggression, General Offending, and Substance Use.

Given conduct disorder is a relatively stable syndrome, often leading to antisocial behaviors or antisocial personality disorder in adulthood (Robins, 1978), its absence here may be considered a significant limitation. Individuals with conduct disorder and/or oppositional defiant disorder are not a homogeneous group. As Kazdin (1993) points out, the types may vary in their onset, specific behavioral problems, intensity, repetitiveness, etc. Further, the prevalence of conduct disorder when examining the occurrence of specific externalizing behaviors, appears quite high. Kazdin notes that for treatment purposes, it is more useful to target the specific behaviors as opposed to the syndrome. In addition, research has been able to delineate the specific childhood behaviors predictive of more recalcitrant outcomes. Therefore, in the risk factors under this domain, we have delineated the behaviors that in any combination would justify a conduct disorder diagnosis.

General Behavioral Problems. According to Lipsey and Derzon (1998), "prior antisocial behavior was the best overall predictor of subsequent antisocial behavior" (p. 92) in adolescence. As Farrington (this book) summarized, there are strong intercorrelations between conduct problems, delinquency, and physical aggression in samples of multi-problem youth. In the Cambridge Study, childhood misconduct at age 8-10 was among the significant predictors of self-reported violence at age 10-14 (Farrington & West, 1971). During adolescence, Farrington (this book) found that the more behavioral problems a youth had exhibited, the more likely they were to also commit acts of violence and non-violent offences.

This factor refers to antisocial behaviors other than violence and aggression, substance use, or general offending. This involves a wide range of antisocial behaviors, including but not limited to cruelty to animals or people, firesetting, promiscuity or irresponsible sexual practices, prostitution, truancy, cheating, conning, an unstable job record, serious rule violation, serious disruptive classroom behavior, and running away from home. Many researchers have indicated that the earlier the onset of antisocial behaviors, the more likely the individual will become a serious and chronic offender (e.g., Loeber, 1990; Moffitt, 1993). Therefore, this risk factor is considered critical in the earlier age groups. Likewise, the threshold of the severity rating will vary considerably across the developmental stages.

Violence and Aggression. This factor refers to physical or verbal aggression exhibited towards objects or people. In their meta-analysis of 34 longitudinal source studies, Lipsey and Derzon (1998) found that prior aggression and violence during the 6 to 11 and 12 to 14 age groups was among the better predictors of serious and violent delinquency from age 15 to 25. Aggression and violence were operationalized to include physical violence, aggressive disruptive behavior, aggression toward objects, and verbal aggression. Thornberry (1995) found that an onset of violence by age 9 had a positive association with self-reported violence in later adolescent years. In addition, some researchers have demonstrated that aggression in childhood is also predictive of later severity of delinquency (Roff & Witt, 1984).

As a predictor variable, violence and aggression has been coded using many methods, all of which have resulted in a significant association with subsequent violence and/or violent criminal offenses (see Hawkins et al., 1998 for a review). These methods may include parent, teacher, or peer-rated aggression; as well as the youth's self-report. There is evidence that the association between early and subsequent aggression is not as strong in girls (see Hawkins et al., 1998).

Substance Use. Although substance use is an early behavior problem, the robust association between this behavior in itself and later serious and violent offending calls for a risk factor of its own. In their meta-analysis, Lipsey and Derzon (1998) demonstrated substance use during middle childhood was one of the strongest predictors, with those engaging in substance use being 5 to 20 times more likely to exhibit later violence and delinquency. Moreover, the use of substances in younger childhood years necessary for this factor to be highly predictive of later serious and violent offending is fairly minor, involving chiefly the use of tobacco and alcohol. Alternatively, in adolescence, *use* of substances, including illicit drugs, had only a small effect size in the prediction of violence and offending (Lipsey & Derzon, 1998). However, Farrington (this book) found drug use and heavy drinking during adolescence to be highly related to violence during adolescence.

General Offending. Prior offending is among the highest predictors of later offending, particularly in middle childhood (Lipsey & Derzon, 1998). In middle childhood, those with histories of offending were 5-20 times more likely to engage in subsequent violent or serious delinquency. In his summary of research findings, Farrington (this book) noted that adolescent convictions for non-violent crimes actually predicted adult violence better than adolescent violent convictions.

Given the laws in many countries do not allow for official arrests for criminal offending during middle childhood, criminal records are not necessary for this risk factor to be present and highly predictive of later violence. Mitchell and Rosa (1979) for example, discovered that parent and teacher reports of stealing at age 5 to 15 were linked to later violent offending. Thus, this item may be coded on the basis of official, self, or collateral reports.

7. Future Directions

In sum, the clinically and empirically based factors selected for inclusion in this conceptual framework have been divided along five domains. Although this inventory may be deemed over inclusive by some, and too selective by others, the primary goal in designing this framework was to incorporate a broad range of factors from various disciplines and domains. Ideally, this approach will facilitate a more in-depth understanding of the complex interactions, processes and independent effects that previously identified risk factors have on violent behavior across the lifespan. This is not to say that the individual factors will not be transformed into more specific indicators, or amalgamated into broader themes and constructs in the future. On the contrary, care has been taken in the initial stages of the design to allow for the use of this framework to meet an array of clinical, research, and prevention objectives.

The next stage of this process will be to adapt the conceptual framework presented in this chapter into separate case management tools, in accordance with four developmental stages. Although each of these tools will be structured identically with respect to the five domains, the instrument designed for the earliest developmental stage (pre/peri-natal) will contain significantly less factors than the instrument designed for use with adolescents. In other words, the inclusion of risk factors will be cumulative, with certain factors (e.g., school environment) not appearing in the instrument(s) until they are developmentally relevant, while other factors (e.g., parental coping ability) will be present throughout all four instruments. It should also be noted that the operational definition of the factors may vary by developmental stage. For example, although the underlying definition of parental coping ability will remain constant throughout all four instruments, the operational definition will be conceptually redefined in order to account for the unique stressors that coincide with raising a child at each of the four developmental stages.

To increase the utility of this framework, eventually we will follow it with a proposed scoring, or "checklist" system. In brief, this will be achieved in two ways. First, each factor will be scored as present or absent across each developmental stage. Second, we will include detailed guidelines for a severity rating of each, which may differ depending on the developmental stage of the individual. Finally, we will include a method for flagging those items of utmost importance for designing interventions that will meet the unique needs of the youth being assessed.

Admittedly, the creation of this type of conceptual framework is an ambitious endeavor. Each of the factors included in this review has an associated body of research, with a multitude of indicators and previously developed assessment tools. The goal, therefore, is not to diminish the complexity of each individual factor, nor is it to provide an entirely comprehensive review of every factor. Instead, the following review is intended to provide a framework capable of integrating a wide range of measures from biological, psychological, and criminal justice research, with knowledge derived from clinical expertise and intervention experience: a necessary step towards the development of a comprehensive and multi-systems understanding and approach to the problem of youth violence.

Notes

Authors Note. We wish to give special thanks to Rebecca Dempster and Andrea McEachran for their assistance in the early stages of developing this conceptual framework. We also extend our appreciation to the participants of the NATO Advanced Research Workshop, as their input was crucial to the design of this framework.

References

Adams, S. (1970). The PICO project. In N. Johnston, L. Savitz, & M. E. Wolfgang (Eds.), *The sociology of punishment and correction* (pp. 548-561). New York: Wiley.

American Psychiatric Association. (1994). *Diagnostic and statistical manual of mental disorders* (4th ed.). Washington, DC: Author.

Andrews, D. A., & Bonta, J. (1995). *The psychology of criminal conduct.* Cincinnati: Anderson.

Augimeri, L. K., Koegl, C. J., Webster, C. D., & Levene, K.S. (2001). *Early Assessment Risk List for Boys: EARL-20B (Version 2).* Toronto, ON: Earlscourt Child and Family Centre.

Baker, R .L. A., & Mednick, B. R. (1984). *Influences on human development: A longitudinal perspective.* Boston, MA: Kluwer-Nijhoff.

Bartell, P., Borum, R., & Forth, A. (1999). *SAVRY: Structured Assessment of Violence Risk in Youth* (Version 1). Consultation Edition.

Bates, J. E. (1980). The concept of difficult temperament. *Merrill-Palmer Quarterly, 26,* 299-319.

Battin, S. R., Hill, K. G., Abbott, R. D., Catalano, R. F., & Hawkins, J. D. (1998). The contribution of gang membership to delinquency beyond delinquent friends. *Criminology, 36,* 93–115.

Belfrage, H., Lidberg, L., & Oreland, L. (1992). Platelet monoamine oxidase activity in mentally disordered violent offenders. *Acta Psychiatrica Scandinavica, 85,* 218-221.

Bellinger, D., Leviton, A., Waternaux, C., Needleman, H., & Rabinowitz, R. (1987). Longitudinal analyses of prenatal and postnatal lead exposure and early cognitive development. *New England Journal of Medicine, 316,* 1037-1043.

Benson, D. F., & Zaidel, E. (1985). *The dual brain.* New York: Guilford.

Berman, M. E., Kavoussi, R. J., & Coccaro, E. F. (1997). Neurotransmitter correlates of human aggression. In D. M. Stoff, J. Breiling, & J. D. Master (Eds.), *Handbook of Antisocial Behavior* (pp. 305-313). New York: Wiley.

Born, M., Chevalier, V., & Humblet, I. (1997). Resilience, desistance and delinquent career of adolescent offenders. *Journal of Adolescence, 20,* 679-694.

Bowden, C. L., Deutsch, C. K., & Swanson, J. M. (1987). Plasma dopamine-β-hydroxylase and platelet monoamine oxidase in attention deficit disorder and conduct disorder. *Journal of the American Academy of Child and Adolescent Psychiatry, 27,* 171-174.

Brennan, P. A., & Mednick, S. A. (1997). Medical histories of antisocial individuals. In D. M. Stoff, J. Breiling, & J. D. Master (Eds.), *Handbook of Antisocial Behavior* (pp. 269-279). New York: John Wiley & Sons.

Brewer, D.D., Hawkins, J.D., Catalano, R.F., & Neckermen, H.J. (1995). Preventing serious, violence, and chronic juvenile offenders: A review of evaluations of selected strategies in childhood, adolescence, and the community. In J.C Howell, B. Krisberg, J.D. Hawkins, & J.J. Wilson (Eds.), *A sourcebook: Serious, violent, and chronic offenders* (pp. 61-141). Thousand Oaks, CA: Sage Publications.

Bursik, R. J., & Grasmick, H. G. (1993). *Neighborhoods and crime: The dimensions of effective community control.* New York: Lexington Books.

Carey, G., & Goldman, D. (1997). The genetics of antisocial behavior. In D. M. Stoff, J. Breiling, & J. D. Master (Eds.), *Handbook of Antisocial Behavior* (pp. 243-254). New York: Wiley.

Catalano, R. F., Arthur, M. W., Hawkins, J.D., Berglund, L., & Olson, J. J. (1998). Comprehensive community and school-based interventions to prevent antisocial behavior. In R. Loeber, & D.P., Farrington (Eds.), *Serious and violent juvenile offenders: Risk factors and successful interventions* (pp. 248-283). Thousand Oaks, CA: Sage.

Catalano, R. F., & Hawkins, J. D. (1996). The social development model: A theory of antisocial behavior. In J.D. Hawkins (Ed.), *Delinquency and Crime: Current Theories* (pp. 149–197). New York, NY: Cambridge University Press.

Conry, J., & Fast, D. K. (2000). *Fetal Alcohol Syndrome and the Criminal Justice System.* Vancouver: The

Law Foundation of British Columbia.

Corrado, R., Odgers, C., & Cohen, I. (2001). Girls in jail: Punishment or protection? In R. Roesch, R. R. Corrado, & R. Dempster (Eds.), *Psychology in the courts: International advances in knowledge* (pp. 41-52). Routledge: London.

Cunningham, P. B., & Henggeler, S. W. (1999). Engaging multiproblem families in treatment: Lessons learned throughout the development of multisystemic therapy. *Family Process, 38,* 265-286.

Conry, J., & Fast, D. K. (2000). *Fetal alcohol syndrome and the criminal justice system.* Vancouver: The Law Foundation of British Columbia.

Deater-Deckard, K., Dodge, K. A., Bates, J.E., & Pettie, G. S. (1998). Multiple risk factors in the development of externalizing behavior problems: Group and individual differences. *Development and Psychopathology, 10,* 469-493.

Dembo, R., Williams, L., Wothke, W., Schneidler, J., & Brown, C. (1992). The role of family factors, physical abuse, and sexual victimization experiences in high risk youths' alcohol and other drug use and delinquency: longitudinal model. *Violence and Victims, 7,* 233-246.

Denno, D. W. (1990). *Biology and violence: From birth to adulthood.* Cambridge, UK: Cambridge University Press.

Depue, R. A., & Spoont, M. R. (1986). Conceptualizing a serotonin trait: A behavioral dimension of constraint. *Annals of the New York Academy of Sciences, 487,* 47-62.

Dodge, K. A., & Schwartz, D. (1997). Social information processing mechanisms in aggressive behavior. In D. M. Stoff, J. Breiling, & J. D. Master (Eds.), *Handbook of antisocial behavior* (pp. 171-180). New York: Wiley.

Dubow, E. F., Edwards, S., & Ippolito, M. F. (1997). Life stressors, neighborhood disadvantage, and resources: A focus on intercity children's adjustment. *Journal of Clinical Child Psychology, 26,* 130-144.

DuRant, R. H., Cadenhead, C., Pendergrast, R. A., & Slavens, G. (1994). Factors associated with violence among urban black adolescents. *American Journal of Public Health, 84,* 612-617.

Ellingson, R. J. (1954). The incidence of EEG abnormality among patients with mental disorders of apparently nonorganic origin: A critical review. *American Journal of Psychiatry, 111,* 262-275.

Elliott, D. S., Huizinga, D., & Menard, S. (1989). *Multiple problem youth: Delinquency, substance use and mental health problems.* New York, NY: Springer-Verlag.

Elliot, D. S. (1994). Serious violent offenders: Onset, developmental course, and termination – The American Society of Criminology 1993 presidential address. *Criminology, 32,* 1-21.

Erickson, M. F., Sroufe, L. A., & Egland, B. (1985). The relationship between quality of attachment and behavior problems in preschool in a high risk sample, *Monographs of the Society for Research in Child Development, 50*(1-2), 147-166.

Eron, L. D., Huesmann, L. R., & Zelli, A. (1991). The role of parental variables in the learning of aggression. In D.J. Pepler & K. H. Rubin (Eds.), *The development and treatment of childhood aggression* (pp. 169-188). Hillsdale, NJ: Erlbaum.

Esbensen, F., & Huizinga, D. (1991). Juvenile victimization and delinquency. *Youth and Society, 23,* 202-228.

Eysenck, H. J. (1964). *Crime and personality.* London: Methuen.

Farrington, D. P, & West, D.J. (1971). A comparison between early delinquents and young aggressives. *British Journal of Criminology, 11,* 341-358.

Farrington, D. P. (1989). Early predictors of adolescent aggression and adult violence. Violence and Victims, 4, 79–100.

Farrington, D.P. (1991). Childhood aggression and adult violence: Early precursors and later-life outcomes. In D.J. Pepler & K.H. Rubin (Eds.), *The development and treatment of childhood aggression* (pp. 5-29). NJ: Erlbaum.

Fergusson, D.M., & Woodward, L.J. (1999). Maternal age and educational and psychosocial outcomes in early adulthood. *Journal of Child Psychology and Psychiatry, 40,* 479-489.

Firestone, P., & Peters, S. (1983). Minor physical anomalies and behavior in children: A review. *Journal of Autism and Developmental disorders, 13,* 411-425.

Fitzhardinge, P. M., & Steven, E. M. (1972). The small for date infant: II. Neurological and intellectual sequelae. *Pediatrics, 70,* 50-57.

Frick, P. J., Lilienfeld, S. O., Ellis, M., Loney, B., & Silverthorn, P. (1999). The association between anxiety and psychopathy dimensions in children. *Journal of Abnormal Child Psychology, 27,* 383-392.

Garbarino, J. (1997). Growing up in a socially toxic environment. In D. Cicchetti & S. L. Toth (Eds.), *Developmental perspectives on trauma: Theory, research, and intervention* (pp. 141-154). Rochester, NY: University of Rochester Press.

Garmezy, N. (1993). Children in poverty: Resilience despite risk. *Psychiatry, 56,* 127-136.

Gibson, C. L., & Tibbetts, S. G. (1998). Interaction between maternal cigarette smoking and apgar scores in predicting offending behavior. *Psychological Reports, 83,* 579-586.

Gorman-Smith, D., & Tolan, P. H., Zelli, A., & Huesmann, L. R. (1996). The relation of family functioning to violence among inner-city minority youth. *Journal of Family Psychology, 10,* 115-129.

Graham, P., Rutter, M., & George, S. (1973). Temperamental characteristics as predictors of behavior disorders in children. *American Journal of Orthopsychiatry, 43,* 328-339.

Haapasalo, J., & Tremblay, R. E. (1994). Physically aggressive boys from ages 6 to 12: Family background, parenting behavior, and prediction of delinquency. *Journal of Consulting and Clinical Psychology, 27,* 281-298.

Halstead, S. (1997). Risk assessment and management in psychiatric practice: Inferring predictors of risk. A view from learning disability. *International Review of Psychiatry, 9,* 225-231.

Hare, R. D. (1993). *Without conscience: The disturbing world of the psychopaths among us.* New York: Pocket Books.

Hare, R. D., & Hart, S. D. (1993). Psychopathy, mental disorder, and crime. In S. Hodgins (Ed.), *Mental disorder and crime* (pp. 104-115). London: Sage.

Harris, G. T., Rice, M. E., & Quinsey, V. L. (1993). Violent recidivism of mentally disordered offenders: The development of a statistical prediction instrument. *Criminal Justice and Behavior, 20,* 315-335.

Hawkins, J.D., Herrenkohl, T. L., Farrington, D. P., Brewer, D., Catalano, R. F., & Harachi, T. W. (1998). A review of predictors of youth violence. In R. Loeber & D. P., Farrington (Eds.), *Serious and violent juvenile offenders: Risk factors and successful interventions* (pp. 106-146). Thousand Oaks, CA: Sage.

Henry, B., Avshalom, C., Moffitt, T. E., & Silva, P. A. (1996). Temperamental and familial predictors of violent and non-violent criminal convictions: Age 3 to age 18. *Developmental Psychology, 32,* 614–623.

Henry, B. A., Capsi, A., Moffitt, T. E., & Silva, P. A. (1996). Temperamental and familial predictors of violent and nonviolent criminal convictions: Age 3 to age 18. *Developmental Psychology, 32,* 614-623.

Henry, B., & Moffitt, T. E. (1997). Neuropsychological and neuroimaging studies of juvenile delinquency and adult criminal behavior. In D. M. Stoff, J. Breiling, & J. D. Master (Eds.), *Handbook of Antisocial Behavior* (pp. 280-288). New York: Wiley.

Herrenkohl, T. I., Maguin, E., Hill, K. G., Hawkins, J. D., Abbott, R. D., & Catalano, R. F. (2000). Developmental risk factors for youth violence. *Journal of Adolescent Health, 26,* 176-186.

Hertzig, M. E. (1982). Neurological "soft" signs in low-birthweight children. *Annual Progress in Child Psychiatry and Child Development* (pp 509-524). New York: Brunner/Mazel.

Hill, H. M, Soriano, F. I., Chen S. A., & LaFromboise, T. D. (1994). Sociocultural factors in the etiology and prevention of violence among ethnic minority youth. In L .D. Eron & J. H. Gentry (Eds.), *Reason to hope: A psychosocial perspective on violence & youth* (pp. 59-97). Washington, DC: American Psychological Association.

Hirschi, T. (1969). *Causes of delinquency.* Berkeley, CA: University of California Press.

Hodgins, S. (1992). Mental disorder, intellectual deficiency, and crime: Evidence from a birth cohort. *Archives of General Psychiatry, 49,* 476-483.

Jenkins, J., & Keating, D. (1998). Risk and resilience in six-and ten-year-old children. *Working Paper* (W-98-23E). Applied Research Branch, Strategic Policy, Human Resources Development Canada.

Kandel, E., Brennan, P. A., Mednick, S. A., & Michelsen, N. M. (1989). Minor physical anomalies and recidivistic adult violent offending. *Acta Psychiatrica Scandinavica, 79,* 103-107.

Kazdin, A. E. (1987). Treatment of antisocial behavior in children: Current status and future directions. *Psychological Bulletin, 102,* 187-203.

Kazdin, A. E. (1993). Adolescent mental health: Prevention and treatment programs. *American Psychologist, 48,* 127-141.

Kazdin, A. E., Mazurick, J. L., & Siegel, T. C. (1994). Treatment outcome among children with externalizing disorder who terminate prematurely versus those who complete psychotherapy. *Journal of American Academy of Child and Adolescent Psychiatry, 33,* 549-557.

af Klinteberg, B., Oreland, L., Hallman, J., Wirsen, A., Levander, S., & Schalling, D. (1991). Exploring the connections between platelet monoamine oxidase activity and behavior: Relationships with performance in neuropsychological tasks. *Neuropsychobiology, 23,* 188-196.

Kropp, P. R., Hart, S. D., Webster, C. D., & Eaves, D. (1995). *Manual for the spousal assault risk assessment guide* (2nd ed.). Vancouver: British Columbia Institute Against Family Violence.

Kruh, I. P., & Brodsky, S. L. (1997). Clinical evaluations for transfer of juveniles to criminal court: Current practices and future research. *Behavioral Sciences and the Law, 15,* 151-165.

Laub, J. H., & Lauristen, J. L. (1998). The interdependence of school violence with neighborhood and family conditions. In D. S. Elliot, B. A. Hamburg, & K. R. Williams (Eds.), *Violence in American schools: A new perspective* (pp. 127-155). New York: Cambridge University Press.

Linnoila, M. (1997). On the psychobiology of antisocial behavior. In D. M. Stoff, J. Breiling, & J. D. Master (Eds.), *Handbook of antisocial behavior* (pp. 336-340). New York: Wiley.

Lipman, E. L., Boyle, M.H, Dooley, M.D., & Offord, D. L. (1998). *Children and lone-mother families: An investigation of factors influencing child well-being.* Report to Applied Research Branch, Strategies Policy, Human Resources Development. Ottawa: Human Resources Development.

Lipsey, M. W., & Derzon, J. H. (1998). Predictors of violence and serious delinquency in adolescence and early adulthood: A synthesis of longitudinal research. In R. Loeber & D.P., Farrington (Eds.), *Serious and violent juvenile offenders: Risk factors and successful interventions* (pp. 86-105). Thousand Oaks, CA: Sage.

Litt, S. M. (1972). Perinatal complications and criminality. *Dissertation Abstracts International, 35*(5-B), 2349-2350.

Loeber, R. (1990). Development of risk factors of juvenile antisocial behavior and delinquency. *Clinical Psychology Review, 10,* 1-41.

Lorion, R. (1998). Exposure to urban violence: Contamination of the school environment. In D. S. Elliot, B. A. Hamburg, & K. R. Williams (Eds.), *Violence in American schools: A new perspective* (pp. 293-311). New York: Cambridge University Press.

Lykken, D. T. (1957). A study of anxiety in the sociopathic personality. *Journal of Abnormal and Social Psychology, 55,* 6-10.

Lynam, D., Moffitt, T. E., & Stouthamer-Loeber, J. (1993). Explaining the relation between IQ and delinquency: Race, class, test motivation, school failure, or self-control. *Journal of Abnormal Psychology, 102,* 187-196.

Maguin, E., Hawkins, J.D., Catalano, R.F., Hill, K., Abbott, R., & Herrenkohl, T. (1995, November). Risk factors measured at three ages for violence at age 17-18. *Paper presented at the American Society of Criminology,* Boston.

Marlowe, M., Stellern, J., Moon, C. & Errera, J. (1985). Main and interactive effects of metallic toxins on aggressive classroom behavior. *Aggressive Behavior, 11,* 41-48.

McCord, J. (1979). Some child-rearing antecedents of criminal behavior in adult men. *Journal of Personality and Social Psychology, 37,* 1477-1486.

McCord, J., & Ensminger, M. (1995, November). *Pathways from aggressive childhood to criminality.* Paper presented at the American Society of Criminology, Boston.

McCormick, M.C. (1985). The contribution of low birth weight to infant mortality and childhood morbidity. *New England Journal of Medicine, 312,* 82-90.

McNeil, T. F. (1988). Obstetric factors and perinatal injuries. In M. T. Tsuang & J. C. Simpson (Eds.), *Handbook of schizophrenia, Vol. 3: Nosology, epidemiology and genetics.* New York: Elsevier.

McGuire, J. (1997). Psychosocial approaches to the understanding and reduction of violence in young people. In V. Varma (Ed.), *Violence in children and adolescents* (pp.65-83). Bristol, PA: Jessica Kingsley.

Mitchell, S., & Rosa, P. (1979). Boyhood behavior problems as precursors of criminality: A fifteen year follow up study. *Journal of Child Psychology & Psychiatry, 22,* 19-33.

Moffitt, T. E. (1987). Parental mental disorder and offspring criminal behavior: An adoption study. *Psychiatry, 50,* 346-360.

Moffitt, T. E. (1990). Juvenile delinquency and attention-deficit disorder: Developmental trajectories from age 3 to 15. *Child Development, 61,* 893-910.

Moffitt, T. E. (1993a). Adolescence-limited and life course-persistent antisocial behavior: A developmental taxonomy. *Psychological Review, 100,* 674-701.

Moffitt, T. E. (1993b). The neuropsychology of conduct disorder. *Development and Psychopathology, 5,* 135-151.

Moffitt, T. E., & Silva, P. A. (1988). Self-reported delinquency, neuropsychological deficit, and history of attention deficit disorder. *Journal of Abnormal Child Psychology, 16,* 553-569.

Monahan, J. (1992). Mental disorder and violent behavior. *American Psychologist, 47,* 511-521.

Morash, M., & Rucker, L. (1989). An exploratory study of the connection of mothers age at childbearing to her children's delinquency in four data sets. *Crime and Delinquency, 35,* 45-93.

Mulvey, E. (1984). Resource availability, agency type, and amenability to treatment judgments of juvenile offenders. *Dissertation Abstracts International, 44*(12-B), 3939.

Needleman, H. L., Schell, A., Bellinger, D., Leviton, A., & Allred, A. (1990). The long-term effects of exposure to low doses of lead in childhood: An 11-year follow-up report. *New England Journal of Medicine, 322,* 83-88.

Needleman, H. L., Riess, J. A., Tobin, M. J., Biesecker, G. E., & Greenhouse, J. B. (1996). Bone lead levels and delinquent behavior. *Journal of the American Medical Association, 275,* 363-369.

Osgood, D. W., & Chambers, J. M. (2000). Social disorganization outside the metropolis: An analysis of rural youth violence. *Criminology, 38,* 81-111.

Osofsky, J. D. (1999). The impact of violence on children. *Future of Children, 9,* 33-49.

Paschall, M. J. (1996, June). *Exposure to violence and the onset of violent behavior and substance abuse among black male youth: An assessment of independent effects and psychological mediators.* Paper presented at the Society for Prevention Research, San Juan, PR.

Patrick, C. J., Bradley, M. M., Lang, P. J. (1993). Emotion in the criminal psychopath: Startle reflex modulation. *Journal of Abnormal Psychology, 102,* 82-92.

Patterson, G. R., Ried, J. B., & Dishion, T. J. (1992). *Antisocial boys: A social interactional approach* (Vol. 4). Eugene, OR: Castalia.

Peterson, P.L., Hawkins, J.D., Abbott, R.D., & Catalano, R.F. (1994). Disentangling the effects of parental drinking, family management, and parental alcohol norms on current drinking by black and white adolescents. *Journal of Research on Adolescence, 4,* 203-227.

Peterson, P. L., Matousek, M., Mednick, S. A., Volavka, J., & Pollock, V. (1982). EEG antecedents of thievery. *Criminology, 19,* 219-229.

Plomin, R., Owen, M. J., & McGuffin, P. (1994). The genetic basis of complex human behaviors. *Science, 264,* 1733-1739.

Pontius, A.A. (1972). Neurological aspects in some types of delinquency, especially among juveniles: Toward a neurological model of ethical action. *Adolescence, 7,* 289-308.

Raine, A. (1997). Antisocial behavior and psychophysiology: A biosocial perspective and a prefrontal dysfunction hypothesis. In D. M. Stoff, J. Breiling, & J. D. Master (Eds.), *Handbook of Antisocial Behavior* (pp. 289-304). New York: Wiley.

Raine, A., Brennan, P., & Mednick, S. A. (1994). Birth complications combined with early maternal rejection at age 1 year predispose to violent crime at age 18 years. *Archives of General Psychiatry, 94,* 984-988.

Raine, A., & Venables, P. (1992). Antisocial behavior: Evolution, genetics, neuropsychology, and psychophysiology. In A. Gale & M. Eysenck (Eds.), *Handbook for individual differences: Biological perspectives.* London: Wiley.

Ramsey, E., & Walker, H. M. (1998). Family management correlates of antisocial behavior among middle school boys. *Behavioral Disorders, 13,* 187-201.

Rasanen, P., Hakko, H., Isohanni, M., Hodgins, S., Jarvelin, M. R., & Tiihonen, J. (1999). Maternal smoking during pregnancy and risk of criminal behavior among adult male offspring in the Northern Finland 1966 birth cohort. *American Journal of Psychiatry, 156,* 857-862.

Rasanen, P., Hakko, H., Jarvelin, M. R., & Tiihonen, J. (1999). Is a large body size during childhood a risk factor for later aggression? *Archives of General Psychiatry, 56,* 518-527.

Reiss, A. J., & Roth, J. A. (1993). *Understanding and preventing violence.* National Academy Press: Washington, DC.

Robins, L. N. (1978). Sturdy childhood predictors of adult antisocial behavior: Replications from longitudinal studies. *Psychological Medicine, 8,* 611-622.

Rogers, R., & Webster, C. D. (1989). Assessing treatability in mentally disordered offenders. *Law and Human Behavior, 13,* 19-29.

Roff, J. D., & Witt, R. D. (1984). Childhood social adjustment, adolescent status, and young adult mental health. *American Journal of Orthopsychiatry, 54,* 595-602.

Roussey, S., & Toupin, J. (2000). Behavioral inhibition deficits in juvenile psychopaths. *Aggressive Behavior, 26,* 413-424.

Rutter, M. (1985). Resilience in the face of adversity: Protective factors and resistance to psychiatric disorder. *British Journal of Psychiatry, 147,* 598-611.

Rutter, M. (1987). Psychosocial resilience and protective mechanisms. *American Journal of Orthopsychiatry, 57,* 316-331.

Rutter, M., Giller, H., & Hagell, A. (1998). *Antisocial behavior by young people.* New York: Cambridge University Press.

Sampson, R. J., & Lauritsen, J. L. (1994). Violent victimization and offending: Individual-, situational-, and community-level risk factors. In A. J. Reiss, Jr. & J. A. Roth (Eds.), *Understanding and preventing violence: Social influences* (Vol. 3, pp. 1-114). Washington, DC: National Academy Press.

Saner, H., & Ellickson, P. (1996). Concurrent risk factors for adolescent violence. *Journal of Adolescent Health, 19,* 94-103.

Shaw, D. S., & Winslow, E. B. (1997). Precursors and correlates of antisocial behavior from infancy to preschool. In D. M. Stoff, J. Breiling, & J. D. Master (Eds.), *Handbook of antisocial behavior* (pp. 148-158). New York: Wiley.

Silverman, A. B., Reinherz, H. Z., & Ginconia, R. M. (1996). The long-term sequelae of child and adolescent abuse: A longitudinal community study. *Child Abuse & Neglect, 20,* 709-724.

Singer, M., Anglin, T. M., Song, L., & Lunghofer, L. (1995). Adolescents' exposure to violence and associated symptoms of psychological trauma. *Journal of the American Medical Association, 273,* 477-482.

Smith, C., & Thornberry, T. P. (1995). The relationship between childhood maltreatment and adolescent involvement in delinquency. *Criminology, 33,* 451-481.

Thornberry, T. P. (1998). Membership in youth gangs and involvement in serious, violent offending. In R. Loeber & D.P. Farrington (Eds.), *Serious and violent juvenile offenders: Risk factors and successful interventions* (pp. 147-166). Thousand Oaks, CA: Sage.

Tolan, P. H., & Thomas, P. (1995). The implications of age of onset for delinquency risk II: Longitudinal data. *Journal of Abnormal Child Psychology, 23*, 157-181.

U.S. Department of Health and Human Services. (2001). *Youth violence: A report of the Surgeon General.* Rockville, MD: U.S Department of Health and Human Services, Centers for Disease Control and Prevention, National Center for Injury Prevention and Control; Substance Abuse and Mental Health Services Administration, Center for Mental Health Services; and National Institutes of Health, National Institute of Mental Health.

Vuchinich, S., Bank, L. & Patterson, G. R. (1992). Parenting, peers, and the stability of antisocial behavior in preadolescent boys. *Developmental Psychology, 28*, 510-521.

Wadsworth, M. E. J. (1978). Delinquency prediction and its uses: The experience of a 21 year follow-up study. *International Journal of Mental Health, 7*, 43-62.

Walhler, R. G. (1980). The insular mother: Her problems in parent-child treatment. *Journal of Applied Behavioral Analysis, 13*, 207-219.

Wasserman, G. A., & Miller, L. S. (1998). The prevention of serious and violent juvenile offending. In R. Loeber, & D.P., Farrington (Eds.), *Serious and violent juvenile offenders: Risk factors and successful interventions* (pp. 197-247). Thousand Oaks, CA: Sage Publications.

Webster, C. D, Douglas, K. S., Eaves, D., & Hart, S. D. (1997). HCR-20 assessing risk for violence: Version 2. Burnaby, British Columbia: Mental Health, Law, and Policy Institute, Simon Fraser University.

Webster-Stratton, C. (1990). Enhancing the effectiveness of self-administered videotape parent training for families with conduct problem children. *Journal of Abnormal Child Psychology and Psychiatry, 32*, 1047-1062.

Weiss, B., Dodge, K. A., Bates, J. E., & Pettit, G. S. (1992). Some consequences of early harsh discipline: Child aggression and a maladaptive social information processing style. *Child Development, 63*, 1321-1335.

Wells, L. E., & Rankin, J. H. (1988). Direct parental controls and delinquency. *Criminology, 26*, 263–285.

Werner, E. E., & Smith, R. S. (1992). *Overcoming the odds: High risk children from birth to adulthood.* Ithaca, NY: Cornell University Press.

Widom, C. S. (1989). Child abuse, neglect, and violent criminal behavior. *Criminology, 27*, 251-271.

Widom, C. S. (1997). Child abuse, neglect, and witnessing violence. In D. M. Stoff, J. Breiling, & J. D. Master (Eds.), *Handbook of antisocial behavior* (pp. 159-170). New York: Wiley.

Williams, J. H. (1994). Understanding sub-stance use, delinquency involvement, and juvenile justice system involvement among African-American and European-American adolescents. *Unpublished dissertation*, University of Washington, Seattle, WA.

Woodward, L. J., & Fergusson, D. M. (1999). Early conduct problems and later risk of teenage pregnancy in girls. *Development and Psychopathology, 11*, 127-141.

Author Index